Modeling Dynamic Transportation Networks

Springer
Berlin
Heidelberg
New York
Barcelona
Budapest
Hong Kong
London
Milan
Paris
Santa Clara
Singapore
Tokyo

Bin Ran · David Boyce

Modeling Dynamic Transportation Networks

An Intelligent Transportation System Oriented Approach

Second Revised Edition

With 51 Figures

Springer

Dr. Bin Ran
University of Wisconsin at Madison
Department of Civil and Environmental Engineering
2256 Engineering Hall
1415 Engineering Drive
Madison, WI 53706, USA

Dr. David Boyce
University of Illinois at Chicago
Urban Transportation Center
1033 West Van Buren Street
Suite 700 South
Chicago, IL 60607, USA

Ran, Bin:
Modeling dynamic transportation networks : an intelligent transportation system oriented approach / Bin Ran ; David Boyce. - 2. ed. - Berlin ; Heidelberg ; New York ; Barcelona ; Budapest ; Hong Kong ; London ; Milan ; Paris ; Santa Clara ; Singapore ; Tokyo : Springer, 1996

1. Aufl. u.d.T.: Ran, Bin: Dynamic urban transportation network models

NE: Boyce, David E.:

ISBN-13:978-3-642-80232-4 e-ISBN-13:978-3-642-80230-0
DOI: 10.1007/978-3-642-80230-0

Softcover reprint of the hardcover 2nd edition 1996

SPIN 10537415 42/2202-5 4 3 2 1 0 – Printed on acid-free paper

To our parents

Preface

This book seeks to summarize our recent progress in dynamic transportation network modeling. It concentrates on ideal dynamic network models based on actual travel times and their corresponding solution algorithms. In contrast, our first book *Dynamic Urban Transportation Network Models - Theory and Implications for Intelligent Vehicle-Highway Systems* (Springer-Verlag, 1994) focused on instantaneous dynamic network models. Comparing the two books, the major differences can be summarized as follows:

1. This book uses the variational inequality problem as the basic formulation approach and considers the optimal control problem as a subproblem for solution purposes. The former book used optimal control theory as the basic formulation approach, which caused critical problems in some circumstances.

2. This book focuses on ideal dynamic network models based on actual travel times. The former book focused on instantaneous dynamic network models based on currently prevailing travel times.

3. This book formulates a stochastic dynamic route choice model which can utilize any possible route choice distribution function instead of only the logit function.

4. This book reformulates the bilevel problem of combined departure time/ route choice as a one-level variational inequality.

5. Finally, a set of problems is provided for classroom use.

In addition, this book offers comprehensive insights into the complexity and challenge of applying these dynamic network models to Intelligent Transportation Systems (ITS). Nevertheless, the models in this text are not yet fully evaluated and are subject to revision based on future research.

A summary of the necessary mathematical background, including variational inequalities, optimal control and constrained optimization, provides a reference for transportation engineers involved in ITS projects. By understanding the mathematical requirements of dynamic transportation network models, the professional community can appreciate the intensified research effort required for elaborating this new dimension of transportation science.

Our first book was used as a major reference in a summer course on "Dynamic Transportation Networks and Intelligent Vehicle-Highway Systems" at the National Central University, Taiwan, in 1994. Later, it was used in the course, "Transportation Network Equilibrium Analysis," at Massachusetts Institute of Technology in 1995. Through this classroom usage of our first book, many comments and problem sets have been developed, which provide a basis for the problem sets in this book.

Many researchers have contributed their ideas and comments to this work. In particular, Professor Larry LeBlanc, Vanderbilt University, collaborated with the authors in writing several of the papers on which this book is based. We wish to acknowledge fully his important contributions to this effort. Dr. Hong K. Lo and Mr. Bruce Hongola, University of California at Berkeley, have collaborated with the first author on several papers on multi-class dynamic route choice models and computational experiments. Professors Nagui Rouphail, North Carolina State University, and Andrzej Tarko, Purdue University, have collaborated with the authors on developing dynamic link travel time functions.

Professor Floyd Hanson, University of Illinois at Chicago, taught the first author numerous skills from supercomputing techniques to advanced optimal control theory. Professors Hani Mahmassani, University of Texas at Austin, Anna Nagurney, University of Massachusetts at Amherst, and Martin Beckmann, Brown University, provided continuing encouragement for our research during the past six years. Professor Huey-Kuo Chen and Ms. Yali Chieu at the National Central University at Taiwan constructively commented on the manuscript. Professor Bruce Janson, University of Colorado at Denver, wrote a comprehensive review of the previous book. Professor Norbert Oppenheim, City University of New York, also provided extensive comments. In addition, Professor Piet Bovy and Mr. Serge Hoogendoorn, Delft University of Technology, the Netherlands, offered many detailed comments.

Professors Toshikazu Shimazaki, Masao Kuwahara and Takashi Akamatsu at the University of Tokyo helped the first author to begin his research on one of the most challenging topics in transportation engineering. Their continuing support and constructive comments are appreciated. Appreciation is also extended to Professors Carlos Daganzo and Pravin Varaiya, and Dr. Steven Shladover, University of California at Berkeley, for their valuable advice and encouragement. Special thanks go to Professors Robert L. Smith, Jr., Erhard F. Joeres, Jeffrey Russell and William D. Berg, University of Wisconsin at Madison, for their encouragement to the first author.

Appreciation is extended to Professors Randolph Hall, University of Southern California, Moshe Ben-Akiva, Massachusetts Institute of Technology, Warren Powell and David Bernstein, Princeton University, Gang-Len Chang, University of Maryland at College Park, Michael Florian, University of Montreal, Andre de Palma, Universite de Cergy Pontoise, Malachy Carey, University of Ulster, Mark Daskin, Northwestern University, Terry Friesz, George Mason University, B.-W. Wie, University of Hawaii, Tony Smith, University of Pennsylvania, Kumares Sinha, Purdue University, Chris Hendrickson and Haris Koutsopoulos, Carnegie-Mellon University, Pitu Mirchandani, University of Arizona, Dr. Jia-Hao Wu, University of Montreal, Dr. Stanislaw Berka and Dr. Der-Horng Lee, University of Illinois at Chicago, and many anonymous referees for their constructive and helpful comments.

Special thanks go to other friends and colleagues at the Urban Transportation Center, University of Illinois at Chicago, the Institute of Transportation Studies, University of California at Berkeley, Center for Transportation Studies, Massachusetts Institute of Technology, and the Department of Civil and Environmental Engineering, University of Wisconsin at Madison.

Partial support of this research was provided by the Federal Highway Administration and the Illinois Department of Transportation through the ADVANCE Project and the California Department of Transportation through the PATH Program. The support of the National Science Foundation through its grants to the National Institute of Statistical Sciences and the National Center for Supercomputing Applications is also gratefully acknowledged. Appreciation also goes to Mr. Alberto J. Santiago, Mr. James Clark, Dr. Henry Lieu, Federal Highway Administration, and Dr. Ajay Rathi, Dr. Rekha Pillai and Dr. Shaw-Pin Miaou, Oak Ridge National Laboratory.

Bin Ran
David Boyce

Madison, Wisconsin
February, 1996

Contents

Summary of Notation

Network Flow Constraints and Definitions of Travel Times

$x_a(t)$	=	number of vehicles on link a at time t (main problem variable) *
$x_a(k)$	=	number of vehicles on link a at beginning of interval k *
$x_{ap}^{rs}(t)$	=	number of vehicles on link a and route p between O-D pair rs at time t *
$x_{bp}^{s}(t)$	=	number of vehicles on dummy link b and route p toward destination s at time t *
$x_{a1}(t)$	=	number of vehicles traveling on sublink a_1 at time t
$x_{a2}(t)$	=	number of vehicles traveling on sublink a_2 at time t
$u_a(t)$	=	inflow into link a at time t (main problem variable) **
$u_a(k)$	=	inflow into link a during interval k **
$u_{ap}^{rs}(t)$	=	inflow into link a and route p between O-D pair rs at time t **
$u_{a1}(t)$	=	inflow into sublink a_1 at time t
$u_{a2}(t)$	=	inflow into sublink a_2 at time t
$v_a(t)$	=	exit flow from link a at time t (main problem variable) **
$v_a(k)$	=	exit flow from link a during interval k **
$v_{ap}^{rs}(t)$	=	exit flow from link a and route p between O-D pair rs at time t **
$v_{bp}^{s}(t)$	=	exit flow from dummy link b and route p toward destination s at time t **
$v_{a1}(t)$	=	exit flow from sublink a_1 at time t
$v_{a2}(t)$	=	exit flow from sublink a_2 at time t
l_a	=	length of link a
e_{am}	=	maximum traffic density of link a
v_{am}	=	flow capacity at the exit of link a
$y_a(k)$	=	number of vehicles on link a at the beginning of time interval k (subproblem variable)
$p_a(k)$	=	inflow into link a during interval k (subproblem variable)
$q_a(k)$	=	exit flow from link a during interval k (subproblem variable)
$f^{rs}(t)$	=	departure flow rate from origin r toward destination s at time t
$f^{rs}(k)$	=	departure rate from origin r toward destination s during interval k
$f_p^{rs}(t)$	=	departure flow rate from origin r toward destination s on route p at time t
$f_{mp}^{rs}(t)$	=	departure flow rate of class m vehicles from origin r toward destination s on route p at time t

$F^{rs}(t)$	=	cumulative number of departing vehicles from origin r to destination s by time t
$F^{rs}(k)$	=	cumulative number of departing vehicles from origin r to destination s at the beginning of interval k
$F_p^{rs}(t)$	=	cumulative number of departing vehicles from origin r to destination s on route p by time t
$e^{rs}(t)$	=	arrival flow at destination s from origin r at time t **
$e^{rs}(k)$	=	arrival flow at destination s from origin r during interval k **
$E^{rs}(t)$	=	cumulative number of vehicles arriving at destination s from origin r by time t (main problem variable) *
$E^{rs}(k)$	=	cumulative number of vehicles arriving at destination s from origin r at the beginning of interval k *
$\bar{E}^{rs}(k)$	=	cumulative number of vehicles arriving at s from r at the beginning of interval k (subproblem variable)
$U_{ap}^{rs}(t)$	=	cumulative number of vehicles entering link a on route p between O-D pair rs by time t *
$V_{ap}^{rs}(t)$	=	cumulative number of vehicles exiting link a on route p between O-D pair rs by time t *
$c_a(t)$	=	instantaneous travel time for link a at time t
$\psi_p^{rs}(t)$	=	instantaneous travel time for route p between O-D pair rs at time t
$\sigma^{rs}(t)$	=	minimal instantaneous route travel time between O-D pair rs at time t
$A(j)$	=	set of links whose tail node is j (after j)
$B(j)$	=	set of links whose head node is j (before j)
$\tau_a(t)$	=	actual travel time over link a for flows entering link a at time t
$\bar{\tau}_a(t)$	=	estimated actual travel time over link a for flows entering link a at time t
$\dot{\tau}_a(t)$	=	changing rate of actual link travel time $\tau_a(t)$ over time t
$\tilde{\tau}_{ma}(t)$	=	actual disutility function for travelers of class m entering link a at time t
$\eta_p^{rs}(t)$	=	actual travel time for route p between O-D pair rs for flows departing origin r at time t
$\eta_p^{ri}(t)$	=	actual travel time for route p between origin r and node i for flows departing origin r at time t
$\tilde{\eta}_{mp}^{rj}(t)$	=	actual travel disutility for route p from origin r to node j by class m travelers departing origin r at time t
$\pi^{rs}(t)$	=	minimal actual route travel time between O-D pair rs for flows departing origin r at time t
δ_{ap}^{rs}	=	1, if link a is on route p from r to s; = 0, otherwise

* state variable
** control variable

Redefined Notation for Stochastic Dynamic Network Models

$T_a(t)$ = perceived actual travel time over link a for flows entering link a at time t

$\tau_a(t)$ = mean actual travel time over link a for flows entering link a at time t

$\bar{\tau}_a(t)$ = estimated mean actual travel time over link a for flows entering link a at time t

$\Omega_p^{rs}(t)$ = perceived actual travel time for flows departing origin r toward destination s over route p at time t

$\eta_p^{rs}(t)$ = mean actual travel time experienced by flows departing origin r toward destination s over route p at time t

σ_o^{is} = minimal free flow travel time from node i to destination node s

θ = route choice dispersion parameter

p, q = indexes for routes

$\tau_a(k)$ = mean actual travel time over link a for flows entering link a during interval k

$\bar{\tau}_a(k)$ = mean estimated actual travel time over link a for flows entering link a during interval k

$\eta_p^{rs}(k)$ = mean actual travel time for flows departing from origin r to destination s over route p during interval k

$\pi^{ri}(k)$ = minimal mean actual travel times for flows departing from origin r toward node i during interval k

σ_o^{is} = minimal free flow travel time from node I to destination s

θ = route choice dispersion parameter

$\xi_p^{rs}(t)$ = random error of perceived travel time for route p and O-D pair rs for flows departing origin r at time t

$C_a(t)$ = perceived instantaneous travel time for link a at time t

$c_a(t)$ = mean instantaneous travel time for link a at time t

$\Psi_p^{rs}(t)$ = perceived instantaneous route travel time for route p between rs at time t

$\psi_p^{rs}(t)$ = mean instantaneous route travel time for route p between rs at time t

$\sigma^{ri}(k)$ = minimal mean instantaneous travel times for flows departing from origin r to node i during interval k

Chapter 1

Introduction

Intelligent Transportation Systems (ITS) seek to apply advanced computer, telecommunication, and information technologies to vehicles, transportation networks and operational plans, in order to relieve traffic congestion, reduce travelers' journey times, improve safety, reduce atmospheric emissions and energy consumption and increase the productivity of transportation investment. Using ITS technologies, vehicles and the infrastructure exchange vast amounts of data back and forth, making possible the warning and avoidance of congestion or hazardous conditions, the automatic collection of tolls, the efficient dispatching of trucks and buses, dramatic improvements in safety and other benefits.

Within the framework of ITS, Advanced Traveler Information Systems (ATIS) provide historical, real-time and predictive information to support travel decisions; this in turn influences the travel choices of individuals and consequently improve the time and quality of travel. Successive generations of advanced route guidance systems should improve utilization of the overall capacity of highway and transit systems so as to reduce travel times, congestion and accidents.

By providing early detection of incidents and congestion in the transportation network, route guidance systems redistribute traffic among the available modes and routes when there is excess capacity in some parts of the road network or shift the departure times of travelers to avoid peak-hour congestion when no additional road capacity is available. Furthermore, route guidance systems provide travelers with accurate, current information on both transit and road networks so that some motorists can make their own time-cost tradeoffs and shift to transit, if appropriate.

Likewise, Advanced Traffic Management Systems (ATMS) utilize these new information sources to optimize signal systems, freeway ramp controls and changeable message signs. Such an integrated traffic management and information system is the principal motivation for this book.

1.1 Requirements for Dynamic Modeling

Traditional static network equilibrium models were developed for long range transportation planning. They are not suitable for analyzing and evaluating dynamic route guidance systems which need the capability to solve transportation problems in real time (Boyce, 1989). Friesz et al (1989) analyzed some of the fundamental properties of dynamic models which are pertinent to such route guidance systems. At present, there exists little operational capability to solve large-scale, dynamic network equilibrium models corresponding to the above technological concepts in ITS. The broad goal of current research in this area is to formulate dynamic models which do correspond to the above objectives, and have some reasonable prospects for solution for large urban transportation networks on the present or next generation of computers.

This book aims to present a new generation of dynamic network equilibrium models, incorporating dynamic travel choice problems including traveler's destination choice, mode choice, departure/arrival time choice and route choice. These models are expected to be able to function as off-line dynamic travel forecasting and evaluation tools, and eventually as real-time on-line models of urban transportation networks. Research on dynamic transportation network models is evolving very rapidly; a rich set of new formulations and solution algorithms are presented in this book. The full evaluation of these new models and solution algorithms is a matter for future research. Moreover, extensions to problems of location choice remain to be tackled.

We first describe the general dynamic travel choice problem. Travelers seeking to travel from their current locations to their specified destinations, and to depart *or* arrive at specified times, require best modes, departure times and routings for their trips. These needs could be provided in accordance with one of several objectives, such as the following:

1. each driver seeks to minimize his or her *actual* travel time, given a specified departure time;

2. each driver seeks the current best route at each intersection for current traffic conditions, given a specified departure time (we refer to this type of route choice as minimizing *instantaneous* travel time);

3. each driver agrees to accept a route at a specified departure time that minimizes the travel time of all vehicles traveling during a longer time period;

4. each traveler agrees to accept a mode, destination, departure time and route that minimizes the travel time of all travelers, but makes his or her arrival time as close to a specified target as possible.

Each of these objectives corresponds conceptually to a proposed route guidance system.

In addition to ATIS, Advanced Traffic Management Systems predict traffic congestion and provides real time optimal control strategies for freeways and arterials. As one important component of ATMS, dynamic traffic control systems respond to changing traffic conditions to control ramp flows to improve the efficiency of freeways, to maintain priorities for high-occupancy vehicles and to coordinate signal timing strategies across regional arterial network.

In conjunction with ATIS and ATMS, Advanced Public Transportation Systems (APTS) are another important component of ITS. APTS applies advanced navigation, information and communication technologies that most benefit public transportation. The application of these technologies attracts travelers to transit and ridesharing modes, thereby reducing traffic congestion, atmospheric emissions and energy consumption.

Figure 1.1 shows the framework of an integrated APTS/ATMS/ATIS system. The analysis of such systems is also summarized by Kaysi et al (1993).

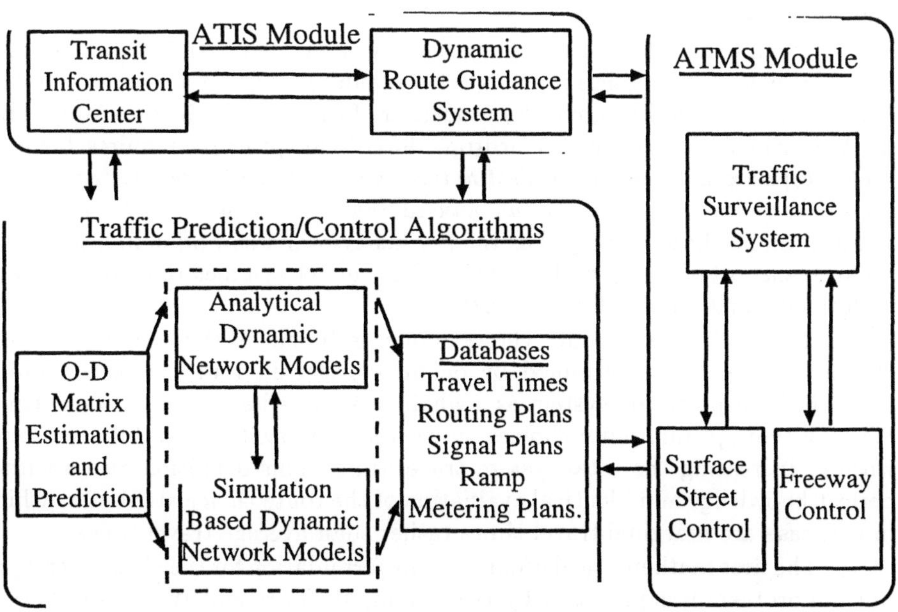

Figure 1.1: Structure of an Integrated APTS/ATMS/ATIS System

This book seeks to present our recent research findings on formulating dynamic models which bear a reasonable correspondence to the above objectives. The problems of dynamic travel choice (destination, mode, departure time, route choice, enroute diversion and parking) are described in a combined modeling framework. Using the variational inequality and optimal control theory approaches, various dynamic travel choice models are formulated for a congested transportation network. If the analysis time period is discretized, these problems reduce to nonlinear programming problems. Solution algorithms are presented for solving some of these problems. Finally, some numerical examples illustrate the properties of these models. In the following, we first analyze transportation network models in general. Then, we review the development of dynamic network models and describe the hierarchy of models in this book.

1.2 Urban Transportation Network Analysis

Existing urban transportation network models or travel choice models can be classified according to the application purpose: long term and short term. Urban travel choice modeling activities to support the planning of urban freeway and transit networks began in the mid-1950s. Initially, models were oriented towards the task of evaluating alternative land use patterns and transportation network proposals. Later, some efforts to consider the optimal design and extension of land use and transportation networks were undertaken with modest success. (Boyce et al, 1970).

Urban travel choice models are usually associated with network equilibrium concepts. The models for long-term planning purpose are usually static. Most static user equilibrium (UE) models are formulated to be consistent with the Wardrop's first principle (Wardrop, 1952). This principle requires *for used routes between a given origin-destination pair that the route cost equals the minimum route cost, and that no unused route has a lower cost.* This user equilibrium has been employed as the key behavioral assumption in most static urban transportation network models. This model assumes that drivers have perfect information about route costs.

A second type of route choice model is based on the second principle of Wardrop, and is determined by defining preferred routes as those which allow the transportation system as a whole to achieve its minimum travel time. With system-optimal route choices, no traveler can switch to a different route without increasing the total system travel time. This definition also assumes perfect knowledge and identical evaluation of the route choice attributes, which in this case are marginal travel times rather than average travel times.

The first mathematical programming formulation for the static user equilibrium problem was proposed by Beckmann, McGuire and Winsten (1956) as an equivalent optimization problem. Their formulation considered the general case in which origin-destination flows are determined by a demand function. The fixed demand case follows immediately from their result. This formulation allows the derivation of existence, uniqueness (in terms of link flows) and optimality conditions of the solution, satisfying Wardrop's UE principle. This model was studied extensively by Dafermos and Sparrow (1969) and efficiently solved by LeBlanc et al (1975) who used the minimum-cost route algorithm to implement the Frank-Wolfe algorithm. A comprehensive account of the static UE problem is given by Sheffi (1985). A more general problem with asymmetric link interactions was studied by Smith (1979) and Dafermos (1980, 1982), as have other variants of the basic formulation. The asymmetric link interactions require the model to be formulated as a variational inequality (VI) problem (Nagurney, 1993).

In the category of static user equilibrium models, there are many extensions which incorporate joint trip distribution/mode/route choice (Florian and Nguyen, 1978) and residential location (Boyce, 1980). The above deterministic models are based on the assumption that drivers have perfect information and comply with the UE travel choice criterion.

Another important extension is stochastic route choice modeling. The initial stochastic route choice model was proposed by Dial (1971) who described a flow-independent logit model for network loading. Daganzo and Sheffi (1977) formulated a stochastic user-equilibrium (SUE) route choice model which is a generalization of the UE criterion, defined as follows: *at SUE, no traveler can improve his or her perceived travel time by unilaterally changing routes.* The SUE models are more realistic than the deterministic UE models since SUE assumes drivers to have less than perfect information when choosing routes. Detailed reviews of user-equilibrium models have been prepared by Friesz (1985), Florian (1986), Boyce et al (1988), and Florian and Hearn (1995).

In addition to models for long term analyses, there are models for short term purposes, which can be termed dynamic models. Previously, most network equilibrium models focused on the static description of traffic flows on the network, implying that flows and travel times are invariant over the duration of the peak period. In response to the untenable assumption of static traffic flow over the entire peak period, several models of dynamic route choice have been proposed. These short term or dynamic models can predict large time-dependent variations of traffic in a road network and should be able to predict the travel times of vehicles during their journey. Thus they are appropriate for the assessment of the impact of ITS on highway network performance, and eventually they should be useful in managing the real time operation of ITS. A comprehensive review of dynamic models is given in Section 1.3.

In dynamic transportation networks, the traditional traveler behavior assumption of the static models needs significant revision to consider short term variations of traffic. Thus, we are no longer considering a day-to-day traffic equilibrium. Instead, we are trying to influence or control traffic and travel patterns of travelers optimally by providing accurate traffic information (such as travel time) and effective traffic control measures. Therefore, we will no longer use the term "user-equilibrium" for dynamic traffic. Instead, we use *dynamic user-optimal* (DUO) to represent the objective we seek to achieve.

In a transportation system, travelers experiment with alternative routes, listen to traffic reports on radios, and compare experiences with other travelers. Based on this evaluation, travelers settle on routes which are most preferred. The underlying assumption of DUO models is that the optimal flow rates are representative of the transportation system state attained when all travelers have determined routes which satisfy some definition of dynamic route choice. Network equilibrium analysis thus solves for these DUO conditions for a transportation network and travel demand pattern.

In an ATIS, two kinds of information are available to travelers: current and future travel time information. Current travel time information can be obtained using the currently prevailing instantaneous link travel times. Future travel time information can be obtained using predicted link travel times. We discuss these two kinds of travel times in more detail in Chapter 4. More generally, both time and cost are considered.

For a dynamic transportation network, the types of control might be classified as follows: total control; partial control; no control (Lo et al, 1994). *Total control* mandates full compliance by travelers; that is, travelers are not given any choice as to whether they would like to comply or not. Some examples include traffic signal control at intersections and ramp metering at an entrance to a freeway. *Partial control* tries to actively influence traffic patterns, but compliance is not mandatory; for example, consider route guidance advice provided by changeable message signs. By providing information on best routes to travelers, traffic patterns may be altered and more travelers can be diverted to less congested routes or time periods. It is up to the travelers themselves to decide whether to follow the advice. This kind of control is rather flexible. *No control* provides real-time traffic information to the travelers, but specific routing is not proposed for individual travelers. It is entirely up to the travelers themselves to decide what to do with the information. Current radio broadcasts on traffic congestion and accident information in metropolitan areas is a simple example of no control.

Each type of control in the above classification requires a substantial modeling effort for dynamic traffic analysis and evaluation. The development of models and algorithms for optimizing flows in real time on transportation networks is fundamental to the success of the ATMS/ATIS component of ITS. The near term objective of ATMS/ATIS is to provide accurate traffic information, mainly travel time information so that travelers may adjust their travel patterns individually. In the long run, when more and more travelers are equipped with more mature systems, the coordination of route and departure time choices will become crucial. It is in this context that some of our dynamic network models are proposed.

1.3 Overview of Dynamic Network Models

Dynamic network models can be classified as flow-based models and vehicle-based models. Flow-based models are based on macroscopic flow equations; vehicle-based models are based on microscopic movement of vehicles. Flow-based models are more applicable for large-scale transportation networks, since mathematical representations and capabilities appropriate for the corresponding dynamic network problems are more suitable.

Vehicle-based models comprise both simulation and optimization models. INTEGRATION (Van Aerde, 1992) and DYNASMART (Mahmassani et al, 1992) are examples of simulation models. More recently, Ben-Akiva et al

(1994) presented a simulation-based dynamic traffic assignment model in which some interactions with flow-based models are proposed. Ghali and Smith (1993) described a set of dynamic network models using packets to represent traffic flows on links. Their models are basically vehicle-based. Lafortune et al (1991) also presented an integer-based dynamic system-optimal (DSO) traffic assignment model.

There have been two stages of development of flow-based dynamic network models. Yagar (1971), Hurdle (1974) and Merchant and Nemhauser (1978a) were among the first to consider dynamic models for congested traffic networks. But the assumptions of these models are very limiting and they are unsuitable for application to general large-scale networks. Important breakthroughs began to occur in the late 1980s when ITS ignited the potential applicability of such models to the next generation of surface transportation systems.

The study of dynamic route choice models over a general road network was begun by Merchant and Nemhauser (1978a, 1978b) who formulated a dynamic system-optimal (DSO) route choice model for a many-to-one network. Subsequently, Carey (1987) reformulated the Merchant-Nemhauser problem as a convex nonlinear program which has analytical and computational advantages over the original formulation. DSO route choice models over a multiple origin-destination (O-D) network were established by using optimal control theory (Friesz et al, 1989; Ran and Shimazaki, 1989a). Recently, many simulation-based DSO route choice models have also been proposed by various researchers, especially for freeway corridor problems (Chang et al 1993). In related studies, Carey and Srinivasan (1993) and Carey (1995) derived system marginal costs, user perceived costs, user externality costs and a set of optimal congestion tolls on congested networks by using the Kuhn-Tucker conditions.

An important dynamic generalization of the static UE concept is called dynamic user-optimal (DUO) route choice. One dynamic user-optimal route choice problem is to determine vehicle flows at each instant of time on each link resulting from drivers using minimal-time routes. Friesz et al (1989) proposed a DUO route choice model by considering the equilibration of instantaneous route costs. Furthermore, a generalized DUO route choice model over a multiple origin-destination network was presented by Wie, Friesz and Tobin (1990). In the formulation of some dynamic models (Friesz et al, 1989; Wie, 1989; Ran and Shimazaki, 1989a), only the inflow into each link at a given time is defined as a *control variable*; the exit flow from each link is considered to be a *function* of the number of vehicles on that link. If the exit flow function is nonlinear, it is impossible to establish an optimization model of DUO route choice for a network for multiple origin-destination pairs.

By defining the exit flow as a *control variable*, Ran and Shimazaki (1989b) presented a DUO route choice model which considered the equilibration of instantaneous travel times. Subsequently, Ran, Boyce and LeBlanc (1993) formulated a set of new instantaneous DUO route choice models with flow propagation constraints. Some of the basic constraints for a dynamic network model were also discussed in Ran et al (1992a). Wie (1993, 1995) examined the dynamic mixed behavior traffic equilibrium problem by applying the optimal control theory and non-cooperative N-person nonzero-sum differential game theory. Mahmassani et al (1993) explored the dynamic user equilibrium assignment within a simulation framework. Among other dynamic network models, Janson (1991) presented a set of dynamic network models using average link travel time/flow relationships and proposed a heuristic solution algorithm. Janson and Robles (1995) also described a quasi-continuous dynamic traffic assignment model. Jayakrishnan et al (1995) examined dynamic traffic assignment by using fundamental traffic flow relationships.

Further studies concern the extension of deterministic dynamic network models to *stochastic* dynamic network models. Vythoulkas (1990) developed a logit-type stochastic dynamic route choice model for a general network. However, some of the key constraints, such as the flow propagation, are missing, which results in unrealistic traffic flows under general network conditions. Cascetta (1991) and Cascetta et al (1993) presented a dynamic stochastic route choice model for day-to-day route choices using a stochastic process theory approach. Ran et al (1992) formulated two logit-type stochastic dynamic user-optimal (SDUO) route choice models considering both instantaneous and actual travel times for general transportation networks.

Computational issues of dynamic route choice problems have received increasing attention in recent years. Following Merchant and Nemhauser's (1978b) proposal of a conceptual algorithm for solving a single-destination DSO route choice model, Ho (1980) solved the same model by successively optimizing a sequence of linear programs. Later, Ho (1990) designed a nested decomposition algorithm for the same problem and tested this algorithm on a hypercube computer. Zawack and Thompson (1987) examined the time-space network application in dynamic network model. Ran and Shimazaki (1989b) proposed a time decomposition algorithm to solve a multiple-destination DSO assignment model. Using the time-space expansion technique, Boyce et al (1991) presented a Frank-Wolfe algorithm to solve an instantaneous DUO route choice model. Codina and Barcelo (1991) also applied a time decomposition algorithm to solve a preliminary DUO route choice model. Drissi and Hameda (1992) solved a heuristic dynamic traffic assignment model on a time-space network. Janson (1993) solved a dynamic route choice problem and combined departure time/route choice problem using a heuristic based on the Frank-Wolfe algorithm. Wie et al (1994) used an augmented Lagrangian method to solve a discrete dynamic network traffic assignment problem. Wie et al (1995) also presented a discrete nested cost operator approach for solving a dynamic network user equilibrium model. Boyce et al (1995) solved an instantaneous DUO route choice model by using Frank-Wolfe algorithm on an expanded time-space network. Recently, Boyce et al (1996) implemented Janson's model on a large-scale network for the northwest suburban area of Chicago, which provided the first implementation of a dynamic network model on a large-scale network.

The choice of departure time has been addressed by several researchers, including Abkowitz (1981) and Hendrickson and Plank (1984), who developed work trip scheduling models. De Palma et al (1983) and Ben-Akiva et al (1984) modeled departure time choice over a simple network with one bottleneck using the general continuous logit model. Mahmassani and Herman (1984) used a traffic flow model to derive the equilibrium joint departure time and route choice pattern over a parallel route network. Mahmassani and Chang (1987) further developed the concept of equilibrium departure time choice and presented the boundedly-rational user equilibrium concept under which all drivers in the system are satisfied with their current travel choices, and thus feel no need to improve their outcome by changing to an alternate choice. Friesz et al (1993) formulated a simultaneous departure time/route choice model using the variational inequality approach. An overview of our modeling activity in dynamic departure time, mode and route choices was presented in LeBlanc et al (1992); in addition, papers presented at various conferences formed the basis for several chapters of this book.

1.4 Hierarchy of Dynamic Network Models

In this section, we give an overview of the chapters of the book and their interrelationships. In Chapter 2, an introduction to variational inequality and continuous optimal control theory is provided first. This background knowledge provides a basis for the formulation of dynamic network models in this book. Then, the equivalency of variational inequality and constrained optimization problem is discussed. The relaxation algorithm is presented for solving a general variational inequality. Subsequently, we summarize the basic concepts and principles of continuous time optimal control theory. This knowledge provides a basis for the formulation and solution of dynamic network models in this book. Discrete time optimal control problems are introduced in Chapter 3. Then, nonlinear programming (NLP) problems and their similarity to the discrete time optimal control problems are discussed. The conventional Frank-Wolfe algorithm is presented for solving NLP problems.

We began our dynamic network model research with the dynamic route choice problem, also known as the dynamic traffic assignment problem. Based on the two types of travel times, instantaneous link travel times and actual link travel times, two route choice models can be formulated. The relation of these dynamic route choice problems is depicted in Figure 1.2. The discussion of the two types of travel times is given in Chapter 4 together with all the constraint conditions necessary for dynamic network models. In addition to flow conservation conditions, flow propagation constraints are especially emphasized. Other important constraints include link capacity and spillback constraints.

In Chapter 5, we explore the concept of *ideal* DUO route choice behavior of travelers. The formulation of the ideal DUO route choice problem is based on the underlying choice criterion that each traveler uses the route that minimizes his/her future (actual) travel time when departing from the origin to his/her

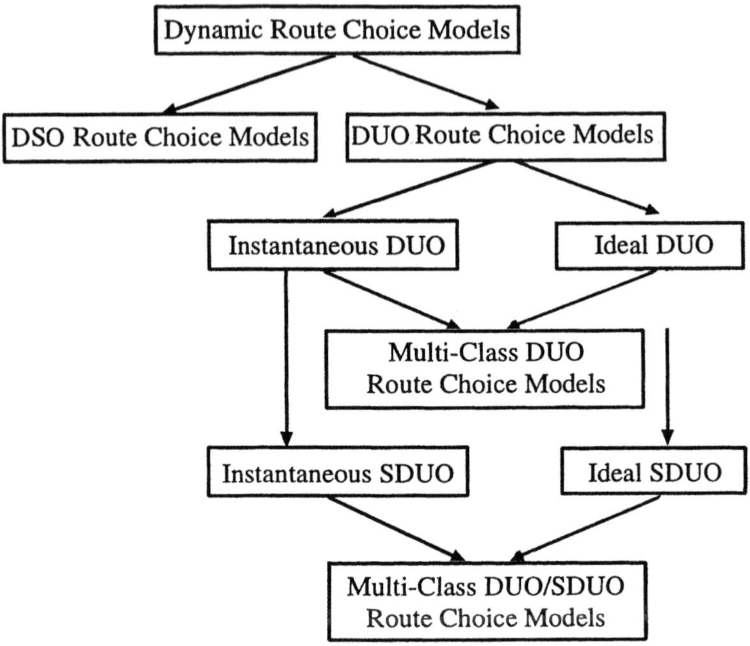

Figure 1.2: A Hierarchy of Dynamic Route Choice Models

destination. Thus, for any O-D pair, under ideal DUO route choice conditions vehicles departing the origin at the same time must arrive at the destination at the same time. The necessary and sufficiency conditions for the equivalence of the VI models to the ideal DUO conditions are proven. The link-based models are computationally more tractable than the route-based models. These models are fundamentally different from earlier DSO and DUO route choice models by:

1. using variational inequality theory;

2. employing exit flows as a set of control variables rather than functions;

3. including new flow propagation constraints.

Furthermore, multi-class route-based and link-based VI models for ideal DUO route choice problems are presented to allow more flexible capability for dynamic route choice modeling. Finally, a numerical example is provided to illustrate the ideal DUO route choice models.

In Chapter 6, the continuous time formulation of the DUO route choice program is transformed into a discrete time VI formulation. The VI model is reformulated as a discrete optimal control model under relaxation and solved using the relaxation/Frank-Wolfe algorithm. Since the model is convex, the discrete version should be efficient for applications to large networks. We present a new algorithm for solving this NLP in Chapter 6. It is solved by the Frank-Wolfe technique embedded in a relaxation procedure. In the relaxation procedure, the estimated link travel times are updated iteratively and the Frank-Wolfe technique is applied in each iteration to solve the resulting NLP. For the linearized Frank-Wolfe subproblem, an expanded time-space network is constructed so that the subproblem can be decomposed according to O-D pairs and can be viewed as a set of minimal-cost route problems. Flow propagation constraints representing the relationship between link flows and travel times are satisfied in the minimal-cost route search so that only flow conservation constraints for links and nodes remain. The proposed formulation has computational advantages since the gradient vector of the objective function with respect to the control and state variables is always nonnegative which allows for much more efficient minimal-cost route calculations. Preliminary computational results from applying the algorithm to a test network are reported.

The trip pattern is assumed to be known *a priori* in the dynamic route choice problem. We define a decision node for each route between each O-D pair as any node on that route including the origin. The instantaneous route travel time between a decision node and the destination node is calculated by using the currently prevailing link times. The instantaneous DUO route choice problem is to determine vehicle flows at each instant of time on each link resulting from drivers using minimal-time routes under currently prevailing travel times. The instantaneous DUO route choice model is formulated using the optimal control theory approach. In Chapter 7, we discuss instantaneous DUO route choice problems and formulate route-based and link-based VI models for single-class travelers. The necessity and sufficiency of the equivalence of the VI models with the instantaneous DUO route choice conditions are demonstrated. As an example, the link-based continuous time VI is transformed into a discrete time VI. The VI model is then reformulated as discrete optimal control model under relaxation and solved using the relaxation/Frank-Wolfe algorithm. Finally, a numerical example is used to illustrate the solution of the instantaneous DUO model. In Ran and Shimazaki (1989b), an instantaneous DUO route choice model that avoids the use of a link exit flow function was presented. Using the same approach, two instantaneous DUO optimal control models are presented in Chapter 8. Furthermore, multi-class VI models are presented, considering both route-based and link-based instantaneous DUO route choice conditions.

In Chapters 9 and 10, we consider two stochastic dynamic user-optimal (SDUO) route choice models which are extensions of our previous deterministic DUO route choice models. We present an ideal SDUO route choice model based on a probabilistic distribution function of future travel times perceived by drivers over alternative routes for each O-D pair. In particular, probit and logit distribution functions are discussed. In parallel, the formulation of the instantaneous SDUO route choice problem is based on the underlying choice criterion that each traveler uses the route that minimizes his/her perceived instantaneous travel time when departing from any decision node to his/her destination. The solution of this instantaneous SDUO model results in instantaneous stochastic network flows at each decision node based on a probabilistic distribution function of perceived instantaneous travel times of alternative routes. Furthermore, we use the relaxation technique, the Method of Successive Averages, and Monte Carlo sampling to solve our ideal SDUO and instantaneous SDUO route choice problems. In particular, we discuss solution algorithms (DYNASTOCH) for logit-type flow-independent ideal SDUO and instantaneous SDUO route choice problems. These algorithms are very similar to STOCH algorithm proposed by Dial (1971) for static logit-type assignment. One important advantage of these algorithms is that route enumeration is avoided.

Next, we study a joint dynamic departure time and route choice problem in Chapter 11. In this problem, travelers' departure times can be shifted based on origin-destination travel times corresponding to each possible departure time. This problem, then, is to determine travelers' departure times and choose their best routes at each instant of time. General VI formulations for joint departure time/route choice problems are presented; both route-based and link-based VI models are discussed for this joint dynamic travel choice problem. The model extends our previous DUO route choice models to the case where both departure times and routes over a general road network must be chosen. The necessity and sufficiency proofs of the variational inequalities demonstrate that our variational inequalities are consistent with the desired DUO departure time and route choice properties. As an example, the link-based continuous time VI problem is transformed into a discrete time VI formulation. The VI model is reformulated as discrete optimization problem model under relaxation and solved using the relaxation/Frank-Wolfe method. Note that the time-space network expansion should include dummy links for departure flows. A numerical example illustrates that total travel time can be decreased by choosing appropriate departure times.

In Chapter 12, a combined DUO mode/departure time/route choice model with multi-class travelers is presented for a general transportation network. Our model extends the dynamic user-optimal departure time/route choice conditions to include mode choice as well. The model extends the earlier combined departure time/route choice model to the case where dynamic traffic flows by different modes affect other modes' costs. By formally stratifying travelers into different groups, an accurate analysis of the time-cost tradeoff in mode choice is possible. This model presents a VI formulation of this DUO mode/departure time/route choice problem. By proving the necessity and sufficiency, it is demonstrated that our variational inequality is consistent with the DUO mode/departure time/route choice conditions. The hierarchy of dynamic travel choice models is summarized in Figure 1.3.

In order to implement the above dynamic network models in realistic urban transportation networks, we investigate time-dependent travel time functions for signalized arterial and freeway segment links in Chapter 13. Dynamic link travel times are first classified according to various applications. Subsequently, travel time functions for arterial links with long and short time horizons are discussed separately, and two sets of functions are recommended

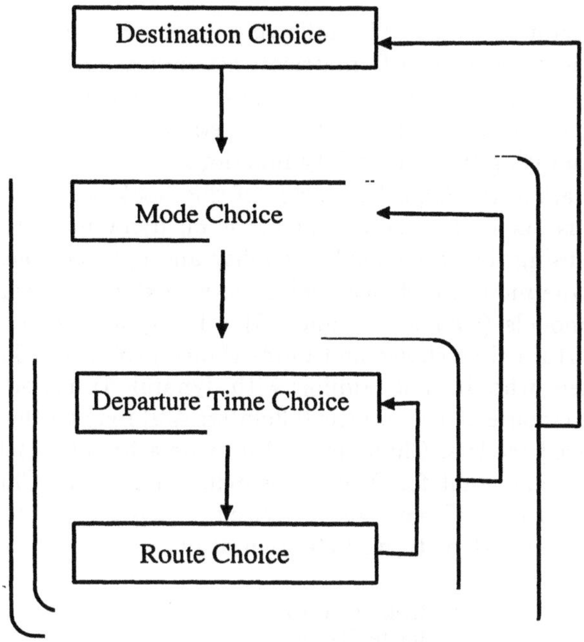

Figure 1.3: A Hierarchy of Dynamic Travel Choice Models

for dynamic transportation network problems. Implications of those functional forms are analyzed and some modifications for dynamic network models are suggested. In addition, based on dynamic link travel time functions, we discuss how many independent variables are necessary to describe the temporal traffic flow and properly estimate the time-dependent travel time over an arterial link. As a result, six link flow variables and corresponding link state equations are proposed as the basis for formulating dynamic transportation network models. Finally, time-dependent travel time functions for freeway segment links are recommended.

Several practical considerations of implementing dynamic network models are presented in Chapter 14, including rolling horizon method, travelers' knowledge of and response to traffic information, and travelers stratification. Furthermore, various implementation issues are discussed. Among those issues, we focus on the following items: dynamic traffic prediction; dynamic route guidance; integrated traffic control and information system; incident management; dynamic congestion pricing; and automated highway systems analysis; and dynamic transportation planning. The application of dynamic network models to these tasks is discussed specifically. In addition, we analyze how these models can serve operating and evaluation functions in ITS. The detailed requirements of ITS for dynamic network models are also identified. Subsequently, we discuss data needs of dynamic network models to accomplish the above implementation tasks. The following items are identified as the most important: time-dependent O-D matrices; network geometry data and intersection/ramp control data; traffic flow data for calibrating various dynamic link travel time functions; traveler information for stratifying travelers into multiple groups and calibrating travel disutility functions.

A flowchart of the logical sequence of this book is shown in Figure 1.4. Throughout this book, the main emphasis is on dynamic travel choice models formulated using the variational inequality and optimal control theory approach. For dynamic route choice problems, we extend deterministic models to stochastic models (Chapters 9 and 10). Then, we discuss more general combined departure time choice and mode choice problems (Chapters 11-12). Thus, for readers who are not familiar with dynamic transportation network models, dynamic route choice models (Chapters 4-9) may be the most interesting. For advanced readers, Chapters 9-12 provide a broader picture of models and a good starting point for future research. In general, Chapters 1-8 can be used as a classroom reference for advanced transportation network courses, especially for those with a strong ITS orientation.

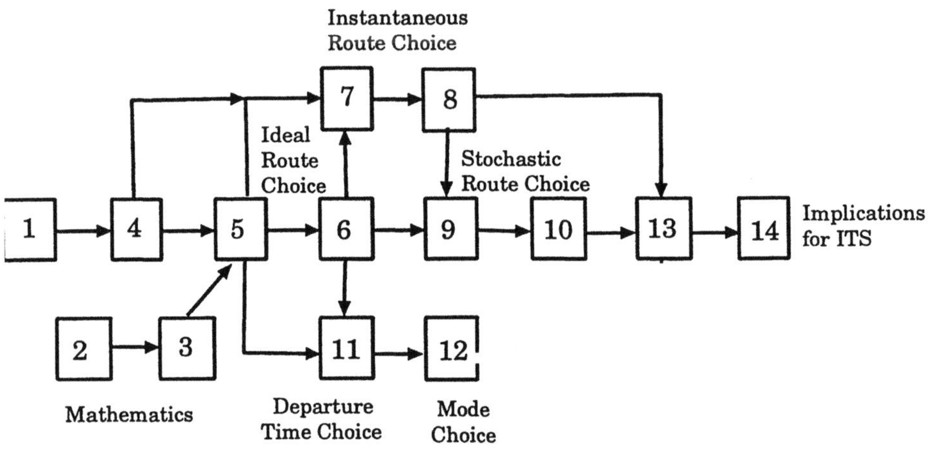

Figure 1.4: Logical Flowchart of the Chapters

1.5 Notes

The success of ATMS/ATIS depends on the development and successful applications of advanced transportation network models and control models. This field needs to develop models and algorithms that use real-time data to determine optimal control strategies of traffic. They must enable real-time management of traffic while accommodating both pre-trip planning and en route travel plan modification. They must also provide the means of evaluating the benefits of various aspects of ITS. Technologies that are likely to be useful include artificial intelligence, expert systems and parallel computing. Traffic modeling for ATMS/ATIS includes:

1. travel forecasting models;
2. optimal routing methods;
3. support systems for traffic management centers;
4. dynamic route choice models;
5. traffic simulation models;
6. network-wide optimization programs;
7. driver/traveler behavior models (human factors).

In addition, developments in traffic modeling should take into account multi-modal travel requirements, such as HOV or public transit, and should also include the development of the following capabilities:

1. real-time, traffic-adaptive logic for signal control;
2. real-time, traffic-adaptive logic for freeway control, including ramp and possibly mainline metering;
3. real-time integration of freeway and surface street control;
4. transit and emergency vehicle priorities.

Research into each of the above tasks is a long term effort. This book seeks to provide a framework and some background knowledge to study and investigate the above complicated application problems. We developed these dynamic network models bearing the above tasks in mind. Nevertheless, application of dynamic network models in ITS is still premature. The main concerns for dynamic network models which prevent their large-scale application include:

1. accurate representation of traffic propagation and travel time functions for links;
2. accurate representation of travel choice behavior;
3. validation of models using real-time data.

However, extensive potential applications of dynamic network models can be expected in future ATIS/ATMS systems, including dynamic route guidance, freeway ramp control, arterial signal optimization/coordination and automated highway, etc. We expect to summarize these applications in a subsequent book.

Problems

1.1 What are the major functional requirements of ITS for dynamic network modeling?

1.2 When generalizing static network models into dynamic network models, what are the major factors to be considered?

1.3 Static user-equilibrium is based on the assumptions of day-to-day route choice adjustment and the availability of perfect traffic information. In a time-dependent transportation network, the above assumptions do not hold any more. Therefore, is there a dynamic user-equilibrium in a dynamic transportation network? If so, what should be the corresponding conditions required to achieve the dynamic user-equilibrium state?

1.4 Compare the stock market and a dynamic traffic network. Many similarities can be found. It is generally true that the prediction of a stock market is unreliable. Could we conclude that the prediction of a dynamic traffic network is also not reliable? In other words, dynamic network models may not be able to predict the dynamic traffic flows with a reasonable accuracy. Give your thinking with regard to the above argument.

1.5 Traffic information is important for travelers to make their travel choice decisions. What kind of information can a traveler obtain to improve his/her travel choices? Should he/she obtain the information from friends and neighbors or from a Traffic Information Center?

1.6 Discuss the limitations, advantages and disadvantages of vehicle-based and flow-based dynamic network models. Also discuss how both models can be integrated.

1.7 Compare the similarities and major differences of dynamic network models and static network models.

1.8 Consider the traditional four-step travel forecasting procedure (i.e., trip generation, trip distribution, mode choice and traffic assignment) in transportation planning. What are the major difficulties and changes for each of the steps in a dynamic generalization of the procedure?

Chapter 2

Variational Inequalities and Continuous Optimal Control

Variational inequality (VI) problems are being used extensively in formulating economic and engineering problems in new ways that lead to tractable solution algorithms. Applications in transportation engineering are relatively recent, beginning about 1980. Because of its general capability to formulate transportation problems, VI has received increasing attention from transportation network modelers during the last decade. In this chapter, we introduce the basic concepts of variational inequality theory which is capable of formulating and analyzing more general problems than the constrained optimization approach. In Section 2.1, we first define variational inequality problems for both static and dynamic problems. We then introduce some fundamental definitions, along with qualitative results for variational inequality problems, such as conditions for existence and uniqueness of solutions. Subsequently, the relaxation method for solving a variational inequality is discussed.

Optimal control theory has been extensively used in solving many engineering problems, such as in mechanical and aeronautics engineering. Its application in transportation engineering has been limited to traffic signal control on surface streets and ramp metering control on freeways. Recently, with the rapid advance of supercomputing facilities and techniques, solution of optimal control theory formulations of large-scale problems has become feasible. Therefore, the application of this approach to dynamic transportation network modeling is attractive. The objective of conventional optimal control theory is to determine control strategies that cause a process to satisfy the physical constraints and at the same time minimize or maximize some performance criterion. In this book, we use optimal control theory to formulate and analyze time-dependent transportation network problems. These optimal control models have many similarities with the optimization models for solving static counterparts of these problems which are formulated and solved using nonlinear programming theory.

In this chapter, we review some basic concepts of continuous optimal control theory to provide a basis for formulating dynamic transportation network models and analyzing the conditions necessary for the existence of an optimal solution. In Section 2.2, we first present definitions which are associated with any optimal control problem. Then, we discuss optimal control problems assuming that the admissible controls are not bounded (no constraints) and derive the corresponding necessary conditions. These necessary conditions are then employed to find the optimal control law for the important linear regulator problem. Furthermore, Pontryagin's minimum principle is introduced as a generalization of the fundamental theorem of the calculus of variations.

Subsequently, optimal control problems with general bounded control and state variables are discussed. For problems with bounded control and state variables, we discuss a generalization of Pontryagin's Minimum Principle and the corresponding partial differential equations of the necessary (optimality) conditions derived using a general Hamiltonian function. Then, we discuss some optimal control problems with simple constraints on control and state variables. These optimal control problems will be used in the analysis of dynamic transportation network models in later chapters.

The main applications of both variational inequality and optimal control problems in this book concern modeling of traveler choices. Such models are essential for the operation and evaluation of large-scale Intelligent Transportation Systems. A natural way to represent the inter-relationships between an ITS and traveler behavior is a hierachical optimal control problem. This concept is introduced in Section 2.3, and illustrated by an application to congestion pricing in Chapter 14.

2.1 Variational Inequality Problems

The variational inequality problem is a general problem formulation that encompasses a set of mathematical problems, including nonlinear equations, optimization problems, complementarity problems and fixed point problems. Variational inequalities were originally developed as a tool for the study of certain classes of partial differential equations such as those that arise in mechanics. The focus of this section is on variational inequality problems suitable for the analysis of dynamic network equilibrium models.

2.1.1 Definitions

In this section, we present several types of variational inequality problems. First, we discuss the variational inequality for static problems. Here, we are dealing with a vector of decision variables $\mathbf{x} = (x_1, x_2, \cdots, x_n)$ and a vector of cost functions $\mathbf{f}(\mathbf{x}) = [f_1(\mathbf{x}), f_2(\mathbf{x}), \cdots, f_n(\mathbf{x})]$. Define G as a given closed convex set of the decision variables $\mathbf{x}$; $\mathbf{f}$ is a vector of given continuous functions defined on R^n. Then, we define the static case as follows.

Definition 2.1.1. *The infinite-dimensional variational inequality problem is to determine a vector* $\mathbf{x}^* \in G \subset R^n$, *such that*

$$\mathbf{f}[\mathbf{x}^*] \cdot [\mathbf{x} - \mathbf{x}^*] \geq 0, \qquad \forall\, \mathbf{x} \in G \tag{2.1}$$

In geometric terms, variational inequality (2.1) states that $\mathbf{f}(\mathbf{x}^*)$ is orthogonal to the feasible set G at the point $\mathbf{x}^*$.

Next, we discuss the variational inequality for dynamic problems. Unlike the static problem, we are concerned with a vector of control variables $\mathbf{u}(t) = [u_1(t), u_2(t), \cdots, u_m(t)]$ and their dynamic processes

$$\dot{\mathbf{x}}(t) = \mathbf{h}[\mathbf{x}(t), \mathbf{u}(t)]$$

where the state variables $\mathbf{x}(t) = [x_1(t), x_2(t), \cdots, x_n(t)]$ and state equations $\mathbf{h} = [h_1(t), h_2(t), \cdots, h_n(t)]$. Associated with the dynamic processes, there is a vector of cost functions $\mathbf{F}(t) = [F_1(t), F_2(t), \cdots, F_m(t)]$. Each element of the cost function vector is a function of state and control variables, i.e.,

$$F_i(t) = F_i[\mathbf{x}(t), \mathbf{u}(t)] \qquad i = 1, 2, \cdots, m$$

Since the state variables $\mathbf{x}(t)$ can be determined by the state equations when the control variables $\mathbf{u}(t)$ are given, the vector of cost functions can be further simplified as $\mathbf{F}(t) = \mathbf{F}[\mathbf{u}(t)]$. Define $G(t)$ as a given closed convex set of the control variables $\mathbf{u}(t)$. We assume $\mathbf{F}(t)$ is a set of given continuous functions from $G(t)$ to $R^n(t)$. Then, we give the following definition of the dynamic variational inequality problem.

Definition 2.1.2. *The infinite-dimensional variational inequality problem is to determine a control vector* $\mathbf{u}^*(t) \in G(t) \subset R^n(t)$, *such that*

$$\mathbf{F}[\mathbf{u}^*(t)] \cdot [\mathbf{u}(t) - \mathbf{u}^*(t)] \geq 0, \qquad \forall\, \mathbf{u}(t) \in G(t) \tag{2.2}$$

This definition facilitates the formulation of dynamic network equilibrium problems as variational inequality problems. However, the following definition is also useful where continuous time problems need to be transformed to discrete time problems and comparisons need to be made with static problems.

Definition 2.1.3. *The infinite-dimensional variational inequality problem is to determine a control vector* $\mathbf{u}^*(t) \in G(t) \subset R^n(t)$, *such that*

$$\int_0^T \mathbf{F}^T[\mathbf{u}^*(t)] \cdot [\mathbf{u}(t) - \mathbf{u}^*(t)]\, dt \geq 0, \qquad \forall\, \mathbf{u}(t) \in G(t) \tag{2.3}$$

Many dynamic transportation network equilibrium problems can be formulated as systems of equations. The systems of equations can be written as

$$\mathbf{F}[\mathbf{u}^*(t), \mathbf{x}^*(t)] = 0 \tag{2.4}$$

This problem can be regarded as a special case of a variational inequality.

We next discuss several problems related to the variational inequality. These problems include optimization problems and complementarity problems. We first discuss the relationship between an optimization problem and a variational inequality problem. In this discussion, we mainly consider the static case; the analysis can be readily extended to their dynamic counterparts.

Optimization Problems

A general optimization problem is to maximize or minimize an objective function, and in the case of a constrained problem, subject to a given set of constraints. Both unconstrained and constrained mathematical programming problems can be formulated as variational inequality problems. The following two theorems describe the relationship between an optimization problem and a variational inequality problem.

Theorem 2.1.1. *Let* $\mathbf{x}^*$ *be a solution to the minimization problem:*

$$\min_{\mathbf{x}} \quad Z(\mathbf{x}) \tag{2.5}$$

$$\text{s.t.} \quad \mathbf{x} \in G,$$

where Z *is continuously differentiable and* G *is closed and convex. Then,* $\mathbf{x}^*$ *is a solution to the variational inequality problem:*

$$\nabla Z(\mathbf{x}^*) \cdot (\mathbf{x} - \mathbf{x}^*) \geq 0, \qquad \forall\, \mathbf{x} \in G. \tag{2.6}$$

Proof: Denote an auxiliary function $Y(\alpha) = Z[\mathbf{x}^* + \alpha(\mathbf{x} - \mathbf{x}^*)]$, where $\alpha \in [0,1]$ is the decision variable. Note that $Y(\alpha)$ achieves its minimum at $\alpha = 0$, since

$$[\mathbf{x}^* + \alpha(\mathbf{x} - \mathbf{x}^*)]|_{\alpha=0} = \mathbf{x}^*$$

Thus, the derivative $dY/d\alpha$ must be nonnegative within the interval $\alpha \in [0,1]$:

$$\frac{dY}{d\alpha} = \nabla Z(\mathbf{x}^*) \cdot (\mathbf{x} - \mathbf{x}^*) \geq 0$$

Therefore, we obtain variational inequality (2.6) and $\mathbf{x}^*$ is a solution to (2.6).

Theorem 2.1.2. *If* $Z(\mathbf{x})$ *is a convex function and* $\mathbf{x}^*$ *is a solution to variational inequality (2.6), then* $\mathbf{x}^*$ *is a solution to minimization problem (2.5).*

Proof: Since $Z(\mathbf{x})$ is convex, we have

$$Z(\mathbf{x}) \geq Z(\mathbf{x}^*) + \nabla Z(\mathbf{x}^*) \cdot (\mathbf{x} - \mathbf{x}^*) \geq 0, \qquad \forall\, \mathbf{x} \in G. \tag{2.7}$$

Note that $(\nabla Z(\mathbf{x}^*)\cdot(\mathbf{x}-\mathbf{x}^*) \geq 0)$ is true because $\mathbf{x}^*$ is a solution to variational inequality (2.6). Therefore, from inequality (2.7) we have

$$Z(\mathbf{x}) \geq Z(\mathbf{x}^*), \forall\, \mathbf{x} \in G \tag{2.8}$$

In other words, $\mathbf{x}^*$ is a minimum of mathematical program (2.5).

Note that the above two theorems apply to both constrained and unconstrained optimization problems, because the feasible set G may or may not be constrained. In addition, the variational inequality problem can be formulated as an optimization problem when a certain symmetry condition holds. More specifically, if the variational inequality formulation of the optimality conditions is characterized by a function with a symmetric Jacobian, then the solution of the optimality conditions and the solution of a particular optimization problem are the same. We have the following theorem to depict this relationship.

Theorem 2.1.3. *If $\mathbf{F}(\mathbf{x})$ is a set of continuously differentiable functions on G and the Jacobian matrix*

$$\nabla F(\mathbf{x}) = \begin{bmatrix} \frac{\partial F_1}{\partial x_1} & \frac{\partial F_1}{\partial x_2} & \cdots & \frac{\partial F_1}{\partial x_n} \\ \frac{\partial F_2}{\partial x_1} & \frac{\partial F_2}{\partial x_2} & \cdots & \frac{\partial F_2}{\partial x_n} \\ \vdots & \vdots & & \vdots \\ \frac{\partial F_n}{\partial x_1} & \frac{\partial F_n}{\partial x_2} & \cdots & \frac{\partial F_n}{\partial x_n} \end{bmatrix}$$

is symmetric and positive semidefinite. Then, there is a real-valued convex function $Z(\mathbf{x})$ satisfying

$$\nabla Z(\mathbf{x}) = \mathbf{F}(\mathbf{x})$$

where $\nabla Z(\mathbf{x})$ is the gradient vector of function $Z(\mathbf{x})$. The solution $\mathbf{x}^*$ to variational inequality (2.6) is also the solution to the optimization problem:

$$\begin{aligned} \min_{\mathbf{x}} \quad & Z(\mathbf{x}) \\ \text{s.t.} \quad & \mathbf{x} \in G. \end{aligned} \tag{2.9}$$

Proof: Since $\nabla \mathbf{F}(\mathbf{x})$ is symmetric and positive semidefinite, it follows from Green's Theorem that

$$Z(\mathbf{x}) = \oint \mathbf{F}(\mathbf{x})dx$$

where $\oint$ is a line integral. Since $\mathbf{x}^*$ is the solution to variational inequality (2.6), using Theorem 2.1.2, we conclude that $\mathbf{x}^*$ is also the solution to optimization problem (2.9).

By Theorem 2.1.3, a variational inequality problem can be reformulated as a convex optimization problem only when the cost function $\mathbf{F}(\mathbf{x})$ is symmetric and positive semidefinite. Thus, the variational inequality problem encompasses the optimization problem. Therefore, the variational inequality is a more general problem in that it can also accomodate a function $\mathbf{F}(\mathbf{x})$ with an asymmetric Jacobian matrix. Historically, some static transportation network equilibrium problems which cannot be formulated as optimization problems were formulated successfully and solved as variational inequality problems. A similar observation can be made for dynamic transportation network equilibrium problems. Only certain types of dynamic problems can be formulated as optimal control problems. Many dynamic problems do not satisfy the symmetry condition and have to be formulated as variational inequalities directly.

We note that although the above theorems are only proven for static problems, they apply to dynamic problems as well. To avoid repetition, we omit the proofs. Note that in dynamic problems, the corresponding optimization problems have to be replaced by optimal control problems.

Complementarity Problems

Complementarity problems are defined on the nonnegative orthant. The nonlinear complementarity problem is a system of equations and inequalities stated as follows:

Find $\mathbf{x}^* \geq 0$ such that

$$\mathbf{f}(\mathbf{x}^*) \geq \mathbf{0} \quad \text{and} \quad \mathbf{f}^T(\mathbf{x}^*) \cdot \mathbf{x}^* = 0 \tag{2.10}$$

When $\mathbf{f}(\mathbf{x})$ is a set of linear functions, that is, $\mathbf{f}(\mathbf{x}) = \mathbf{A}\mathbf{x} + \mathbf{B}$, where $\mathbf{A}$ is an $n \times n$ matrix and $\mathbf{B}$ an $n \times 1$ vector, problem (2.10) is a linear complementarity problem. In general, the complementarity problem is a special case of the variational inequality problem. The relationship between the complementarity problem and the variational inequality problem is as follows.

Theorem 2.1.4. *The variational inequality (2.1) and the complementarity problem (2.10) have precisely the same solutions, if any solutions exist.*

Proof: First, we need to prove that if $\mathbf{x}^*$ satisfies variational inequality (2.1), then it also satisfies complementarity problem (2.10). Denote $\mathbf{e}_i$ as an n-dimensional vector with 1 in the ith location and 0 elsewhere, i.e.,

$$\mathbf{e}_i = \begin{bmatrix} 0 & \cdots & 0 & 1 & 0 & \cdots & 0 \end{bmatrix}$$

Substituting $\mathbf{x} = \mathbf{x}^* + \mathbf{e}_i$ into variational inequality (2.1), we have $f_i(\mathbf{x}^*) \geq 0$. We can choose any $\mathbf{e}_i$ with 1 at any i-th location so that each component of $\mathbf{f}_i(\mathbf{x}^*)$ is nonnegative. Thus, $\mathbf{f}(\mathbf{x}^*) \geq \mathbf{0}$.

Now substituting $\mathbf{x} = 2\mathbf{x}^*$ into variational inequality (2.1), we obtain

$$\mathbf{f}(\mathbf{x}^*) \cdot (\mathbf{x}^*) \geq 0. \tag{2.11}$$

Then, substituting $\mathbf{x} = \mathbf{0}$ into variational inequality (2.1), we obtain

$$\mathbf{f}(\mathbf{x}^*) \cdot (-\mathbf{x}^*) \geq 0. \tag{2.12}$$

Equations (2.11) and (2.12) together imply that $\mathbf{f}(\mathbf{x}^*)\cdot\mathbf{x}^* = 0$. Thus, we obtain complementarity problem (2.10).

Second, if $\mathbf{x}^*$ satisfies complementarity problem (2.10), then

$$\mathbf{f}(\mathbf{x}^*)\cdot\mathbf{x}^* = 0 \tag{2.13}$$

Since we can find any feasible $\mathbf{x} \geq \mathbf{0}$ and $\mathbf{f}(\mathbf{x}^*) \geq \mathbf{0}$, we obtain

$$\mathbf{f}(\mathbf{x}^*)\cdot\mathbf{x} \geq 0 \tag{2.14}$$

Subtracting equation (2.13) from equation (2.14), we obtain the variational inequality

$$\mathbf{f}(\mathbf{x}^*)\cdot(\mathbf{x}-\mathbf{x}^*) \geq 0 \tag{2.15}$$

2.1.2 Existence and Uniqueness

Next we discuss the existence and uniqueness of the variational inequality problem for static problems. The conclusions will also apply to variational inequalities for the dynamic problems. Existence of a solution to a variational inequality problem follows from continuity of the function $\mathbf{f}$ entering the variational inequality, provided that the feasible set G is compact. In general, we have the following existence theorem.

Theorem 2.1.5. *If G is a compact convex set and $\mathbf{f}(\mathbf{x})$ is continuous on G, then the variational inequality problem has at least one solution $\mathbf{x}^*$.*

The proof of this theorem requires the use of Brouwer's Fixed Point Theorem and is not given here (Nagurney, 1993). Qualitative properties of existence and uniqueness are easily obtained under certain monotonicity conditions. First, we present the following definitions.

Definition 2.1.4. *A vector of functions $\mathbf{f}(\mathbf{x})$ is monotone on G if*

$$[\mathbf{f}(\mathbf{x}^1)-\mathbf{f}(\mathbf{x}^2)]\cdot(\mathbf{x}^1-\mathbf{x}^2) \geq 0 \qquad \forall\, \mathbf{x}^1,\mathbf{x}^2 \in G \tag{2.16}$$

where $\mathbf{x}^1$ and $\mathbf{x}^2$ are any two points on G.

Definition 2.1.5. *A vector of functions $\mathbf{f}(\mathbf{x})$ is* strictly *monotone on G if*

$$[\mathbf{f}(\mathbf{x}^1)-\mathbf{f}(\mathbf{x}^2)]\cdot(\mathbf{x}^1-\mathbf{x}^2) > 0 \qquad \forall\, \mathbf{x}^1,\mathbf{x}^2 \in G, \qquad \mathbf{x}^1 \neq \mathbf{x}^2 \tag{2.17}$$

Definition 2.1.6. *A vector of functions $\mathbf{f}(\mathbf{x})$ is* strongly *monotone on G if for some $\alpha > 0$*

$$[\mathbf{f}(\mathbf{x}^1) - \mathbf{f}(\mathbf{x}^2)] \cdot (\mathbf{x}^1 - \mathbf{x}^2) \geq \alpha \, ||\mathbf{x}^1 - \mathbf{x}^2||^2, \quad \forall \, \mathbf{x}^1, \mathbf{x}^2 \in G \tag{2.18}$$

Assume that $\mathbf{f}(\mathbf{x})$ is continuously differentiable on G and $\nabla\mathbf{f}(\mathbf{x})$ is strongly positive definite. Then $\mathbf{f}(\mathbf{x})$ is strongly monotone. Then, we have the following theorem for uniqueness.

Theorem 2.1.6. *Suppose that* $\mathbf{f}(\mathbf{x})$ *is strictly monotone on* G. *Then, the solution is unique, if one exists.*

Next, we present some easier methods for checking the monotonicity of functions.

Theorem 2.1.7. *Suppose that* $\mathbf{f}(\mathbf{x})$ *is continuously differentiable on* G *and the Jacobian matrix*

$$\nabla f(\mathbf{x}) = \begin{bmatrix} \frac{\partial f_1}{\partial x_1} & \frac{\partial f_1}{\partial x_2} & \cdots & \frac{\partial f_1}{\partial x_n} \\ \frac{\partial f_2}{\partial x_1} & \frac{\partial f_2}{\partial x_2} & \cdots & \frac{\partial f_2}{\partial x_n} \\ \vdots & \vdots & & \vdots \\ \frac{\partial f_n}{\partial x_1} & \frac{\partial f_n}{\partial x_2} & \cdots & \frac{\partial f_n}{\partial x_n} \end{bmatrix}$$

is positive semidefinite (or positive definite), then $\mathbf{f}(\mathbf{x})$ *is monotone (or strictly monotone).*

Theorem 2.1.8. *Assume that* $\mathbf{f}$ *is continuously differentiable at some* $\bar{\mathbf{x}}$. *Then* $\mathbf{f}(x)$ *is locally strictly (or strongly) monotone at* $\bar{\mathbf{x}}$ *if* $\nabla\mathbf{f}(\bar{\mathbf{x}})$ *is positive definite (or strongly positive definite), that is,*

$$\mathbf{v}\mathbf{f}(\bar{\mathbf{x}})\mathbf{v} > 0, \qquad \forall \, \mathbf{v} \in R^n, \mathbf{v} \neq 0 \tag{2.19}$$

$$\mathbf{v}\nabla\mathbf{f}(\bar{\mathbf{x}})\mathbf{v} \geq \alpha||\mathbf{v}||^2, \qquad \text{for some} \quad \alpha > 0, \quad \forall \, \mathbf{v} \in R^n \tag{2.20}$$

where $\mathbf{v}$ *is an arbitrary vector with components of real values.*

Given the above two theorems for monotonicity, we have the following theorem for uniqueness.

Theorem 2.1.9. *Assume that* $\mathbf{f}(\mathbf{x})$ *is continuously differentiable on* G *and that* $\nabla\mathbf{f}(\mathbf{x})$ *is strongly positive definite, then* $\mathbf{f}(\mathbf{x})$ *is strongly monotone.*

The following theorem provides a condition under which both existence and uniqueness of the solution to the variational inequality problem are guaranteed. No assumption on the compactness of the feasible set G is made. This is important for very complicated dynamic problems when convexity of the feasible set is difficult to prove.

Theorem 2.1.10. *If* $\mathbf{f}(\mathbf{x})$ *is strongly monotone, then there exists precisely one solution* $\mathbf{x}^*$ *to the variational inequality (2.1).*

The proof of existence follows from the fact that strong monotonicity implies coercivity, whereas uniqueness follows from the fact that strong monotonicity implies strict monotonicity. In conclusion, in the case of an unbounded feasible set G, strong monotonicity of the function $\mathbf{f}$ guarantees both existence and uniqueness. If G is compact, then existence is guaranteed if $\mathbf{f}$ is continuous, and only the strict monotonicity condition is needed for uniqueness to be guaranteed. The first conclusion is important for some complicated dynamic problems.

2.1.3 Relaxation Algorithm

In this section, we present an iterative method for the solution of the variational inequality problem. For a static problem, the variational inequality problem is to determine a vector $\mathbf{x}^* \in G \subset R^n$, such that

$$\mathbf{f}(\mathbf{x}^*) \cdot (\mathbf{x} - \mathbf{x}^*) \geq 0, \qquad \forall\, \mathbf{x} \in G \tag{2.21}$$

where $\mathbf{f}$ is a vector of continuous functions. Assume that there exists a vector of auxiliary smooth functions $\mathbf{g}(\mathbf{x}, \mathbf{y})$ on $G \times G$ and $\mathbf{g} \in R^n$. The function $\mathbf{g}(\mathbf{x}, \mathbf{y})$ has the following properties:

1. $\mathbf{g}(\mathbf{x}, \mathbf{x}) = \mathbf{f}(\mathbf{x})$ for all $\mathbf{x}$ on G;
2. for every fixea $\mathbf{x}, \mathbf{y} \in G$, the Jacobian matrix $\nabla \mathbf{g}(\mathbf{x}, \mathbf{y})$ is symmetric and positive definite.

In other words, the decision variables $\mathbf{x}$ for the functions $\mathbf{f}$ are partitioned into two groups $\mathbf{x}$ and $\mathbf{y}$. Since the Jacobian matrix $\nabla \mathbf{g}(\mathbf{x}, \mathbf{y})$ is symmetric and positive definite, the line integral $\oint \mathbf{g}(\mathbf{x}, \mathbf{y})\, dx$ creates a new function $Z(\mathbf{x}, \mathbf{y})$ on $G \times G$ and $Z \in R$. For any fixed $\mathbf{y} \in G$, function $Z(\mathbf{x}, \mathbf{y})$ is strictly convex and

$$\mathbf{g}(\mathbf{x}, \mathbf{y}) = \nabla_{\mathbf{x}} Z(\mathbf{x}, \mathbf{y}) \tag{2.22}$$

At each iteration n, we solve the following variational inequality subproblem:

$$\mathbf{g}(\mathbf{x}^{(n)}, \mathbf{x}^{(n-1)}) \cdot (\mathbf{x} - \mathbf{x}^{(n)}) \geq 0, \qquad \forall\, \mathbf{x} \in G \tag{2.23}$$

or an equivalent mathematical programming problem:

$$\min_{\mathbf{x} \in G} \quad Z(\mathbf{x}, \mathbf{x}^{(n-1)}) \tag{2.24}$$

for which a unique solution $\mathbf{x}^{(n)}$ exists. The solution to program (2.24) may be computed using any appropriate mathematical programming algorithm. For transportation equilibrium network problems, the F-W algorithm is most widely used for such purpose. We note that a variational inequality subproblem can be constructed in various ways, some easier to solve than others.

If the sequence of solution $\mathbf{x}^{(n)}$ is convergent, i.e., $\mathbf{x}^{(n)} \to \mathbf{x}^*$ as $n \to \infty$, variational inequality subproblem (2.23) yields

$$\mathbf{f}(\mathbf{x}^*) \cdot (\mathbf{x} - \mathbf{x}^*) = \mathbf{g}(\mathbf{x}^*, \mathbf{x}^*) \cdot (\mathbf{x} - \mathbf{x}^*) \geq 0, \qquad \forall\, \mathbf{x} \in G \tag{2.25}$$

Thus, $\mathbf{x}^*$ is a solution to variational inequality problem (2.21). The relaxation method is stated as follows:

Step 0: Initialization.
Find a set of feasible decision variables $\mathbf{x}^{(n)}$. Set $n := 0$.

Step 1: Relaxation.
Solve the mathematical programming subproblem:

$$\min_{\mathbf{x} \in G} \quad Z(\mathbf{x}^{(n)}, \mathbf{x}^{(n-1)}) \tag{2.26}$$

obtaining solution $\mathbf{x}^{(n)}$.

Step 2: Convergence Test.
If $|\mathbf{x}^{(n)} - \mathbf{x}^{(n-1)}| \leq \epsilon$, for a prespecified small value ϵ, then stop. Otherwise, set $n := n + 1$, and go to Step 1.

We note that in the relaxation method, each component $g_i(\mathbf{x}, \mathbf{y})$ of function $\mathbf{g}(\mathbf{x}, \mathbf{y})$ should correspond to the relaxation of variable x_i

$$g_i(\mathbf{x}, \mathbf{y}) = f_i(y_1, \cdots, y_{i-1}, x_i, y_{i+1}, \cdots, y_n) \qquad i = 1, \cdots, n \tag{2.27}$$

This relaxation method is sometimes termed the *diagonalization method* because the Hessian matrix of Z is diagonal, since all cross-link effects have been fixed. The mathematical programming subproblem is also known as the *diagonalized* problem. Therefore, the relaxation method resolves variational inequality (2.21) into a sequence of variational inequality subproblems (2.23) or mathematical programming subproblems (2.24).

2.2 Continuous Optimal Control Problems

The basic theory of optimal control is derived from the calculus of variations. Compared with the calculus of variations, optimal control theory has many advantages in solving time-dependent optimization problems. It also has many similarities with static optimization problems, such as nonlinear programming theory. We discuss these relationships in more detail in Chapter 3.

2.2.1 Definitions

Before presenting the optimality conditions for the optimal control problem, we introduce some basic definitions. The formulation of an optimal control problem is associated with a time-dependent process or system and requires the following:

1. a mathematical representation of the process to be controlled or to be optimized;

2. a statement of the physical constraints;

3. specification of a performance criterion or an objective function.

Dynamic Process/Dynamic System

The modeling of a dynamic process seeks to obtain the simplest mathematical description that adequately predicts the response of the physical system to all anticipated inputs. We restrict our discussion to systems described by ordinary differential equations. We define

$$x_1(t), x_2(t), \cdots, x_n(t)$$

as the *state variables* (or simply the *states*) of the process at time t, and

$$u_1(t), u_2(t), \cdots, u_m(t)$$

as the *control variables* (or the *controls*) to the process at time t. The system or process may be described by n first-order differential equations, where $\dot{x}_i(t) = dx_i(t)/dt$:

$$\begin{aligned}
\dot{x}_1(t) &= f_1[x_1(t), x_2(t), \cdots, x_n(t), u_1(t), u_2(t), \cdots, u_m(t), t] \\
\dot{x}_2(t) &= f_2[x_1(t), x_2(t), \cdots, x_n(t), u_1(t), u_2(t), \cdots, u_m(t), t] \\
&\vdots \\
\dot{x}_n(t) &= f_n[x_1(t), x_2(t), \cdots, x_n(t), u_1(t), u_2(t), \cdots, u_m(t), t]
\end{aligned} \tag{2.28}$$

We define the *state vector* of the system as $\mathbf{x}(t) = [x_1(t), x_2(t), \cdots, x_n(t)]$, and the *control vector* as $\mathbf{u}(t) = [u_1(t), u_2(t), \cdots, u_m(t)]$. The state equations can then be written in vector notation as

$$\dot{\mathbf{x}}(t) = \mathbf{f}[\mathbf{x}(t), \mathbf{u}(t), t] \tag{2.29}$$

where $\mathbf{f}$ is also a vector.

State Variables

The starting point for optimal control investigations is a mathematical model in state variable form. Why are state variables used in control problems? Formulating a mathematical model in state variable form is convenient because:

1. the concept of state has a strong physical motivation;
2. the state variable form is easy to use in theoretical investigations and the resulting differential equations are suitable for digital or analog solution;
3. the state form provides a unified framework for the study of nonlinear and linear systems.

Referring to the state of a system, we have the following definition:

Definition 2.2.1 *The state of a system is a set of quantities $x_1(t)$, $x_2(t)$, $\cdots$, $x_n(t)$ which, if known at $t = 0$, are determined for $t \geq 0$ by specifying the inputs to the system for $t \geq 0$.*

System Classification

Systems are described by the terms linear, nonlinear, time-invariant and time-variant (explicit function of time). We shall classify systems according to the form of their state equations. For example, if a system is nonlinear and time-variant, the state equations are written as

$$\dot{\mathbf{x}}(t) = \mathbf{f}[\mathbf{x}(t), \mathbf{u}(t), t] \tag{2.30}$$

When the system does not depend on time t explicitly, a special form of the above nonlinear dynamic systems is represented by equations of the form

$$\dot{\mathbf{x}}(t) = \mathbf{f}[\mathbf{x}(t), \mathbf{u}(t)] \tag{2.31}$$

which is time-invariant. Many dynamic transportation network problems have time-invariant state equations. If a system is linear and time-invariant, its state equations are

$$\dot{\mathbf{x}}(t) = \mathbf{A}(t)\mathbf{x}(t) + \mathbf{B}(t)\mathbf{u}(t) \tag{2.32}$$

where $\mathbf{A}(t)$ and $\mathbf{B}(t)$ are $n \times n$ and $n \times m$ matrices with time-dependent elements. State equations for one special type of linear dynamic systems have the form

$$\dot{\mathbf{x}}(t) = \mathbf{A}\mathbf{x}(t) + \mathbf{B}\mathbf{u}(t) \tag{2.33}$$

where $\mathbf{A}$ and $\mathbf{B}$ are constant matrices.

Performance Measure/Objective Function

In order to evaluate the performance of a system quantitatively, the system controller needs to select a performance measure or an objective function. An optimal control is defined as one that minimizes (or maximizes) this objective function. In certain cases, the objective function may be self-explanatory, whereas in other problems the objective function may be artificial. For example, the statement,"To minimize the total travel time over a traffic network," clearly indicates that the objective function is defined from a system point of view. On the other hand, the statement, "To achieve equal minimal travel times over each used route for each O-D pair," does not directly suggest an objective function. In such a problem, the modeler may need to design an artificial objective function so that the resulting optimization conditions of the optimal control program represent desired physical properties. However, such an objective function may not have any direct physical interpretation. We shall discuss the definition of an objective function in more detail in the following chapters for various dynamic network problems.

In the following, we assume that the *performance* of a system can be evaluated by a measure J of the form

$$J = \int_0^T F[\mathbf{x}(t), \mathbf{u}(t), t]\, dt + S[\mathbf{x}(T), T] \tag{2.34}$$

where 0 and T are initial and final times; F and S are scalar functions. T may be pre-specified or changeable according to the problem requirement. If T is changeable, it is termed "free". F is a cost term that depends on the state and control variables at any time t. S is associated with final state $\mathbf{x}(T)$ and the final time T. It is sometimes termed a salvage cost in economics.

Starting from the initial state $x(0) = x_0$, assigning values to control variables $u(t)$ for $t \in [0, T]$ will cause a system to achieve some value for each state variable. The sequence of values achieved by each state variable through time is termed a *state trajectory*. Minimization of the objective function with respect to the control variables assigns unique real values to each trajectory of the system so that some optimal controls can be found.

We now present an explicit statement of the *optimal control problem* (OCP). First, some definitions of various forms of optimal controls are presented.

Definition 2.2.2 *If a functional relationship of the form*

$$\mathbf{u}^*(t) = \mathbf{f}[\mathbf{x}(t), t] \tag{2.35}$$

can be found for the optimal control at each instant of time t, then the function **f** *is called an optimal control law.*

Notice that equation (2.35) implies that **f** is a rule which determines the optimal control at time t for any admissible state value at time t. For example, if

$$\mathbf{u}^*(t) = \mathbf{G}\,\mathbf{x}(t) \tag{2.36}$$

where **G** is an $m \times n$ matrix of real constants, then we say that the optimal control law is linear and time-invariant.

Definition 2.2.3 *If the optimal control is determined as a function of time for a specified initial state value, i.e.,*

$$\mathbf{u}^*(t) = \mathbf{e}[\mathbf{x}(0), t] \tag{2.37}$$

then the optimal control is said to be in open-loop form.

Thus, the optimal open-loop control is optimal only for a particular initial state value. In other words, if the optimal control law is known, the optimal control history starting from any initial state value can be generated.

Conceptually, it is helpful to imagine the difference between a closed-loop optimal control law and an open-loop optimal control law as shown in Figure 2.1. Although engineers normally prefer closed-loop solutions to optimal control problems, there are cases when an open-loop control may be desirable. For example, in radar tracking of a satellite, once the orbit is set, very little can happen to cause an undesired change in the trajectory parameters. In this situation a pre-programmed control for the radar antenna might well be used. In dynamic transportation network analysis, it is hard to find an example of an open-loop control. Thus, open-loop optimal control is not our major focus.

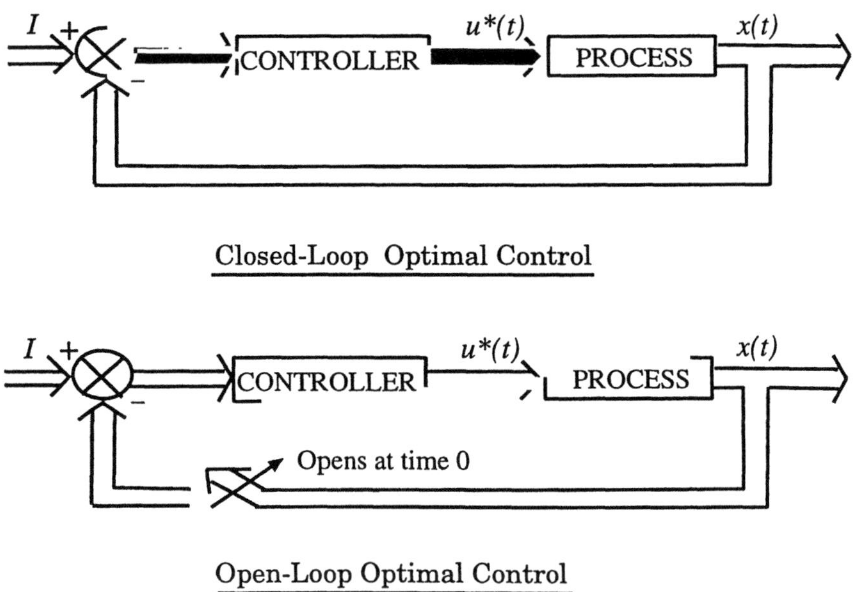

Figure 2.1: Closed-Loop and Open-Loop Optimal Controls

Definition 2.2.4 *If the optimal control is determined as a function of the state variable only, i.e.,*

$$\mathbf{u}^*(t) = \mathbf{f}[\mathbf{x}(t)] \tag{2.38}$$

then the optimal control is an optimal feedback control.

A typical example of feedback control is the classic servomechanism problem where actual and desired outputs are compared and any deviation produces a control signal that attempts to reduce the discrepancy to zero. In dynamic transportation network problems, there are many applications of such a control law.

With the above basic definitions in mind, we begin our discussion of optimal control problems. First, we present the simplest form of optimal control problem which has no constraints on state and control variables.

2.2.2 No Constraints

In this section, we mainly present an optimal control problem with a fixed end time and no constraints. We now consider a dynamic system having a cost function $F[\mathbf{x}(t), \mathbf{u}(t), t]$ for a fixed time period $[0, T]$. The optimal control problem is to seek a control function $\mathbf{u}$ which is feasible, or admissible, in order to minimize the objective function as follows:

$$\min_{\mathbf{x},\mathbf{u}} \ J = \int_0^T F[\mathbf{x}(t), \mathbf{u}(t), t]\, dt + S[\mathbf{x}(T), T] \tag{2.39}$$

s.t.

$$\dot{\mathbf{x}}(t) = \mathbf{f}[\mathbf{x}(t), \mathbf{u}(t), t] \tag{2.40}$$

$$T \quad \text{and} \quad \mathbf{x}(0) = \mathbf{x}_o \ \text{fixed}; \quad \mathbf{x}(T) \ \text{free} \tag{2.41}$$

where $F[\mathbf{x}(t), \mathbf{u}(t), t]$ has continuous partial derivatives with respect to $\mathbf{x}(t)$ and $S[\mathbf{x}(T), T]$ has continuous partial derivatives with respect to $\mathbf{x}(T)$.

We use the method of Lagrange multipliers to adjoin the system differential state equations to the objective function, which gives

$$\begin{aligned} I &= \int_0^T \{F[\mathbf{x}(t), \mathbf{u}(t), t] + \boldsymbol{\lambda}(t)\ [\mathbf{f}[\mathbf{x}(t), \mathbf{u}(t), t] - \dot{\mathbf{x}}(t)]\}\ dt \\ &+ S[\mathbf{x}(T), T] \end{aligned} \tag{2.42}$$

where $\boldsymbol{\lambda}(t)$ is a vector of Lagrange multipliers associated with the dynamic state equations. It is standard in optimal control theory to designate the major part of the integrand of equation (2.42) as the Hamiltonian,

$$\mathcal{H} = F[\mathbf{x}(t), \mathbf{u}(t), t] + \boldsymbol{\lambda}(t)\, \mathbf{f}[\mathbf{x}(t), \mathbf{u}(t), t] \tag{2.43}$$

Then, the first-order necessary conditions for the optimal control problem may be summarized as follows:

$$\frac{\partial \mathcal{H}}{\partial \mathbf{u}(t)} = \frac{\partial F[\mathbf{x}(t), \mathbf{u}(t), t]}{\partial \mathbf{u}(t)} + \boldsymbol{\lambda}(t) \frac{\partial \mathbf{f}[\mathbf{x}(t), \mathbf{u}(t), t]}{\partial \mathbf{u}(t)} = \mathbf{0} \tag{2.44}$$

$$-\dot{\boldsymbol{\lambda}}(t) = \frac{\partial \mathcal{H}}{\partial \mathbf{x}(t)} = \frac{\partial F[\mathbf{x}(t), \mathbf{u}(t), t]}{\partial \mathbf{x}(t)} + \boldsymbol{\lambda}(t) \frac{\partial \mathbf{f}[\mathbf{x}(t), \mathbf{u}(t), t]}{\partial \mathbf{x}(t)} \tag{2.45}$$

$$\dot{\mathbf{x}}(t) = \frac{\partial \mathcal{H}}{\partial \boldsymbol{\lambda}(t)} = \mathbf{f}[\mathbf{x}(t), \mathbf{u}(t), t] \tag{2.46}$$

$$\mathbf{x}(0) = \mathbf{x}_0 \tag{2.47}$$

$$\boldsymbol{\lambda}(T) = \frac{\partial S[\mathbf{x}(T), T]}{\partial \mathbf{x}(T)} \tag{2.48}$$

Equations (2.46) and (2.47) are simply the restatement of state equation (2.40) and boundary condition (2.41). Equations (2.47) and (2.48) are termed the transversality conditions which constitute two-point boundary conditions for the set of differential equations. The above first-order necessary conditions are derived using Pontryagin's Minimum (or Maximum) Principle or the Hamilton-Jacobi equation. The Pontryagin's Minimum Principle can simply be stated as follows. The inequality

$$\mathcal{H}[\mathbf{u}^*(t)] \leq \mathcal{H}[\mathbf{u}(t)] \tag{2.49}$$

is valid for all admissible $\mathbf{u}(t)$, where $*$ represents that the solution is optimal. Sometimes, we directly use this principle to analyze complicated optimal control problems. It is also interesting to compute the total derivative with respect to time t as

$$\begin{aligned} \frac{d\mathcal{H}}{dt} &= \frac{\partial F}{\partial t} + \dot{\mathbf{x}}(t)\left[\frac{\partial F}{\partial \mathbf{x}(t)} + \boldsymbol{\lambda}(t)\frac{\partial \mathbf{f}}{\partial \mathbf{x}(t)}\right] + \dot{\mathbf{u}}(t)\left[\frac{\partial F}{\partial \mathbf{u}(t)} + \boldsymbol{\lambda}(t)\frac{\partial \mathbf{f}}{\partial \mathbf{u}(t)}\right] \\ &+ \dot{\boldsymbol{\lambda}}(t)\,\mathbf{f} + \boldsymbol{\lambda}(t)\frac{\partial \mathbf{f}}{\partial t} \end{aligned} \tag{2.50}$$

Now, we consider that equation (2.44) may or may not hold. Substituting equations (2.44) and (2.46) into equation (2.50), we obtain

$$\frac{d\mathcal{H}}{dt} = \frac{\partial F}{\partial t} + \dot{\mathbf{u}}(t)\frac{\partial \mathcal{H}}{\partial \mathbf{u}(t)} + \boldsymbol{\lambda}(t)\frac{\partial \mathbf{f}}{\partial t} \tag{2.51}$$

Thus, if F and f are not explicit functions of time t, the Hamiltonian is constant along an optimal trajectory where $\partial\mathcal{H}/\partial\mathbf{u} = 0$. It can be shown that this is true along an optimal trajectory, even if we cannot require $\partial\mathcal{H}/\partial\mathbf{u} = 0$. Therefore, the following additional condition holds

$$\frac{d\mathcal{H}}{dt} = 0 \tag{2.52}$$

when F and f are not explicit functions of time. This is an important result which we will use in later developments. In the following, we discuss a special problem using the above first-order necessary conditions.

The Linear Regulator

We now study a particular optimal control problem which has its solution as a linear feedback control law. The typical problem is stated as follows. We have a linear differential system

$$\dot{\mathbf{x}}(t) = \mathbf{A}(t)\,\mathbf{x}(t) + \mathbf{B}(t)\,\mathbf{u}(t), \qquad \mathbf{x}(0) = \mathbf{x}_0 \tag{2.53}$$

and wish to find the control which minimizes the objective function

$$\mathbf{J} = \frac{1}{2}\int_0^T [\mathbf{x}(t)\,\mathbf{Q}(t)\,\mathbf{x}(t) + \mathbf{u}(t)\,\mathbf{R}(t)\,\mathbf{u}(t)]\,dt + \frac{1}{2}\mathbf{x}(T)\,\mathbf{W}\,\mathbf{x}(T) \tag{2.54}$$

Without loss of generality, we assume $\mathbf{Q}(t)$, $\mathbf{R}(t)$ and $\mathbf{W}$ are symmetric matrices. We may solve this problem using the first-order necessary conditions or the Minimum Principle. The Hamiltonian is

$$\begin{aligned}\mathcal{H}[\mathbf{x}(t), \mathbf{u}(t), \boldsymbol{\lambda}(t), t] &= \frac{1}{2}\mathbf{x}(t)\,\mathbf{Q}(t)\,\mathbf{x}(t) + \frac{1}{2}\mathbf{u}(t)\,\mathbf{R}(t)\,\mathbf{u}(t) \\ &+ \boldsymbol{\lambda}(t)\,\mathbf{A}(t)\,\mathbf{x}(t) + \boldsymbol{\lambda}(t)\,\mathbf{B}(t)\,\mathbf{u}(t) \end{aligned} \tag{2.55}$$

Application of the first-order necessary conditions requires, for an optimal control, that

$$\frac{\partial \mathcal{H}}{\partial \mathbf{u}(t)} = \mathbf{0} = \mathbf{R}(t)\,\mathbf{u}(t) + \boldsymbol{\lambda}(t)\,\mathbf{B}(t) \tag{2.56}$$

and

$$\frac{\partial \mathcal{H}}{\partial \mathbf{x}(t)} = -\dot{\boldsymbol{\lambda}}(t) = \mathbf{Q}(t)\,\mathbf{x}(t) + \boldsymbol{\lambda}(t)\,\mathbf{A}(t) \tag{2.57}$$

with the boundary condition

$$\boldsymbol{\lambda}(T) = \frac{\partial \mathbf{S}[\mathbf{x}(T)]}{\partial \mathbf{x}(T)} = \mathbf{W}\,\mathbf{x}(T) \tag{2.58}$$

Thus, from equation (2.56), we require that

$$\mathbf{u}(t) = -\mathbf{R}^{-1}(t)\,\mathbf{B}(t)\,\boldsymbol{\lambda}(t) \tag{2.59}$$

We shall inquire whether we can convert this to a closed-loop control by assuming that the solution for the Lagrange multiplier is similar to equation (2.58)

$$\boldsymbol{\lambda}(t) = \mathbf{P}(t)\,\mathbf{x}(t) \tag{2.60}$$

where $\mathbf{P}(t)$ is a symmetric matrix, with $n \times n$ time-dependent elements. If we substitute equations (2.60) and (2.59) into equations (2.53), we must require

$$\dot{\mathbf{x}}(t) = \mathbf{A}(t)\,\mathbf{x}(t) - \mathbf{B}(t)\,\mathbf{R}^{-1}(t)\,\mathbf{B}(t)\,\mathbf{P}(t)\,\mathbf{x}(t) \tag{2.61}$$

Also, from equations (2.60) and (2.57), we require that

$$\dot{\boldsymbol{\lambda}}(t) = \dot{\mathbf{P}}(t)\,\mathbf{x}(t) + \mathbf{P}(t)\,\dot{\mathbf{x}}(t) = -\mathbf{Q}(t)\,\mathbf{x}(t) - \mathbf{A}(t)\,\mathbf{P}(t)\,\mathbf{x}(t) \tag{2.62}$$

Combining equations (2.61) and (2.62), we obtain

$$\begin{aligned} &\Big[\dot{\mathbf{P}}(t) + \mathbf{P}(t)\,\mathbf{A}(t) + \mathbf{A}(t)'\,\mathbf{P}(t) \\ - \;&\mathbf{P}(t)\,\mathbf{B}(t)\,\mathbf{R}^{-1}(t)\,\mathbf{B}(t)\,\mathbf{P}(t) + \mathbf{Q}(t)\Big]\;\mathbf{x}(t) = \mathbf{0} \end{aligned} \tag{2.63}$$

Since this equation must hold for all nonzero $\mathbf{x}(t)$, the term premultiplying $\mathbf{x}(t)$ must be zero. Thus, the $\mathbf{P}(t)$ matrix, which must satisfy equation (2.63), is called the Riccati matrix equation, and must be positive definite, so that

$$\dot{\mathbf{P}}(t) = -\mathbf{P}(t)\mathbf{A}(t) - \mathbf{A}(t)\mathbf{P}(t) + \mathbf{P}(t)\mathbf{B}(t)\mathbf{R}^{-1}(t)\mathbf{B}(t)\mathbf{P}(t) - \mathbf{Q}(t) \tag{2.64}$$

with a boundary condition given by equations (2.58) and (2.60), stated as

$$\mathbf{P}(T) = \mathbf{W} \tag{2.65}$$

Thus, we may solve the Riccati matrix equation backward in time from T to 0 and obtain $\mathbf{P}(t)$. The solution for $\mathbf{P}(t)$ will be illustrated in Example 2.2.1 below. Define the matrix $\mathbf{K}(t)$ as

$$\mathbf{K}(t) = -\mathbf{R}^{-1}(t)\,\mathbf{B}(t)\,\mathbf{P}(t) \tag{2.66}$$

Combining equations (2.59), (2.60) and (2.66), we obtain a closed-loop control of the form

$$\mathbf{u}(t) = \mathbf{K}(t)\,\mathbf{x}(t) \tag{2.67}$$

$\mathbf{Q}(t)$, $\mathbf{R}(t)$, and $\mathbf{W}$ must be at least positive semidefinite in order to establish the sufficient condition for a minimum. In addition, we know from equation (2.59) that $\mathbf{R}(t)$ must have an inverse; therefore, it is sufficient that $\mathbf{R}(t)$ be positive definite and the $\mathbf{Q}(t)$ and $\mathbf{W}$ be at least positive semidefinite. A detailed discussion on sufficient conditions is not given in this book. Interested readers should refer to the notes at the end of this chapter.

Example. Find the feedback optimal control law for the scalar system

$$\dot{x} = x(t) + u(t), \qquad x(0) = x_0 \tag{2.68}$$

to minimize the objective function

$$J = \frac{1}{2}\int_0^T u^2(t)dt + 2\,x^2(T) \tag{2.69}$$

The state and control variables are unconstrained. The final time T is specified and $x(T)$ is free. Riccati equation (2.64) and boundary condition (2.65) become

$$\dot{p}(t) = -2p(t) + p^2(t), \qquad p(T) = 4 \tag{2.70}$$

Solving the above differential equation, we obtain the solution

$$p(t) = \frac{2}{1 - e^{(2t-2T-\ln 2)}} \tag{2.71}$$

Thus, using equation (2.66), we have

$$K(t) = -R^{-1}(t)\ B(t)\ P(t) = -\frac{2}{1 - e^{(2t-2T-\ln 2)}} \tag{2.72}$$

The optimal feedback control law is

$$u(t) = K(t)\ x(t) \tag{2.73}$$

2.2.3 Equality and Inequality Constraints

In this section, we consider optimal control problems with nonlinear equality and inequality constraints on state and control variables. Those constraints can represent most practical constraints in realistic applications. The set of equality constraints for control and state variables are denoted as $\mathbf{G}[\mathbf{u}(t), \mathbf{x}(t), t] = \mathbf{0}$ and the set of inequality constraints are denoted as $\mathbf{K}[\mathbf{u}(t), \mathbf{x}(t), t] \geq \mathbf{0}$. We assume $\mathbf{G}$ and $\mathbf{K}$ are continuous and differentiable with respect to $\mathbf{u}$, $\mathbf{x}$ and t.

We first consider an optimal control problem with a fixed beginning $\mathbf{x}_0$ and fixed end time T. The optimal control problem is formulated as follows

$$\min_{\mathbf{x},\mathbf{u}} \quad J = \int_0^T F[\mathbf{x}(t), \mathbf{u}(t), t]dt + S[\mathbf{x}(T), T] \tag{2.74}$$

s.t.

$$\frac{d\mathbf{x}(t)}{dt} = \mathbf{f}[\mathbf{u}(t), \mathbf{x}(t), t] \tag{2.75}$$

$$\mathbf{G}[\mathbf{u}(t), \mathbf{x}(t), t] = \mathbf{0} \tag{2.76}$$

$$\mathbf{K}[\mathbf{u}(t), \mathbf{x}(t), t] \leq \mathbf{0} \tag{2.77}$$

$$\mathbf{x}(0) = \mathbf{x}_0, \qquad T \text{ fixed.} \tag{2.78}$$

Denote the Lagrangian multipliers associated with equations (2.75)-(2.77) as $\boldsymbol{\lambda}(t)$, $\boldsymbol{\sigma}(t)$ and $\boldsymbol{\eta}(t)$, respectively. We construct the augmented Hamiltonian $\mathcal{H}$ for the above optimal control problem as

$$\begin{aligned} \mathcal{H} &= F[\mathbf{x}(t), \mathbf{u}(t), t] + \boldsymbol{\lambda}(t)\ \mathbf{f}[\mathbf{u}(t), \mathbf{x}(t), t] \\ &+ \boldsymbol{\sigma}(t)\ \mathbf{G}[\mathbf{u}(t), \mathbf{x}(t), t] + \boldsymbol{\eta}(t)\ \mathbf{K}[\mathbf{u}(t), \mathbf{x}(t), t] \end{aligned} \tag{2.79}$$

where

$$\boldsymbol{\eta}(t) \begin{cases} > \mathbf{0} & \text{if } \mathbf{K} = \mathbf{0} \\ = \mathbf{0} & \text{if } \mathbf{K} < \mathbf{0} \end{cases}$$

The first order necessary conditions for the optimal control program are

$$\begin{aligned}\frac{\partial \mathcal{H}}{\partial \mathbf{u}(t)} = 0 &= \frac{\partial F[\mathbf{x}(t),\mathbf{u}(t),t]}{\partial \mathbf{u}(t)} + \boldsymbol{\lambda}(t)\frac{\partial \mathbf{f}[\mathbf{u}(t),\mathbf{x}(t),t]}{\partial \mathbf{u}(t)} \\ &+ \boldsymbol{\sigma}(t)\frac{\partial \mathbf{G}[\mathbf{u}(t),\mathbf{x}(t),t]}{\partial \mathbf{u}(t)} + \boldsymbol{\eta}(t)\frac{\partial \mathbf{K}[\mathbf{u}(t),\mathbf{x}(t),t]}{\partial \mathbf{u}(t)} \end{aligned} \tag{2.80}$$

$$\begin{aligned}-\dot{\boldsymbol{\lambda}}(t) = \frac{\partial \mathcal{H}}{\partial \mathbf{x}(t)} &= \frac{\partial F[\mathbf{x}(t),\mathbf{u}(t),t]}{\partial \mathbf{x}(t)} + \boldsymbol{\lambda}(t)\frac{\partial \mathbf{f}[\mathbf{x}(t),\mathbf{u}(t),t]}{\partial \mathbf{x}(t)} \\ &+ \boldsymbol{\sigma}(t)\frac{\partial \mathbf{G}[\mathbf{u}(t),\mathbf{x}(t),t]}{\partial \mathbf{x}(t)} + \boldsymbol{\eta}(t)\frac{\partial \mathbf{K}[\mathbf{u}(t),\mathbf{x}(t),t]}{\partial \mathbf{x}(t)} \end{aligned} \tag{2.81}$$

$$\dot{\mathbf{x}}(t) = \frac{\partial \mathcal{H}}{\partial \boldsymbol{\lambda}(t)} = \mathbf{f}[\mathbf{x}(t),\mathbf{u}(t),t] \tag{2.82}$$

$$\frac{\partial \mathcal{H}}{\partial \boldsymbol{\sigma}(t)} = \mathbf{0} = \mathbf{G}[\mathbf{u}(t),\mathbf{x}(t),t] \tag{2.83}$$

$$\frac{\partial \mathcal{H}}{\partial \boldsymbol{\eta}(t)} = \mathbf{K}[\mathbf{u}(t),\mathbf{x}(t),t] \geq \mathbf{0} \tag{2.84}$$

$$\mathbf{x}(0) = \mathbf{x}_0 \tag{2.85}$$

$$\boldsymbol{\lambda}(T) = \frac{\partial S[\mathbf{x}(T),T]}{\partial \mathbf{x}(T)} \tag{2.86}$$

Equations (2.80)-(2.81) are similar to equations (2.44) and (2.45) except the additional terms resulting from equality and inequality constraints on the control and state variables. Equation (2.82) is a restatement of the state equation (2.75). Equations (2.83)-(2.84) are restatements of the equality and inequality constraints. Equations (2.85)-(2.86) are two-point boundary conditions.

2.2.4 Equality and Nonnegativity Constraints

We now consider a set of special optimal control problems with a fixed end time T. These problems will be widely used in the following chapters for problem formulation and analysis for dynamic transportation networks. The state equations are linear

$$\frac{d\mathbf{x}(t)}{dt} = \mathbf{A}\,\mathbf{u}(t) \tag{2.87}$$

where $\mathbf{A}$ is a matrix of constants. We consider only linear equality constraints for control variables

$$\mathbf{G}\,\mathbf{u}(t) = \mathbf{0} \tag{2.88}$$

and nonlinear equality constraints involving only state variables

$$\mathbf{K}[\mathbf{x}(t)] = \mathbf{0} \tag{2.89}$$

where **G** is a matrix of constants.

The optimal control problem is formulated as follows

$$\min_{\mathbf{x},\mathbf{u}} \quad J = \int_0^T F[\mathbf{x}(t), \mathbf{u}(t)]dt + S[\mathbf{x}(T)] \tag{2.90}$$

s.t. Lagrange Multiplier

Linear State Equations	$\dot{\mathbf{x}}(t) = \mathbf{A}\ \mathbf{u}(t)$	$\boldsymbol{\lambda}(t)$	(2.91)
Linear Control Variable Constraints	$\mathbf{G}\ \mathbf{u}(t) = \mathbf{0}$	$\boldsymbol{\sigma}(t)$	(2.92)
State Variable Constraints	$\mathbf{K}[\mathbf{x}(t)] = \mathbf{0}$	$\boldsymbol{\eta}(t)$	(2.93)

$$\mathbf{x}(t) \geq \mathbf{0}, \qquad \mathbf{u}(t) \geq \mathbf{0}, \tag{2.94}$$

$$\mathbf{x}(0) \quad \text{given.} \tag{2.95}$$

We construct the augmented Hamiltonian $\mathcal{H}$ for the above optimal control problem as

$$\mathcal{H} = F[\mathbf{x}(t), \mathbf{u}(t)] + \boldsymbol{\lambda}(t)\ \mathbf{A}\ \mathbf{u}(t) + \boldsymbol{\sigma}(t)\ \mathbf{G}\ \mathbf{u}(t) + \boldsymbol{\eta}(t)\ \mathbf{K}\ [\mathbf{x}(t)] \tag{2.96}$$

The first order necessary conditions for the optimal control program are

$$\frac{\partial \mathcal{H}}{\partial \mathbf{u}(t)} = \frac{\partial F[\mathbf{x}(t), \mathbf{u}(t)]}{\partial \mathbf{u}(t)} + \boldsymbol{\lambda}(t)\ \mathbf{A} + \boldsymbol{\sigma}(t)\ \mathbf{G} \geq \mathbf{0}, \tag{2.97}$$

$$\text{and} \qquad \mathbf{u}(t)\frac{\partial \mathcal{H}}{\partial \mathbf{u}(t)} = 0, \tag{2.98}$$

$$-\dot{\boldsymbol{\lambda}}(t) = \frac{\partial \mathcal{H}}{\partial \mathbf{x}(t)} = \frac{\partial F[\mathbf{x}(t), \mathbf{u}(t)]}{\partial \mathbf{x}(t)} + \boldsymbol{\eta}(t)\frac{\partial \mathbf{K}[\mathbf{x}(t)]}{\partial \mathbf{x}(t)} \tag{2.99}$$

$$\dot{\mathbf{x}}(t) = \frac{\partial \mathcal{H}}{\partial \boldsymbol{\lambda}(t)} = \mathbf{A}\ \mathbf{u}(t) \tag{2.100}$$

$$\frac{\partial \mathcal{H}}{\partial \boldsymbol{\sigma}(t)} = \mathbf{0} = \mathbf{G}\ \mathbf{u}(t) \tag{2.101}$$

$$\frac{\partial \mathcal{H}}{\partial \boldsymbol{\eta}(t)} = \mathbf{K}[\mathbf{x}(t)] = \mathbf{0} \tag{2.102}$$

$$\boldsymbol{\lambda}(T) = \frac{\partial S[\mathbf{x}(T)]}{\partial \mathbf{x}(T)} \tag{2.103}$$

$$\mathbf{x}(0) = \mathbf{x}_0 \tag{2.104}$$

$$\mathbf{x}(t) \geq \mathbf{0}, \qquad \mathbf{u}(t) \geq \mathbf{0}. \tag{2.105}$$

Equations (2.97)-(2.99) are similar to equations (2.44) and (2.45) except for additional terms resulting from equality constraints on control and state variables. Inequality (2.97) is caused by the nonnegativity constraint on the control variable. Equation (2.100) is a restatement of state equation (2.91). Equations (2.101)-(2.102) are restatements of the equality constraints. Equations (2.103)-(2.105) are two-point boundary conditions.

2.3 Hierarchical Optimal Control Problems

In this section, we discuss some bilevel optimal control problems. We first discuss the utilization of the leader-follower or Stackelberg strategy concept in structuring the control of interconnected systems. These control methods are appropriate for system problems where there are multiple criteria, multiple decision makers, decentralized information and natural hierarchy of decision making levels. In applications of dynamic transportation network models to ATIS/ATMS, the multi-level formulation approach is very important. For example, routing strategies in these systems and travelers' choices constitute a natural hierarchy of decision making. System-optimal and user-optimal strategies are conflicting in general. By imposing dynamic congestion pricing, we can formulate a bilevel model which coordinates the two routing criteria, as shown in Chapter 14. Although we do not explore the application of these concepts further in this book, we include this review as a basis for further research.

The basic leader-follower strategy was originally suggested for static duopoly by von Stackelberg (1952). The generalization of this concept to dynamic nonzero-sum two-person games was given by Cruz. Based on Cruz (1978), we first discuss a static two-person game which will be extended to multi-level optimal control problems.

2.3.1 Static Two-Person Games

The basic idea of a leader-follower strategy for a static two-person game is simple. Players 1 and 2 choose static controls $u_1 \in R$ and $u_2 \in R$, respectively. There are two scalar costs $J_1(u_1, u_2)$ and $J_2(u_1, u_2)$ associated with Players 1 and 2, respectively. Designate Player 1 as the leader and Player 2 as the follower. For each control u_1 chosen by Player 1, Player 2 chooses $u_2 = f_2(u_1)$ where f_2 is determined by u_1 and u_2 such that for Player 2, his/her cost using controls $[u_1, f_2(u_1)]$ is smaller than or equal to a cost using any controls (u_1, u_2). It follows that

$$J_2[u_1, f_2(u_1)] \le J_2(u_1, u_2) \qquad \text{for each } u_1 \text{ and for all } u_2 \tag{2.106}$$

For simplicity, we assume that for each control u_1, $f_2(u_1)$ yields a unique u_2. The leader chooses an optimal control u_1^* such that his/her cost using optimal controls $[u_1^*, f_2(u_1^*)]$ is smaller than or equal to a cost using any controls $[u_1, f_2(u_1)]$. It follows that

$$J_1[u_1^*, f_2(u_1^*)] \le J_1[u_1, f_2(u_1)] \qquad \text{for all } u_1 \tag{2.107}$$

Then, the optimal strategy u_1^* is termed the Stackelberg strategy for Player 1 and $u_2^* = f_2(u_1^*)$ is termed the Stackelberg strategy for Player 2 when the leader is Player 1.

In this problem, we assume that the leader knows the cost mapping of the follower, but the follower doesn't know the cost mapping of the leader.

However, the follower knows the control strategy of the leader and he/she takes this into account in computing his/her strategy. This reaction behavior of the follower is known to the leader who optimizes his/her choice of control u_1.

When Player 1 is the follower and Player 2 is the leader, we have

$$J_1[f_1(u_2), u_2] \leq J_1(u_1, u_2) \qquad \text{for each } u_2 \text{ and for all } u_1 \tag{2.108}$$

and

$$J_2[f_1(u_2^{**}), u_2^{**}] \leq J_2[f_1(u_2), u_2] \qquad \text{for all } u_2 \tag{2.109}$$

where f_1 is determined by u_2 and u_1. The optimal control u_2^{**} is the leader Stackelberg strategy, and optimal control $u_1^{**} = f_1(u_2^{**})$ is the follower Stackelberg strategy.

Sometimes, we may refer to a Nash strategy or Nash control when we use game theory to tackle multi-level optimization problems. A Nash strategy pair $(\tilde{u}_1, \tilde{u}_2)$ is defined by

$$J_1(\tilde{u}_1, \tilde{u}_2) \leq J_1(u_1, \tilde{u}_2) \qquad \text{for all } u_1 \tag{2.110}$$

and

$$J_2(\tilde{u}_1, \tilde{u}_2) \leq J_2(\tilde{u}_1, u_2) \qquad \text{for all } u_2 \tag{2.111}$$

where the tilde ($\sim$) denotes that the control is a Nash control. We note that the Nash strategy may not be unique. As before, we define $\tilde{u}_2 = f_2(\tilde{u}_1)$ where f_2 is determined by $\tilde{u}_1$ and $\tilde{u}_2$. It follows that

$$J_2[\tilde{u}_1, \tilde{u}_2] = J_2[\tilde{u}_1, f_2(\tilde{u}_1)] \tag{2.112}$$

Combining equations (2.107) and (2.112), we have

$$J_1[u_1^*, f_2(u_1^*)] \leq J_1[f_1(\tilde{u}_2), \tilde{u}_2] \tag{2.113}$$

Similarly, we define $\tilde{u}_1 = f_1(\tilde{u}_2)$ where f_1 is determined by $\tilde{u}_1$ and $\tilde{u}_2$. It follows that

$$J_1(\tilde{u}_1, \tilde{u}_2) = J_1[f_1(\tilde{u}_2), \tilde{u}_2] \tag{2.114}$$

Combining equations (2.109) and (2.114), we have

$$J_2[f_1(u_2^{**}), u_2^{**}] \leq J_2(\tilde{u}_1, \tilde{u}_2) \tag{2.115}$$

Thus, for the leader, a Stackelberg strategy is at least as good as any Nash strategy. For the follower, the Stackelberg strategy may or may not be preferable compared to a Nash strategy.

2.3.2 Dynamic Games

Consider a dynamic system

$$\dot{x}(t) = f(x, u_1, u_2) \tag{2.116}$$

where $x \in R^n$ is the state, $u_1 \in R^{m_1}$ and $u_2 \in R^{m_2}$ are the controls, and f is a piecewise continuous function from $R^n \times R^{m_1} \times R^{m_2}$ to R^n. The time interval $[0, T]$ is fixed and the initial state $x(0) = x_0$ is given. In a dynamic system, it is necessary to specify what type of information is available to each player. Suppose no state measurements are available. In this case, we consider open-loop strategies. Associated with each player is a scalar cost function, i.e.,

$$J_i = \int_0^T F_i(x, u_1, u_2)\, dt + S_i[x(T)] \qquad \forall\, i = 1, 2 \tag{2.117}$$

Designate Player 1 as the leader. The dynamic game can be written as a bilevel problem as follows.

The Leader Problem

$$\min \quad J_1 = \int_0^T F_1(x, u_1, u_2)\, dt + S_1[x(T)] \tag{2.118}$$

The Follower Problem

$$\min \quad J_2 = \int_0^T F_2(x, u_1, u_2)\, dt + S_2[x(T)] \tag{2.119}$$

s.t.

$$\dot{x}(t) = f(x, u_1, u_2) \tag{2.120}$$

$$x(0) = x_0 \tag{2.121}$$

The Hamiltonian function for the follower problem is written as

$$\mathcal{H}_2(x, u_1, u_2, p) = F_2(x, u_1, u_2) + p(t)\, f(x, u_1, u_2) \tag{2.122}$$

where $p(t)$ is the Lagrange multiplier. The necessary conditions for the follower problem are

$$\frac{\partial \mathcal{H}_2}{\partial u_2(t)} = 0 \tag{2.123}$$

$$\dot{p}(t) = -\frac{\partial \mathcal{H}_2}{\partial x(t)} \tag{2.124}$$

$$p(T) = \frac{\partial S_2[x(T)]}{\partial x(T)} \tag{2.125}$$

In order to obtain the overall necessary conditions for this dynamic game, we place the necessary conditions (2.123)-(2.125) and constraints (2.120)-(2.121) of the follower problem as constraints for the leader problem. It follows that

$$\min \quad J_1 = \int_0^T F_1(x, u_1, u_2)\, dt + S_1[x(T)] \tag{2.126}$$

s.t.

$$\dot{x}(t) = f(x, u_1, u_2) \tag{2.127}$$

$$\dot{p}(t) = -\frac{\partial \mathcal{H}_2}{\partial x(t)} \tag{2.128}$$

$$\frac{\partial \mathcal{H}_2}{\partial u_2(t)} = 0 \tag{2.129}$$

$$p(T) = \frac{\partial S_2[x(T)]}{\partial x(T)} \tag{2.130}$$

$$x(0) = x_0 \tag{2.131}$$

where equations (2.127)-(2.128) are considered as state equations for the redefined one-level problem. This is an optimal control problem with equality constraint and constraint for final states $x(T)$ and $p(T)$. The Hamiltonian function $\mathcal{H}_1$ for the redefined one-level problem is

$$\begin{aligned} \mathcal{H}_1(x, u_1, u_2, p, \lambda_1, \lambda_2, \beta_2) &= F_1(x, u_1, u_2) + \lambda_1(t)\, f(x, u_1, u_2) \\ &+ \lambda_2(t)\, (-\frac{\partial \mathcal{H}_2}{\partial x}) + \beta_2(t)\, \frac{\partial \mathcal{H}_2}{\partial u_2} \end{aligned} \tag{2.132}$$

where $\lambda_1(t), \lambda_2(t), \beta_2(t)$ are Lagrange multipliers. We define another scalar function as

$$\Theta[x(T), p(t), \nu(T), T] = S_1[x(T)] + \nu(T)\, \{p(T) + \frac{\partial S_2[x(T)]}{\partial x(T)}\} \tag{2.133}$$

where $\nu(T)$ is a Lagrange multiplier. The necessary conditions for the redefined one-level problem are

$$\frac{\partial \mathcal{H}_1}{\partial u_i(t)} = 0 \qquad \forall\, i = 1, 2 \tag{2.134}$$

$$\dot{\lambda}_1(t) = -\frac{\partial \mathcal{H}_1}{\partial x(t)} \tag{2.135}$$

$$\lambda_2(t) = -\frac{\partial \mathcal{H}_1}{\partial p(t)} \tag{2.136}$$

$$\lambda_1(T) = \frac{\partial \Theta}{\partial x(T)} = \frac{\partial S_1[(x(T)]}{\partial x(T)} + \nu(T)\,\frac{\partial^2 S_2[(x(T)]}{\partial x^2(T)} \tag{2.137}$$

$$\lambda_2(T) = \frac{\partial \Theta}{\partial p(T)} = \nu(T) \tag{2.138}$$

Therefore, the necessary conditions for (u_1, u_2) to be an open-loop Stackelberg strategy pair are equations (2.123)-(2.130) and (2.134)-(2.138). Explicit solution in terms of the Riccati matrix equations can be obtained for the linear-quadratic problem. However, necessary conditions for the closed-loop Stackelberg strategy are extremely difficult to characterize. Simplification is possible when the structure of the control law is constrained, e.g., restricting the control law to be linear and the effects of random initial conditions are averaged.

Instead of discussing detailed control strategies, we state some conclusions from the literature (Cruz, 1978). The open-loop strategy for the leader for the entire duration of the game is declared in advance. If the follower minimizes its cost function, it obtains its follower Stackelberg strategy which is the optimal reaction to the declared leader strategy. By declaring his/her strategy in advance, the leader influences the follower to react in a manner which minimizes the follower's cost function, but more importantly, in a manner which is favorable to the leader. This is a direct interpretation of the definition of the leader's strategy. Similarly, for closed-loop strategies where the state is available for measurement, the leader has to declare his control law for the entire duration of the game. In situations where either player might be a leader, both cases should be examined because both players may insist on leader strategies in which case there may be disequilibrium, or both may play follower strategies and a stalemate may occur.

2.3.3 Bilevel Optimal Control Problems

We now generalize the above dynamic games into bilevel optimal control problems. Sometimes, we term this kind of optimization problem as hierarchical optimization, which is contrary to the bilevel optimization problems having less coordination or non-hierarchical properties. To simplify our presentation, we only discuss optimal control problems with equality and nonnegativity constraints which are constrained by fixed beginning and end times. Consider two linear dynamic systems

$$\frac{d\mathbf{x}_i(t)}{dt} = \mathbf{A}_i\, \mathbf{u}_i(t) \qquad \forall\, i = 1, 2 \tag{2.139}$$

The two dynamic systems have two decision makers, each with an objective function as follows

$$\min_{\mathbf{x},\mathbf{u}} J_i = \int_0^T F_i[\mathbf{x}_1(t), \mathbf{u}_1(t), \mathbf{x}_2(t), \mathbf{u}_2(t)]dt + S_i[\mathbf{x}_1(T), \mathbf{x}_2(T)] \qquad (2.140)$$

$$i = 1, 2$$

where each objective function depends on controls and states of both dynamic systems. To make the problem more general, we assume that control variables of both systems are co-related by the following linear constraints

$$\mathbf{G}_1\ \mathbf{u}_1(t) + \mathbf{G}_2\ \mathbf{u}_2(t) = 0 \qquad (2.141)$$

where $\mathbf{G}_1, \mathbf{G}_2$ are vectors of constants. We designate Player (decision maker) 1 as the leader. We then formulate the bilevel optimal control problem as follows.

Upper Level Problem

$$\min_{\mathbf{x}_1,\mathbf{u}_1} \quad J_1 = \int_0^T F_1[\mathbf{x}_1(t), \mathbf{u}_1(t), \mathbf{x}_2(t), \mathbf{u}_2(t)]\ dt + S_1[\mathbf{x}_1(T), \mathbf{x}_2(T)] \qquad (2.142)$$

s.t.

$$\frac{d\mathbf{x}_1(t)}{dt} = \mathbf{A}_1\ \mathbf{u}_1(t) \qquad (2.143)$$

$$\mathbf{G}_1\ \mathbf{u}_1(t) + \mathbf{G}_2\ \mathbf{u}_2(t) = \mathbf{0} \qquad (2.144)$$

$$\mathbf{K}_1[\mathbf{x}_1(t)] = \mathbf{0} \qquad (2.145)$$

$$\mathbf{x}_1(t) \geq \mathbf{0}, \qquad \mathbf{u}_1(t) \geq \mathbf{0} \qquad (2.146)$$

$$\mathbf{x}_1(0) \qquad \text{given.} \qquad (2.147)$$

where $\mathbf{x}_2(t), \mathbf{u}_2(t)$ are determined by the following lower level problem:

Lower Level Problem

$$\min_{\mathbf{x}_2,\mathbf{u}_2} \quad J_2 = \int_0^T F_2[\mathbf{x}_1(t), \mathbf{u}_1(t), \mathbf{x}_2(t), \mathbf{u}_2(t)]\ dt + S_2[\mathbf{x}_1(T), \mathbf{x}_2(T)] \qquad (2.148)$$

s.t.

$$\frac{d\mathbf{x}_2(t)}{dt} = \mathbf{A}_2\ \mathbf{u}_2(t) \qquad (2.149)$$

$$\mathbf{x}_2(t) \geq \mathbf{0}, \qquad \mathbf{u}_2(t) \geq \mathbf{0} \qquad (2.150)$$

$$\mathbf{x}_2(0) \qquad \text{given.} \qquad (2.151)$$

In the lower level problem, $\mathbf{x}_1(t), \mathbf{u}_1(t)$ are determined by the upper level problem. In this bilevel problem, the leader (upper level problem) has the priority to minimize its objective function J_1. We now examine the necessary conditions for the bilevel optimal control problem. We first consider the necessary conditions for the lower level problem which is designated as the follower. The overall necessary conditions for the bilevel problem are constructed by placing the necessary conditions and constraints of the lower level problem as constraints for the upper level problem.

We state the augmented Hamiltonian $\mathcal{H}_2$ for the lower level problem as

$$\mathcal{H}_2 = F_2[\mathbf{x}_1(t), \mathbf{u}_1(t), \mathbf{x}_2(t), \mathbf{u}_2(t)] + \mathbf{p}(t)\ \mathbf{A}_2\ \mathbf{u}_2(t) \tag{2.152}$$

where $\mathbf{p}(t)$ is a vector of Lagrange multipliers. The first order necessary conditions for the lower level optimal control program are

$$\frac{\partial \mathcal{H}_2}{\partial \mathbf{u}_2(t)} = \frac{\partial F_2[\mathbf{x}_1(t), \mathbf{u}_1(t), \mathbf{x}_2(t), \mathbf{u}_2(t)]}{\partial \mathbf{u}_2(t)} + \mathbf{A}_2\ \mathbf{p}(t)\ \geq \mathbf{0} \tag{2.153}$$

$$\mathbf{u}_2(t)\frac{\partial \mathcal{H}_2}{\partial \mathbf{u}_2(t)} = \mathbf{0} \tag{2.154}$$

$$-\ \dot{\mathbf{p}}(t) = \frac{\partial \mathcal{H}_2}{\partial \mathbf{x}_2(t)} = \frac{\partial F_2[\mathbf{x}_1(t), \mathbf{u}_1(t), \mathbf{x}_2(t), \mathbf{u}_2(t)]}{\partial \mathbf{x}_2(t)} \tag{2.155}$$

$$\frac{d\mathbf{x}_2(t)}{dt} = \frac{\partial \mathcal{H}_2}{\partial \mathbf{p}(t)} = \mathbf{A}_2\ \mathbf{u}_2(t) \tag{2.156}$$

$$\mathbf{p}(T) = \frac{\partial S_2[\mathbf{x}_1(T), \mathbf{x}_2(T)]}{\partial \mathbf{x}_2(T)} \tag{2.157}$$

$$\mathbf{x}_2(t) \geq \mathbf{0}, \qquad \mathbf{u}_2(t) \geq \mathbf{0} \tag{2.158}$$

$$\mathbf{x}_2(0) \qquad \text{given} \tag{2.159}$$

Then, we convert the bilevel optimal control problem into a one-level optimal control problem by placing necessary conditions (2.153)-(2.159) of the lower level problem as constraints for the upper level problem. It follows that

$$\min_{\mathbf{x},\mathbf{u}}\ J_1 = \int_0^T F_1[\mathbf{x}_1(t), \mathbf{u}_1(t), \mathbf{x}_2(t), \mathbf{u}_2(t)]\ dt + S_1[\mathbf{x}_1(T), \mathbf{x}_2(T)] \tag{2.160}$$

s.t.

$$\frac{d\mathbf{x}_1(t)}{dt} = \mathbf{A}_1\ \mathbf{u}_1(t) \tag{2.161}$$

$$\frac{d\mathbf{x}_2(t)}{dt} = \mathbf{A}_2\ \mathbf{u}_2(t) \tag{2.162}$$

$$\dot{\mathbf{p}}(t) = -\frac{\partial \mathcal{H}_2}{\partial \mathbf{x}_2(t)} \tag{2.163}$$

$$\mathbf{G}_1\ \mathbf{u}_1(t) + \mathbf{G}_2\ \mathbf{u}_2(t) = \mathbf{0} \tag{2.164}$$

$$\mathbf{K}_1[\mathbf{x}_1(t)] = \mathbf{0} \tag{2.165}$$

$$\frac{\partial \mathcal{H}_2}{\partial \mathbf{u}_2(t)} \geq \mathbf{0} \tag{2.166}$$

$$\mathbf{u}_2(t)\frac{\partial \mathcal{H}_2}{\partial \mathbf{u}_2(t)} = \mathbf{0} \tag{2.167}$$

$$\mathbf{x}_1(t) \geq \mathbf{0}, \qquad \mathbf{u}_1(t) \geq \mathbf{0}, \qquad \mathbf{x}_2(t) \geq \mathbf{0}, \qquad \mathbf{u}_2(t) \geq \mathbf{0} \tag{2.168}$$

$$\mathbf{x}_1(0), \qquad \mathbf{x}_2(0) \qquad \text{given} \tag{2.169}$$

$$\mathbf{p}(T) = \frac{\partial S_2[\mathbf{x}_1(T), \mathbf{x}_2(T)]}{\partial \mathbf{x}_2(T)} \tag{2.170}$$

where equations (2.161)-(2.163) are considered as state equations. We then construct the augmented Hamiltonian $\mathcal{H}_1$ for the converted one-level optimal control program as

$$\begin{aligned} \mathcal{H}_1 &= F_1 + \boldsymbol{\lambda}_1(t)\ \mathbf{A}_1\ \mathbf{u}_1(t) + \boldsymbol{\lambda}_2(t)\ \mathbf{A}_2\ \mathbf{u}_2(t) + \boldsymbol{\lambda}_3(t)\ [-\frac{\partial \mathcal{H}_2}{\partial \mathbf{x}_2(t)}] \\ &+ \boldsymbol{\sigma}(t)\ [\mathbf{G}_1\ \mathbf{u}_1(t) + \mathbf{G}_2\ \mathbf{u}_2(t)] + \boldsymbol{\eta}(t)\ \mathbf{K}_1[\mathbf{x}_1(t)] \\ &+ \boldsymbol{\beta}_1(t)\ \frac{\partial \mathcal{H}_2}{\partial \mathbf{u}_2(t)} + \boldsymbol{\beta}_2(t)\ [\mathbf{u}_2(t)\frac{\partial \mathcal{H}_2}{\partial \mathbf{u}_2(t)}] \end{aligned} \tag{2.171}$$

where $\boldsymbol{\lambda}_1(t), \boldsymbol{\lambda}_2(t), \boldsymbol{\lambda}_3(t), \boldsymbol{\sigma}(t), \boldsymbol{\eta}(t), \boldsymbol{\beta}_1(t), \boldsymbol{\beta}_2(t)$ are Lagrange multipliers and $\boldsymbol{\beta}_1(t) \leq 0$. We define another scalar function as

$$\Theta[\mathbf{x}(T), \mathbf{p}(t), \boldsymbol{\nu}(T), T] = S_1[\mathbf{x}(T)] + \boldsymbol{\nu}(T)\ \{\mathbf{p}(T) - \frac{\partial S_2[\mathbf{x}(T)]}{\partial \mathbf{x}_2(T)}\} \tag{2.172}$$

where $\boldsymbol{\nu}(T)$ is a vector of Lagrange multipliers. The first order necessary conditions for the converted one-level optimal control program are

$$\frac{\partial \mathcal{H}_1}{\partial \mathbf{u}_i(t)} \geq \mathbf{0} \qquad \forall\, i = 1, 2 \tag{2.173}$$

$$\mathbf{u}_i(t)\frac{\partial \mathcal{H}_1}{\partial \mathbf{u}_i(t)} = 0 \qquad \forall\, i = 1, 2 \tag{2.174}$$

$$\dot{\boldsymbol{\lambda}}_i(t) = -\frac{\partial \mathcal{H}_1}{\partial \mathbf{x}_i(t)} \qquad \forall\, i = 1, 2 \tag{2.175}$$

$$\dot{\boldsymbol{\lambda}}_3(t) = -\frac{\partial \mathcal{H}_1}{\partial \mathbf{p}(t)} \tag{2.176}$$

$$\dot{\mathbf{x}}_i(t) = \frac{\partial \mathcal{H}_1}{\partial \boldsymbol{\lambda}_i(t)} = \mathbf{A}_i\ \mathbf{u}_i(t) \qquad \forall\, i = 1, 2 \tag{2.177}$$

$$\dot{\mathbf{p}}(t) = \frac{\partial \mathcal{H}_1}{\partial \boldsymbol{\lambda}_3(t)} = -\frac{\partial \mathcal{H}_2}{\partial \mathbf{x}_2(t)} \tag{2.178}$$

$$\frac{\partial \mathcal{H}_1}{\partial \boldsymbol{\sigma}(t)} = \mathbf{0} = \mathbf{G}_1\, \mathbf{u}_1(t) + \mathbf{G}_2\, \mathbf{u}_2(t) \tag{2.179}$$

$$\frac{\partial \mathcal{H}_1}{\partial \boldsymbol{\eta}(t)} = \mathbf{K}_1[\mathbf{x}_1(t)] = \mathbf{0} \tag{2.180}$$

$$\frac{\partial \mathcal{H}_1}{\partial \boldsymbol{\beta}_1(t)} = \frac{\partial \mathcal{H}_2}{\partial \mathbf{u}_2(t)} \tag{2.181}$$

$$\frac{\partial \mathcal{H}_1}{\partial \boldsymbol{\beta}_2(t)} = \mathbf{u}_2(t)\frac{\partial \mathcal{H}_2}{\partial \mathbf{u}_2(t)} \tag{2.182}$$

$$\boldsymbol{\lambda}_i(T) = \frac{\partial \Theta}{\partial \mathbf{x}_i(T)} \qquad \forall\, i = 1, 2 \tag{2.183}$$

$$\boldsymbol{\lambda}_3(T) = \frac{\partial \Theta}{\partial \mathbf{p}(T)} = \boldsymbol{\nu}(T) \tag{2.184}$$

$$\mathbf{x}_i(0) = \mathbf{x}_{i0}, \qquad \forall\, i = 1, 2 \tag{2.185}$$

$$\mathbf{x}_i(t) \geq \mathbf{0}, \qquad \mathbf{u}_i(t) \geq \mathbf{0} \qquad \forall\, i = 1, 2 \tag{2.186}$$

Therefore, the necessary conditions for the bilevel program are equations (2.153)-(2.159) and (2.173)-(2.186). However, the detailed analysis of the necessary conditions is complicated, especially when the two level problems have many constraints. In general, the analytical solution for the above bilevel program is not possible even for simple cost functions.

On the other hand, if we assume that the above problem has no hierarchical relationship, we can solve the two level problems separately and transmit some shared variables back and forth. In this way, no leader-follower relationship is assumed between the two level problems. Thus, this situation can be termed a coordination problem where the shared variables function as coordinators. This approach can also be extended to a multi-level optimal control problem where the hierarchical relationship is hard to identify.

2.4 Notes

Many complex transportation problems can be formulated and solved as a variational inequality. A comprehensive summary of variational inequality problems is provided by Nagurney (1993). The text by Kinderlehrer and Stampacchia (1980) provides an introduction to some variational inequality problems. For the rigorous proofs of the existence and uniqueness for dynamic problems, especially optimal control problems, refer to Cesari (1983). The relaxation method is described in more detail in Nagurney (1993).

In classical control system design, the objective is to obtain a controller that allows a system to perform in a desirable manner. The objective in control problem formulation for transportation networks is also to obtain a set of control variables which can adjust traffic flows in the network to behave in desired ways. The basic knowledge of the optimal control problem can be found in any standard optimal control text. At the elementary level, readers may consult the text by Kirk (1970). Bryson and Ho (1975) provide many examples which may help readers to understand the theory. Other closely related texts are Kamien and Schwartz (1981) and Sage and White (1977). A proof of optimality conditions for optimal control problems with general constraints for state and control variables is provided by Russak (1970).

Problems

2.1 To formulate a dynamic network equilibrium model, the following two variational inequalities are defined:

$$\mathbf{F}[\mathbf{u}^*(t)] \cdot [\mathbf{u}(t) - \mathbf{u}^*(t)] \geq 0, \qquad \forall\, \mathbf{u}(t) \in G(t)$$

$$\int_0^T \mathbf{F}^T[\mathbf{u}^*(t)] \cdot [\mathbf{u}(t) - \mathbf{u}^*(t)]\, dt \geq 0, \qquad \forall\, \mathbf{u}(t) \in G(t)$$

Show under what conditions these two variational inequalities are equivalent.

2.2 Prove that the systems of equations

$$\mathbf{F}[\mathbf{u}^*(t), \mathbf{x}^*(t)] = 0$$

is a special case of a variational inequality.

2.3 Calculate the Jacobian matrix of the function

$$Z(x_1, x_2) = 5x_1^2 - 3x_1x_2 + 2x_2^2 + 5$$

Is the Jacobian matrix symmetric and positive semidefinite?

2.4 In which of the following cases does the variational inequality

$$\mathbf{F}[\mathbf{u}^*(t)] \cdot [\mathbf{u}(t) - \mathbf{u}^*(t)] \geq 0, \qquad \forall\, \mathbf{u}(t) \in G(t)$$

have a unique solution. Explain your answers.

1. $\mathbf{F}[\mathbf{u}^*(t)]$ is convex on $G(t)$.
2. $\mathbf{F}[\mathbf{u}^*(t)]$ is monotone on $G(t)$.

2.5 Design a relaxation method for solving the variational inequality problem

$$\mathbf{f}(\mathbf{x}^*) \cdot (\mathbf{x} - \mathbf{x}^*) \geq 0, \qquad \forall\, \mathbf{x} \in G$$

where $\mathbf{x} = (x_1, x_2)$, $\mathbf{f}(\mathbf{x}) = c_1(x_1, x_2) + c_2(x_1, x_2)$ and c_1, c_2 are convex functions. State the resulting mathematical programming subproblem.

2.6 Find the optimal control law for the following optimal control problem

$$\min \qquad J = \int_0^T [\frac{1}{2}u^2(t) + x^2(t)]dt + \frac{1}{2}\, x^2(T)$$

s.t.

$$\dot{x} = -\frac{1}{2}x(t) + u(t), \qquad x(0) = x_0.$$

2.7 Solve the following optimal control problem

$$\min \qquad J = \int_0^T [au^2(t) + bu(t) + cx(t) + x^2(t)]dt$$

s.t.

$$\dot{x} = u(t), \qquad x(0) = x_0, \qquad a > 0.$$

2.8 The optimal control problem

$$\min \qquad J = \int_0^T f[u(t), x(t)]dt$$

s.t.

$$\dot{x} = g[u(t), x(t)], \qquad x(0) = x_0$$

has no explicit dependence on time t. Show that the Hamiltonian is a constant function of time along the optimal path. Hint: Using the chain rule, compute

$$\frac{d\mathcal{H}}{dt} = \frac{\partial \mathcal{H}}{\partial x}\frac{dx}{dt} + \frac{\partial \mathcal{H}}{\partial u}\frac{du}{dt} + \frac{\partial \mathcal{H}}{\partial \lambda}\frac{d\lambda}{dt}$$

and substitute from the necessary conditions for the partial derivatives of $\mathcal{H}$.

Chapter 3

Discrete Optimal Control and Nonlinear Programming

In this chapter, we present more mathematical background which is necessary for modeling and solving dynamic transportation network problems. This chapter covers discrete optimal control and nonlinear programming problems. First, we introduce the discrete optimal control problem (OCP). To simplify our presentation, we consider discrete optimal control problems with fixed end times as examples in Section 3.1. The discussion is focused on the analysis of optimality conditions. Then, some nonlinear programming (NLP) problems with equality and nonnegativity constraints are presented in Section 3.2. Similarities between discrete optimal control and nonlinear programming are emphasized.

In Section 3.3, we present algorithms for solving nonlinear programs, including one-dimensional search and the Frank-Wolfe algorithm. Unlike traditional algorithms for solving optimal control problems, we suggest using these algorithms to solve discrete optimal control problems.

3.1 Discrete Optimal Control Problems

In this section, we consider several discrete optimal control problems with fixed end times. These problems are widely used in our formulations and solutions in the following chapters. We first discuss discrete optimal control problems with no constraints. Then, discrete optimal control problems with equality and inequality constraints are analyzed. Finally, we investigate discrete optimal control problems with equality and nonnegativity constraints.

In discrete optimal control problems, we discretize the time period $[0, T]$ into $K + 1$ small time intervals or increments, i.e., $k = 1, 2, \cdots, K + 1$. For simplicity, each interval is assumed to have equal length Δ. In general, we use the first difference approximation to replace $\dot{\mathbf{x}}(t)$. It follows that

$$\dot{\mathbf{x}}(t) \approx \frac{\mathbf{x}(k+1) - \mathbf{x}(k)}{\Delta} \tag{3.1}$$

In most analysis, it is convenient to assume Δ is a unit value. Thus, the above difference approximation can be rewritten as

$$\dot{\mathbf{x}}(t) \approx \mathbf{x}(k+1) - \mathbf{x}(k) \tag{3.2}$$

Thus, the discrete state equation can be written as

$$\mathbf{x}(k+1) = \mathbf{x}(k) + \mathbf{f}[\mathbf{x}(k), \mathbf{u}(k), k] \tag{3.3}$$

In the same way, the discretization of the Lagrange multipliers

$$\dot{\boldsymbol{\lambda}}(t) \approx \frac{\boldsymbol{\lambda}(k+1) - \boldsymbol{\lambda}(k)}{\Delta} \tag{3.4}$$

can be rewritten as

$$\dot{\boldsymbol{\lambda}}(t) \approx \boldsymbol{\lambda}(k+1) - \boldsymbol{\lambda}(k) \tag{3.5}$$

In continuous optimal control problems, we minimize cost functions which are integrals of scalar cost functions. We now consider a dynamic system having a cost function $F[\mathbf{x}(k), \mathbf{u}(k), k]$ for each time interval k. Thus, we are interested in minimization of cost functions which are summations of scalar functions. It follows that

$$\min_{\mathbf{x},\mathbf{u}} \; J = \sum_{k=1}^{K} F[\mathbf{x}(k), \mathbf{u}(k), k] \; + \; S[\mathbf{x}(K+1), K+1] \tag{3.6}$$

3.1.1 No Constraints

The discrete optimal control problem is to seek an admissible control function $\mathbf{u}$ in order to minimize the objective function

$$\min_{\mathbf{x},\mathbf{u}} \; J = \sum_{k=1}^{K} F[\mathbf{x}(k), \mathbf{u}(k), k] \; + \; S[\mathbf{x}(K+1), K+1] \tag{3.7}$$

s.t.

$$\mathbf{x}(k+1) = \mathbf{x}(k) + \mathbf{f}[\mathbf{x}(k), \mathbf{u}(k), k] \tag{3.8}$$

$$K \quad \text{and} \quad \mathbf{x}(1) = \mathbf{x}_1 \text{ fixed}; \quad \mathbf{x}(K+1) \text{ free} \tag{3.9}$$

where $F[\mathbf{x}(k), \mathbf{u}(k), k]$ possesses continuous partial derivatives with respect to $\mathbf{x}(k)$ and $\mathbf{u}(k)$, and $S[\mathbf{x}(K+1), K+1]$ has continuous partial derivatives with respect to $\mathbf{x}(K+1)$. $S[\mathbf{x}(K+1), K+1]$ is associated with the end time $(K+1)$ only, and is termed the salvage cost in many economics problems.

We define the Hamiltonian as

$$\mathcal{H} = F[\mathbf{x}(k), \mathbf{u}(k), k] + \boldsymbol{\lambda}(k)\, \mathbf{f}[\mathbf{x}(k), \mathbf{u}(k), k] \tag{3.10}$$

where $\boldsymbol{\lambda}(k)$ is the vector of Lagrange multipliers associated with the dynamic state equations. Then, the first-order necessary conditions for the optimal control problem are summarized as follows:

$$\frac{\partial \mathcal{H}}{\partial \mathbf{u}(k)} = \frac{\partial F[\mathbf{x}(k), \mathbf{u}(k), k]}{\partial \mathbf{u}(k)} + \boldsymbol{\lambda}(k) \frac{\partial \mathbf{f}[\mathbf{x}(k), \mathbf{u}(k), k]}{\partial \mathbf{u}(k)} = \mathbf{0} \tag{3.11}$$

$$\begin{aligned} -\boldsymbol{\lambda}(k+1) + \boldsymbol{\lambda}(k) &= \frac{\partial \mathcal{H}}{\partial \mathbf{x}(k)} \\ &= \frac{\partial F[\mathbf{x}(k), \mathbf{u}(k), k]}{\partial \mathbf{x}(k)} + \boldsymbol{\lambda}(k) \frac{\partial \mathbf{f}[\mathbf{x}(k), \mathbf{u}(k), k]}{\partial \mathbf{x}(k)} \end{aligned} \tag{3.12}$$

$$\mathbf{x}(k+1) - \mathbf{x}(k) = \frac{\partial \mathcal{H}}{\partial \boldsymbol{\lambda}(k)} = \mathbf{f}[\mathbf{x}(k), \mathbf{u}(k), k] \tag{3.13}$$

$$\mathbf{x}(1) = \mathbf{x}_1 \tag{3.14}$$

$$\boldsymbol{\lambda}(K+1) = \frac{\partial S[\mathbf{x}(K+1), K+1]}{\partial \mathbf{x}(K+1)} \tag{3.15}$$

Equations (3.13) and (3.14) are simply the restatement of state equation (3.8) and boundary condition (3.9). Equations (3.14) and (3.15) are termed transversality conditions which constitute two-point boundary conditions for the set of differential equations. The above first-order necessary conditions are derived using the discrete Minimum (or Maximum) Principle or the Hamilton-Jacobi equation. The discrete Minimum Principle can simply be stated as follows. The inequality

$$\mathcal{H}[\mathbf{u}^*(k)] \leq \mathcal{H}[\mathbf{u}(k)] \tag{3.16}$$

is valid for all admissible $\mathbf{u}(k)$, where $*$ represents that the solution is optimal. Sometimes, we directly use this principle to perform solution analyze of complicated discrete optimal control problems.

3.1.2 Equality and Inequality Constraints

In this section, we consider discrete optimal control problems with nonlinear equality and inequality constraints on state and control variables. Those constraints can represent most practical constraints in realistic applications. The set of equality constraints for control and state variables are denoted as $\mathbf{G}[\mathbf{u}(k), \mathbf{x}(k), k] = 0$ and the set of inequality constraints are denoted as $\mathbf{K}[\mathbf{u}(k), \mathbf{x}(k), k] \leq 0$. We assume both $\mathbf{G}$ and $\mathbf{K}$ are continuous and differentiable with respect to $\mathbf{u}$, $\mathbf{x}$ and k. The optimal control problem is formulated as follows

$$\min_{\mathbf{x},\mathbf{u}} \; J = \sum_{k=1}^{K} F[\mathbf{x}(k), \mathbf{u}(k), k] \; + \; S[\mathbf{x}(K+1), K+1] \tag{3.17}$$

s.t.

$$\mathbf{x}(k+1) = \mathbf{x}(k) + \mathbf{f}[\mathbf{x}(k), \mathbf{u}(k), k] \tag{3.18}$$

$$\mathbf{G}[\mathbf{u}(k), \mathbf{x}(k), k] = \mathbf{0} \tag{3.19}$$

$$\mathbf{K}[\mathbf{u}(k), \mathbf{x}(k), k] \leq \mathbf{0} \tag{3.20}$$

$$\mathbf{x}(1) = \mathbf{x}_1, \qquad K \text{ fixed.} \tag{3.21}$$

Denote the Lagrangian multipliers associated with equations (3.18)-(3.20) as $\boldsymbol{\lambda}(k)$, $\boldsymbol{\sigma}(k)$ and $\boldsymbol{\eta}(k)$, respectively. We construct the augmented Hamiltonian $\mathcal{H}$ for the above optimal control problem as

$$\begin{aligned} \mathcal{H} &= F[\mathbf{x}(k), \mathbf{u}(k), k] + \boldsymbol{\lambda}(k)\, \mathbf{f}[\mathbf{u}(k), \mathbf{x}(k), k] \\ &+ \boldsymbol{\sigma}(k)\, \mathbf{G}[\mathbf{u}(k), \mathbf{x}(k), k] + \boldsymbol{\eta}(k)\, \mathbf{K}[\mathbf{u}(k), \mathbf{x}(k), k] \end{aligned} \tag{3.22}$$

where

$$\boldsymbol{\eta}(k) \begin{cases} > \mathbf{0} & \text{if } \mathbf{K} = \mathbf{0} \\ = \mathbf{0} & \text{if } \mathbf{K} < \mathbf{0} \end{cases}$$

The first order necessary conditions for the optimal control program are

$$\begin{aligned} \frac{\partial \mathcal{H}}{\partial \mathbf{u}(k)} = \mathbf{0} &= \frac{\partial F[\mathbf{x}(k), \mathbf{u}(k), k]}{\partial \mathbf{u}(k)} + \boldsymbol{\lambda}(k) \frac{\partial \mathbf{f}[\mathbf{u}(k), \mathbf{x}(k), k]}{\partial \mathbf{u}(k)} \\ &+ \boldsymbol{\sigma}(k) \frac{\partial \mathbf{G}[\mathbf{u}(k), \mathbf{x}(k), k]}{\partial \mathbf{u}(k)} + \boldsymbol{\eta}(k) \frac{\partial \mathbf{K}[\mathbf{u}(k), \mathbf{x}(k), k]}{\partial \mathbf{u}(k)} \end{aligned} \tag{3.23}$$

$$\begin{aligned} -\boldsymbol{\lambda}(k) + \boldsymbol{\lambda}(k+1) &= \frac{\partial \mathcal{H}}{\partial \mathbf{x}(k)} \\ &= \frac{\partial F[\mathbf{x}(k), \mathbf{u}(k), k]}{\partial \mathbf{x}(k)} + \boldsymbol{\lambda}(k) \frac{\partial \mathbf{f}[\mathbf{x}(k), \mathbf{u}(k), k]}{\partial \mathbf{x}(k)} \\ &+ \boldsymbol{\sigma}(k) \frac{\partial \mathbf{G}[\mathbf{u}(k), \mathbf{x}(k), k]}{\partial \mathbf{x}(k)} + \boldsymbol{\eta}(k) \frac{\partial \mathbf{K}[\mathbf{u}(k), \mathbf{x}(k), k]}{\partial \mathbf{x}(k)} \end{aligned} \tag{3.24}$$

$$\mathbf{x}(k+1) - \mathbf{x}(k) = \frac{\partial \mathcal{H}}{\partial \boldsymbol{\lambda}(k)} = \mathbf{f}[\mathbf{x}(k), \mathbf{u}(k), k] \tag{3.25}$$

$$\frac{\partial \mathcal{H}}{\partial \boldsymbol{\sigma}(k)} = \mathbf{0} = \mathbf{G}[\mathbf{u}(k), \mathbf{x}(k), k] \tag{3.26}$$

$$\frac{\partial \mathcal{H}}{\partial \boldsymbol{\eta}(k)} = \mathbf{K}[\mathbf{u}(k), \mathbf{x}(k), k] \leq \mathbf{0} \tag{3.27}$$

$$\mathbf{x}(1) = \mathbf{x}_1 \tag{3.28}$$

$$\boldsymbol{\lambda}(K+1) = \frac{\partial S[\mathbf{x}(K+1), K+1]}{\partial \mathbf{x}(K+1)} \tag{3.29}$$

Equations (3.23)-(3.24) are similar to equations (3.11) and (3.12) except for additional terms resulting from equality and inequality constraints on the control and state variables. Equation (3.25) is a restatement of state equation (3.18). Equations (3.26)-(3.27) are restatements of the equality and inequality constraints. Equations (3.28)-(3.29) are two-point boundary conditions.

3.1.3 Equality and Nonnegativity Constraints

We now consider a set of special discrete optimal control problems which will be widely used in the following chapters for formulation and analysis of dynamic transportation network models. The state equations are linear

$$\mathbf{x}(k+1) = \mathbf{x}(k) + \mathbf{A}\ \mathbf{u}(k) \tag{3.30}$$

where $\mathbf{A}$ is a matrix of constants. We consider only linear equality constraints for control variables

$$\mathbf{G}\ \mathbf{u}(k) = \mathbf{0} \tag{3.31}$$

and nonlinear equality constraints involving only state variables

$$\mathbf{K}[\mathbf{x}(k)] = \mathbf{0} \tag{3.32}$$

where $\mathbf{G}$ is a matrix of constants. The discrete optimal control problem is formulated as follows

$$\min_{\mathbf{x},\mathbf{u}} \quad J = \sum_{k=1}^{K} F[\mathbf{x}(k), \mathbf{u}(k), k] \ + \ S[\mathbf{x}(K+1), K+1] \tag{3.33}$$

s.t. Lagrange Multiplier

$$\text{Linear State Equations} \qquad \mathbf{x}(k+1) = \mathbf{x}(k) + \mathbf{A}\ \mathbf{u}(k) \qquad \boldsymbol{\lambda}(k) \tag{3.34}$$

$$\text{Linear Control Variable Constraints} \qquad \mathbf{G}\ \mathbf{u}(k) = \mathbf{0} \qquad \boldsymbol{\sigma}(k) \tag{3.35}$$

$$\text{State Variable Constraints} \qquad \mathbf{K}[\mathbf{x}(k)] = \mathbf{0} \qquad \boldsymbol{\eta}(k) \tag{3.36}$$

$$\mathbf{x}(k) \geq \mathbf{0}, \qquad \mathbf{u}(k) \geq \mathbf{0}, \tag{3.37}$$

$$\mathbf{x}(1) \ \text{ given.} \tag{3.38}$$

We construct the augmented Hamiltonian $\mathcal{H}$ for the above discrete optimal control problem as

$$\mathcal{H} = F[\mathbf{x}(k), \mathbf{u}(k), k] + \boldsymbol{\lambda}(k)\ \mathbf{A}\ \mathbf{u}(k) + \boldsymbol{\sigma}(k)\ \mathbf{G}\ \mathbf{u}(k) + \boldsymbol{\eta}(k)\ \mathbf{K}\ [\mathbf{x}(k)] \tag{3.39}$$

First-order necessary conditions for the discrete optimal control program are

$$\frac{\partial \mathcal{H}}{\partial \mathbf{u}(k)} = \frac{\partial F[\mathbf{x}(k), \mathbf{u}(k), k]}{\partial \mathbf{u}(k)} + \boldsymbol{\lambda}(k)\ \mathbf{A} + \boldsymbol{\sigma}(k)\ \mathbf{G} \geq \mathbf{0}, \tag{3.40}$$

$$\text{and} \qquad \mathbf{u}(k) \frac{\partial \mathcal{H}}{\partial \mathbf{u}(k)} = 0, \tag{3.41}$$

$$\begin{aligned} \boldsymbol{\lambda}(k) - \boldsymbol{\lambda}(k+1) &= \frac{\partial \mathcal{H}}{\partial \mathbf{x}(k)} \\ &= \frac{\partial F[\mathbf{x}(k), \mathbf{u}(k), k]}{\partial \mathbf{x}(k)} + \boldsymbol{\eta}(k) \frac{\partial \mathbf{K}[\mathbf{x}(k)]}{\partial \mathbf{x}(k)} \end{aligned} \tag{3.42}$$

$$\mathbf{x}(k+1) - \mathbf{x}(k) = \frac{\partial \mathcal{H}}{\partial \boldsymbol{\lambda}(k)} = \mathbf{A}\ \mathbf{u}(k) \tag{3.43}$$

$$\frac{\partial \mathcal{H}}{\partial \boldsymbol{\sigma}(k)} = \mathbf{0} = \mathbf{G}\ \mathbf{u}(k) \tag{3.44}$$

$$\frac{\partial \mathcal{H}}{\partial \boldsymbol{\eta}(k)} = \mathbf{K}[\mathbf{x}(k)] = \mathbf{0} \tag{3.45}$$

$$\boldsymbol{\lambda}(K+1) = \frac{\partial S[\mathbf{x}(K+1)]}{\partial \mathbf{x}(K+1)} \tag{3.46}$$

$$\mathbf{x}(1) = \mathbf{x}_1 \tag{3.47}$$

$$\mathbf{x}(k) \geq \mathbf{0}, \qquad \mathbf{u}(k) \geq \mathbf{0}. \tag{3.48}$$

Equations (3.40)-(3.42) are similar to equations (3.11) and (3.12) except for additional terms resulting from equality constraints on control and state variables. The inequality sign in equation (3.40) is caused by the nonnegativity constraint for the control variable. Equation (3.43) is a restatement of the state equation (3.34). Equations (3.44)-(3.45) are restatements of the equality constraints. Equations (3.46)-(3.48) are two-point boundary conditions.

3.2 Nonlinear Programming Problems

This section reviews some concepts related to the formulation and solution of nonlinear optimization programs. The focus of the discussion is on the conditions that characterize the solution of such programs and the constraint conditions which are similar to discrete optimal control problems. To have a systematic description of nonlinear programming problems, we first discuss unconstrained minimization problems. Then, we explore minimization problems with general constraints and the Karush-Kuhn-Tucker conditions. The relationship between the discrete optimal control problem and nonlinear programming problem is then investigated.

3.2.1 Unconstrained Minimization

We first discuss minimization problems without constraints. Our problem is to find a set of variables $x_1, x_2, \cdots, x_n$ that minimize an objective function $Z(x_1, x_2, \cdots, x_n)$. Let $\mathbf{x}$ denote the vector of decision variables, i.e. $\mathbf{x} = (x_1, x_2, \cdots, x_n)$. Using vector notation, minimization without constraints can be stated as

$$\min_{\mathbf{x}} \ Z(\mathbf{x}) \tag{3.49}$$

Since the function to be minimized is unconstrained, the first-order necessary condition for a minimum at $\mathbf{x} = \mathbf{x}^*$ is that the gradient of $Z(\mathbf{x})$ vanish at

$\mathbf{x}^*$. The gradient of $Z(\mathbf{x})$ with respect to $\mathbf{x}^*$, $\nabla_{\mathbf{x}} Z(\mathbf{x})$, is the vector of partial derivatives, that is,

$$\nabla_{\mathbf{x}} Z(\mathbf{x}) = \left[\frac{\partial Z(\mathbf{x})}{\partial x_1}, \frac{\partial Z(\mathbf{x})}{\partial x_2}, \cdots, \frac{\partial Z(\mathbf{x})}{\partial x_n}\right] \tag{3.50}$$

At every point $\mathbf{x}$, the gradient points in the direction of the steepest increase in $Z(\mathbf{x})$. The first-order necessary conditions for a minimum are

$$\nabla Z(\mathbf{x}^*) = \mathbf{0} \tag{3.51}$$

In other words, each element of the gradient has to equal to zero. Equivalently,

$$\frac{\partial Z(\mathbf{x}^*)}{\partial x_i} = 0 \qquad \forall\, i = 1, 2, \cdots, n \tag{3.52}$$

The sufficient conditions that x^* is the local minimum of $Z(\mathbf{x})$ depend on establishing that $Z(\mathbf{x})$ is locally convex in the vicinity of x^*. In other words, any line sequent joining two points x_1 and x_2 lies above the surface $Z(\mathbf{x})$. The strict convexity of $Z(\mathbf{x})$ can be established if $Z(\mathbf{x})$ is positive definite, as now defined. In general, suppose we have an $n \times n$ matrix $F(\mathbf{x})$

$$\begin{bmatrix} F_{11}(\mathbf{x}) & \cdots & F_{1n}(\mathbf{x}) \\ \vdots & & \vdots \\ F_{n1}(\mathbf{x}) & \cdots & F_{nn}(\mathbf{x}) \end{bmatrix}$$

and an arbitrary vector $\mathbf{v} = [v_1, \cdots, v_n]$. $F(\mathbf{x})$ is *positive semidefinite* if

$$\mathbf{v}\, F(\mathbf{x})\, \mathbf{v} \geq 0, \qquad \forall\, \mathbf{v} \in R^n$$

The transpose of the left-hand vector is implied. $F(x)$ is *positive definite* if

$$\mathbf{v}\, F(\mathbf{x})\, \mathbf{v} > 0, \qquad \forall\, \mathbf{v} \neq 0, \mathbf{v} \in R^n$$

$F(x)$ is strongly positive definite if

$$\mathbf{v}\, F(\mathbf{x})\, \mathbf{v} > \alpha ||v||^2, \qquad \text{for some } \alpha > 0, \qquad \forall\, \mathbf{v} \in R^n$$

If $\gamma(\mathbf{x})$ is the smallest eigenvalue, which is necessarily real, of the symmetric part of $F(\mathbf{x})$, that is, $\frac{1}{2}[F(\mathbf{x}) + F^T(\mathbf{x})]$, then it follows that:

1. $F(\mathbf{x})$ is positive semidefinite if and only if $\gamma(\mathbf{x}) \geq 0$;
2. $F(\mathbf{x})$ is positive definite if and only if $\gamma(\mathbf{x}) > 0$.

We define the Hessian matrix of the objective function $Z(\mathbf{x})$ as

$$\nabla^2 Z(\mathbf{x}) = \begin{bmatrix} \frac{\partial^2 Z(\mathbf{x})}{\partial x_1^2} & \frac{\partial^2 Z(\mathbf{x})}{\partial x_1\, \partial x_2} & \cdots & \frac{\partial^2 Z(\mathbf{x})}{\partial x_1\, \partial x_n} \\ \frac{\partial^2 Z(\mathbf{x})}{\partial x_2\, \partial x_1} & \frac{\partial^2 Z(\mathbf{x})}{\partial x_2^2} & \cdots & \frac{\partial^2 Z(\mathbf{x})}{\partial x_2\, \partial x_n} \\ \vdots & \vdots & & \vdots \\ \frac{\partial^2 Z(\mathbf{x})}{\partial x_n\, \partial x_1} & \frac{\partial^2 Z(\mathbf{x})}{\partial x_n\, \partial x_2} & \cdots & \frac{\partial^2 Z(\mathbf{x})}{\partial x_n^2} \end{bmatrix}$$

To show that a stationary point $\mathbf{x}^*$ is a local minimum, it is sufficient to demonstrate that $Z(\mathbf{x})$ is positive definite in the vicinity of $\mathbf{x} = \mathbf{x}^*$.

3.2.2 General Constraints

A typical nonlinear programming problem is to choose the values of a set of variables, $x_1, x_2, \cdots, x_n$, which minimize an objective function $Z(x_1, x_2, \cdots, x_n)$, subject to certain constraints. Each constraint can be expressed as an inequality of a function $g(x_1, x_2, \cdots, x_n)$ of the variables. The set of possible values of $x_1, x_2, \cdots, x_n$ that comply with all constraints is termed the *feasible region*. A general minimization problem with M constraints can be written as

$$\min_{\mathbf{x}} \quad Z(x_1, x_2, \cdots, x_n) \tag{3.53}$$

s.t.

$$g_1(x_1, x_2, \cdots, x_n) \geq b_1 \tag{3.54}$$

$$g_2(x_1, x_2, \cdots, x_n) \geq b_2 \tag{3.55}$$

$$\vdots$$

$$g_m(x_1, x_2, \cdots, x_n) \geq b_m \tag{3.56}$$

where $g_i(x_1, x_2, \cdots, x_n) \geq b_i$ denotes constraint i on the variables. This problem can include any type of constraints. Using vector notation, the above problem can be written in a standard form as follows

$$\min_{\mathbf{x}} \quad Z(\mathbf{x}) \tag{3.57}$$

s.t.

$$\mathbf{g}(\mathbf{x}) \geq \mathbf{b} \tag{3.58}$$

where $\mathbf{g}(\mathbf{x}) = [g_1(\mathbf{x}), g_2(\mathbf{x}), \cdots, g_m(\mathbf{x})]$ and $\mathbf{b} = (b_1, b_2, \cdots, b_m)$. If $Z(\mathbf{x})$ is a nonlinear function or the constraints are nonlinear, the above problem is termed a nonlinear programming (NLP) problem.

A generalization of the Lagrangian method can be used to derive the first-order necessary conditions for general mathematical programs. The Lagrangian for this program is given by

$$\mathcal{L}(\mathbf{x}, \boldsymbol{\mu}) = Z(\mathbf{x}) + \sum_j \mu_j \left[b_j - g_j(\mathbf{x})\right] \tag{3.59}$$

where μ_j is the Lagrange multiplier for constraint j and $\mu_j \geq 0$.

The stationary point of the Lagrangian of a convex function is not at a minimum or a maximum of $\mathcal{L}(\mathbf{x}, \boldsymbol{\mu})$, but rather at a saddle point of the Lagrangian. In fact, $\mathcal{L}(\mathbf{x}^*, \mathbf{u}^*)$ minimizes $\mathcal{L}(\mathbf{x}, \boldsymbol{\mu})$ with respect to $\mathbf{x}$ and maximizes it with respect to $\boldsymbol{\mu}$. This condition can be stated as

$$\mathcal{L}(\mathbf{x}^*, \boldsymbol{\mu}) \leq \mathcal{L}(\mathbf{x}^*, \boldsymbol{\mu}^*) \leq \mathcal{L}(\mathbf{x}, \boldsymbol{\mu}^*) \tag{3.60}$$

In order to write the first-order necessary conditions of Lagrangian, note that its minimization is unconstrained with respect to $\mathbf{x}$. Maximization with respect to

$\boldsymbol{\mu}$, however, is subject to the nonnegativity constraints. Therefore, the saddle point of $\mathcal{L}(\mathbf{x}, \boldsymbol{\mu})$ satisfies the following set of first-order necessary conditions:

$$\frac{\partial \mathcal{L}(\mathbf{x}^*, \boldsymbol{\mu}^*)}{\partial x_i} = 0 \qquad \forall\, i \tag{3.61}$$

$$\mu_j \frac{\partial \mathcal{L}(\mathbf{x}^*, \boldsymbol{\mu}^*)}{\partial \mu_j} = 0 \qquad \text{and} \qquad \frac{\partial \mathcal{L}(\mathbf{x}^*, \boldsymbol{\mu}^*)}{\partial \mu_j} \leq 0 \qquad \forall\, j \tag{3.62}$$

In addition, it is required that $\mu_j \geq 0$, $\forall\, j$. Condition (3.61) simply states that the gradient vanishes at the stationary point. Condition (3.62) describes the condition for a maximum of a function subject to nonnegativity constraints. Since $\mathcal{L}(\mathbf{x}, \boldsymbol{\mu})$ has to be maximized with respect to $\boldsymbol{\mu} = (\mu_1, \cdots, \mu_m)$, the maximum of $\mathcal{L}(\mathbf{x}, \boldsymbol{\mu})$ with respect to μ_j can occur either at a point where $\partial \mathcal{L}(\mathbf{x}, \boldsymbol{\mu})/\partial \mu_j = 0$ or at a point where $\mu_j = 0$. In the later case, it must be true that $\partial \mathcal{L}(\mathbf{x}, \boldsymbol{\mu})/\partial \mu_j \leq 0$. This observation gives rise to conditions (3.62). Thus, conditions (3.61)-(3.62) can be written explicitly as

$$\frac{\partial Z(\mathbf{x}^*)}{\partial x_i} - \sum_j \mu_j^* \frac{\partial g_j(\mathbf{x}^*)}{\partial x_i} = 0 \qquad \forall\, j \tag{3.63}$$

$$b_j - g_j(\mathbf{x}^*) \leq 0 \qquad \forall\, j \tag{3.64}$$

$$\mu_j \left[b_j - g_j(\mathbf{x}^*) \right] = 0 \qquad \forall\, j \tag{3.65}$$

$$\mu_j^* \geq 0 \qquad \forall\, j \tag{3.66}$$

These necessary conditions (3.63)-(3.66) are called the Karush-Kuhn-Tucker conditions. They are widely used in the analysis of optimality conditions for nonlinear programming problems.

The Lagrangian approach implies that constrained minimization problems can be solved as unconstrained problems by finding the saddle point of the Lagrangian. This point can be found by minimizing the Lagrangian with respect to $\mathbf{x}$ given $\boldsymbol{\mu}$, and then maximizing over all values of $\boldsymbol{\mu}$. The Lagrangian is widely used as an aid in the formulation of first-order necessary conditions.

The functional form of the Lagrangian demonstrates why the Lagrange multipliers, or dual variables, can be interpreted as a measure of the sensitivity of the optimal solution to a constraint relaxation. At the solution point,

$$\mathcal{L}(\mathbf{x}^*, \boldsymbol{\mu}^*) = Z(\mathbf{x}^*) + \sum_j \mu_j \left[b_j - g_j(\mathbf{x}^*) \right] \tag{3.67}$$

At this point $\mathcal{L}(\mathbf{x}^*, \boldsymbol{\mu}^*) = Z(\mathbf{x}^*)$. If constraint k is relaxed by a small amount, Δb_k, and b_k in (3.67) is replaced by $b_k - \Delta b_k$, the new minimum value of $\mathcal{L}(\mathbf{x}^*, \boldsymbol{\mu}^*)$ will approximately equal the old value (before the relaxation) minus $\mu_k \Delta b_k$. Thus a relaxation of constraint k by Δb_k improves the optimal value of the objective function by approximately $\mu_k \Delta b_k$.

3.2.3 Linear Equality and Nonnegativity Constraints

Constrained minimization problems with nonnegativity and equality constraints are of special interest in the study of static network equilibrium problems. They also have an important role in dynamic network equilibrium problems, especially when continuous dynamic problems are transformed into discrete forms. The general form of these problems is

$$\min_{\mathbf{x}} Z(\mathbf{x}) \tag{3.68}$$

s.t.

$$\sum_i a_{ij}\, x_i = b_j \qquad j = 1, 2, \cdots, J \tag{3.69}$$

$$x_i \geq 0 \qquad i = 1, 2, \cdots, I \tag{3.70}$$

To find the first-order necessary conditions for a minimum for such a problem, write the Lagrangian with respect to the equality constraints:

$$\mathcal{L}(\mathbf{x}, \boldsymbol{\mu}) = Z(\mathbf{x}) + \sum_j \mu_j \, [b_j - \sum_i a_{ij}\, x_i] \tag{3.71}$$

Then the stationary point of this Lagrangian has to be determined, subject to the constraint

$$x_i \geq 0 \qquad i = 1, 2, \cdots, I \tag{3.72}$$

Unlike the case discussed previously, this problem includes nonnegativity constraints. Consequently, the stationary point of the NLP program has to be determined by the following conditions:

$$x_i^* \frac{\partial \mathcal{L}(\mathbf{x}^*, \boldsymbol{\mu}^*)}{\partial x_i} = 0 \qquad \text{and} \qquad \frac{\partial \mathcal{L}(\mathbf{x}^*, \boldsymbol{\mu}^*)}{\partial x_i} \geq 0 \qquad \forall\, i \tag{3.73}$$

$$\frac{\partial \mathcal{L}(\mathbf{x}^*, \boldsymbol{\mu}^*)}{\partial \mu_j} = 0 \qquad \forall\, i \tag{3.74}$$

$$x_i \geq 0 \qquad \forall\, i \tag{3.75}$$

Equation (3.74) requires, simply, that the derivatives of $\mathcal{L}(\mathbf{x}, \boldsymbol{\mu})$ with respect to $\boldsymbol{\mu}$ vanish at the minimum. No other condition is necessary since the values of the $\boldsymbol{\mu}$ are not constrained to be nonnegative. This condition, then, is identical to the original constraints. The first-order conditions for the NLP programs with linear equality and nonnegativity constraints can be written explicitly as follows:

$$x_i^* [\frac{\partial Z(\mathbf{x}^*)}{\partial x_i} - \sum_j \mu_j^* a_{ij}\,] = 0 \qquad \forall\, i \tag{3.76}$$

$$\frac{\partial Z(\mathbf{x}^*)}{\partial x_i} - \sum_j u_j^* a_{ij} \geq 0 \qquad \forall\, i \tag{3.77}$$

$$\sum_i a_{ij}\, x_i^* = b_j \qquad \forall\, j \tag{3.78}$$

$$x_i^* \geq 0 \qquad \forall\, i \tag{3.79}$$

The same conditions can be derived also by applying the Karush-Kuhn-Tucker conditions (3.63)-(3.66) directly.

3.2.4 Discrete Optimal Control and Nonlinear Programs

Section 3.1 described the relationship between discrete and continuous optimal control problems. In this section, we explore the relationship between discrete optimal control and nonlinear programming problems. We recognize both as being multivariate extremization problems subject to various equality and inequality constraints. There are several approaches to reduce discrete optimal control problems to nonlinear programming problems. In the following, we only present a simple transformation to show the analytical relationship. As an example, we discuss how we can transform the following discrete optimal control problem into an NLP. The discrete optimal control problem is as follows:

$$\min_{\mathbf{x},\mathbf{u}} \; J = \sum_{k=1}^{K} F[\mathbf{x}(k), \mathbf{u}(k), k] \; + \; S[\mathbf{x}(K+1), K+1] \tag{3.80}$$

s.t.

$$\mathbf{x}(k+1) = \mathbf{x}(k) + \mathbf{f}[\mathbf{x}(k), \mathbf{u}(k), k] \tag{3.81}$$

$$K \text{ and } \mathbf{x}(1) = \mathbf{x}_1 \text{ fixed; } \; \mathbf{x}(K+1) \text{ free} \tag{3.82}$$

We define a vector $\mathbf{y}$ as follows:

$$\mathbf{y}_1 = \mathbf{u}(1)$$

$$\vdots$$

$$\mathbf{y}_K = \mathbf{u}(K)$$

$$\mathbf{y}_{K+1} = \mathbf{x}(1)$$

$$\vdots$$

$$\mathbf{y}_{2K+1} = \mathbf{x}(K+1)$$

Note that $\mathbf{y}_i$, $i = 1, \cdots, 2K+1$, must satisfy all corresponding constraints and boundary conditions of the discrete optimal control problem. The objective function can be rewritten as

$$\min_{\mathbf{y}} \ Z(\mathbf{y}) = \sum_{i=1}^{K} F[\mathbf{y}_{K+i}, \mathbf{y}_i, i] \ + \ S[\mathbf{y}_{2K+1}, 2K+1] \tag{3.83}$$

The state equation and boundary conditions can be rewritten as

$$\mathbf{y}_{K+i+1} = \mathbf{y}_{K+i} + \mathbf{f}[\mathbf{y}_{K+i}, \mathbf{y}_i, i] \qquad \forall\, i = 1, 2, \cdots, K \tag{3.84}$$

$$\mathbf{y}_{K+1} \text{ given} \tag{3.85}$$

Thus, the discrete optimal control problem can be written in the form of the nonlinear programming problem:

$$\min_{\mathbf{y}} \ Z(\mathbf{y}) \tag{3.86}$$

s.t.

$$\mathbf{g}(\mathbf{y}) \leq \mathbf{0} \tag{3.87}$$

If the discrete optimal control problem was generated from a continuous optimal control problem, clearly the solution to the associated nonlinear programming problem will only approximate the solution to the original continuous control problem. From the discussion of the discrete optimal control problem and the nonlinear programming problem, we can see that if control variables $\mathbf{u}(k)$ and state variables $\mathbf{x}(k)$ are considered as individual variables and the state equations are transformed into corresponding NLP constraints, then the discrete optimal control problem is in fact a kind of nonlinear programming problem. Thus, solution algorithms for nonlinear programming problems can be used for solving discrete optimal control problems.

3.3 Solution Algorithms

In this section, we present several solution algorithms for nonlinear programming problems. These algorithms are also suitable for solving corresponding discrete optimal control problems. To simplify our presentation, we concentrate on the most widely used algorithms for dynamic transportation network equilibrium problems. For other algorithms, we refer readers to mathematical programming texts. We first present the interval reduction algorithm for a one-dimensional minimization problem. Then, the Frank-Wolfe algorithm is discussed. Although the algorithms are presented for static nonlinear programming problems, they are extended to dynamic problems subsequently.

3.3.1 One Dimensional Minimization

In this section, we consider the minimization of a nonlinear function $Z(x)$ of a single variable x. It is well known from elementary calculus that the necessary condition for a differentiable function in one variable, $Z(x)$, to have a minimum at $x = x^*$ is that the derivative of $Z(x)$ evaluated at x^* equals zero.

$$\frac{dZ(x^*)}{dx} = 0 \tag{3.88}$$

This is a first-order condition for a minimum. If there is a minimum at x^*, this condition must hold. To prove that this stationary point is a minimum, we need to prove that it is a global minimum. In other words, the value of $Z(x)$ is lower than $Z(x)$ at any other x.

A sufficient condition for a stationary point to be a global minimum is that the function is strictly convex . The strict convexity condition is equivalent to requiring that the second derivative of $Z(x)$ be positive, that is

$$\frac{d^2Z(x^*)}{dx^2} > 0 \tag{3.89}$$

We now discuss the methods for determining x^*. It is assumed that x lies within some finite interval $[a, b]$ and $Z(x)$ is continuous and uniquely defined everywhere in that interval. These requirements ensure the existence of a finite minimum of $Z(x)$ for some x in the interval. For simplicity, we assume $Z(x)$ is ditonic (has one extreme point) over the interval $[a, b]$, implying that it has only a single, unique minimum in that interval.

The study of one-dimensional optimization methods is important mainly because such an optimization or line search is often a part of an algorithm designed to find a minimum of multivariate functions. We discuss two basic methods, the bisection and golden section methods. Both methods use the interval reduction approach. The interval reduction approach involves iterative procedures in which each iteration is focused on a current interval. The current interval is a portion of $[a, b]$, denoted as $[a^n, b^n]$. This interval must be determined to include the minimal point x^*. At each iteration, this interval is examined and divided into two parts: the part in which the minimum cannot lie and the current interval for the next iteration. The part in which the minimum cannot lie is discarded and the procedure is repeated for the new current interval. The procedure starts by designating $[a, b]$ as the first current interval, i.e., $a^0 = a$ and $b^0 = b$. The interval is then reduced at each successive iteration until a small enough current interval (smaller than a prespecified value) for x^* is obtained. Basically we have two interval reduction methods which differ from each other only in the rules used to examine the current interval and to decide which portion of it can be discarded.

Bisection Method

The bisection method exploits the fact that the function is monotonic on each side of the minimum. In other words, the derivative of the function, $dZ(x)/dx$, is negative for $x < x^*$ and positive for $x > x^*$. The algorithm computes the derivative of $Z(x)$ at the midpoint of the current interval, $[a^n, b^n]$, at iteration n. Denote this point as x^n. If $dZ(x^n)/dx < 0$, then $x^* > x^n$. Thus, the interval $[a^n, x^n]$ can be discarded. The next current interval will be $[x^n, b^n]$. If $dZ(x^n)/dx > 0$, then $x^* < x^n$. Thus, the interval $[x^n, b^n]$ can be discarded. The next current interval will be $[a^n, x^n]$. A prespecified convergence criterion $|a^n - b^n| \leq \epsilon$ (ϵ is a very small value) can be used to terminate the procedure and the middle point of the remaining interval is taken as the estimate of x^*.

Golden Section Method

The golden section method is based on a comparison of the values of $Z(x)$ at two points, x_1^n and x_2^n, where $x_1^n < x_2^n$. The two points are within the current interval, $[a^n, b^n]$, at iteration n. The two interior points are determined by using a reduction ratio of 0.618 or precisely $(\sqrt{5}-1)/2$. The interior points are selected so that x_1^n is 0.618 of the current interval length to the left of b^n and x_2^n is 0.618 of the current interval length to the right of a^n. Since the value of 0.618 is known as the golden section, this one-dimensional search method is termed the golden section method.

At iteration n, if $Z(x_1^n) > Z(x_2^n)$, the optimum must lie to the right of $Z(x_1^n)$ because the function is ditonic. Thus, the interval $[a^n, x_1^n]$ can be discarded. The new interval for iteration $n+1$ is $[a^{n+1}, b^{n+1}]$, where $a^{n+1} = x_1^n$ and $b^{n+1} = b^n$. If $Z(x_1^n) < Z(x_2^n)$, the optimum must lie to the left of $Z(x_2^n)$ because the function is ditonic. Thus, the interval $[x_2^n, b^n]$ can be discarded. The new interval for iteration $(n+1)$ is $[a^{n+1}, b^{n+1}]$, where $a^{n+1} = a^n$ and $b^{n+1} = x_2^n$. The interval reduction process continues with two new interior points x_1^{n+1} and x_2^{n+1} for iteration $n+1$. Thus, at each iteration, the golden section procedure makes use of one of the interior points from the last interval; only one new point needs to be calculated at each iteration.

The bisection method has a faster convergence speed than the golden section method. However, it requires computing the derivatives of $Z(x)$ at each iteration. On the other hand, the golden section method involves computing the function $Z(x)$ itself. Thus, the bisection algorithm can be preferable to the golden section method if calculating the derivatives is easier than calculating the function itself.

3.3.2 Frank-Wolfe Algorithm

The Frank-Wolfe (F-W) algorithm was originally suggested by Frank and Wolfe (1956) as a procedure for solving quadratic programming problems with linear constraints. It is also known as the convex combination method. This method is extensively used in determining equilibrium flows in static transportation network problems. In this book, it is extended to solve the dynamic transportation network equilibrium problems.

We consider a convex minimization program with linear constraints:

$$\min_{\mathbf{x}} \quad Z(\mathbf{x}) \tag{3.90}$$

s.t.

$$\sum_i a_{ij} \, x_i \geq b_j \qquad \forall \, j \tag{3.91}$$

where a_{ij} and b_j are constant coefficients ($i = 1, \cdots, I; j = 1, \cdots, J$). The algorithm is basically a feasible descent direction method. At iteration $(n+1)$, it generates a point $\mathbf{x}^{n+1} = (x_1^{n+1}, \cdots, x_I^{n+1})$ from $\mathbf{x}^n = (x_1^n, \cdots, x_I^n)$ so that $Z(\mathbf{x}^{n+1}) < Z(\mathbf{x}^n)$. Thus, the essence of this algorithm is the calculation of $\mathbf{x}^{n+1}$ from $\mathbf{x}^n$. The algorithmic step can be written in a standard form as

$$\mathbf{x}^{n+1} = \mathbf{x}^n + \alpha^n \, \mathbf{d}^n \tag{3.92}$$

where $\mathbf{d}^n = (d_1^n, \cdots, d_I^n)$ is a descent direction vector and α^n is a nonnegative scalar known as the step size or move size. This equation means that at each point $\mathbf{x}^n$, a direction $\mathbf{d}^n$ is identified along which the function is decreasing. Then, the step size α^n determines how far the next point $\mathbf{x}^{n+1}$ will be along the direction $\mathbf{d}^n$.

The F-W method selects the feasible descent direction not only based on how steep each candidate direction is in the vicinity of $\mathbf{x}^n$, but also according to how far it is possible to move along this direction. It chooses a direction based on the product of the rate of descent in the vicinity of $\mathbf{x}^n$ in a given direction and the length of the feasible region in that direction. This product is the drop or the possible reduction in the objective function value which can be achieved by moving in this direction. Thus, the algorithm uses the direction that maximizes the drop.

To find a descent direction, the algorithm checks the entire feasible region for an auxiliary feasible solution, $\mathbf{y}^n = [y_1^n, \cdots, y_I^n]$, such that the direction from $\mathbf{x}^n$ to $\mathbf{y}^n$ provides the maximum drop. In seeking the feasible direction, the bounding of the move size does not require a separate step of the algorithm. The bounding is accomplished as an integral part of choosing the decent direction. The direction from $\mathbf{x}^n$ to any feasible solution, $\mathbf{y}$, is the vector $(\mathbf{y} - \mathbf{x}^n)$ (or the unit vector $(\mathbf{y} - \mathbf{x}^n)/||\mathbf{y} - \mathbf{x}^n||$). The slope of $Z(\mathbf{x}^n)$ in the direction of $(\mathbf{y} - \mathbf{x}^n)$ is given by the projection of the opposite gradient $[-\nabla Z(\mathbf{x}^n)]$ in this direction, i.e.,

$$-\nabla Z(\mathbf{x}^n) \cdot \frac{(\mathbf{y} - \mathbf{x}^n)}{||\mathbf{y} - \mathbf{x}^n||} \tag{3.93}$$

The drop in the objective function in the direction $(\mathbf{y} - \mathbf{x}^n)$ is obtained by multiplying this slope by the distance from $\mathbf{x}^n$ to $\mathbf{y}$, $||\mathbf{y} - \mathbf{x}^n||$, i.e.,

$$-\nabla Z(\mathbf{x}^n) \cdot (\mathbf{y} - \mathbf{x}^n) \tag{3.94}$$

This term has to be maximized (in $\mathbf{y}$) subject to the feasibility of $\mathbf{y}$. Alternatively, the term can be multiplied by (-1) and minimized. It follows that

$$\min \nabla Z(\mathbf{x}^n) \cdot (\mathbf{y} - \mathbf{x}^n) = \sum_i \frac{\partial Z(\mathbf{x}^n)}{\partial x_i}(y_i - x_i^n) \tag{3.95}$$

s.t.

$$\sum_i a_{ij}\, y_i \geq b_j \qquad \forall\, j \tag{3.96}$$

where constraint set (3.96) is equivalent to the original constraint set (3.91) by replacing $\mathbf{x}$ with $\mathbf{y}$. Thus, finding the descent direction amounts to solving a linear program, in which y_i is the decision variable. Note that $\nabla Z(\mathbf{x}^n)$ is constant at $\mathbf{x}^n$ and the term $\nabla Z^T(\mathbf{x}^n) \cdot (\mathbf{x}^n)$ can be discarded from the objective function. Thus, the linearized problem can be simplified as

$$\min \ F^n(\mathbf{y}) = \nabla Z(\mathbf{x}^n) \cdot \mathbf{y} = \sum_i \frac{\partial Z(\mathbf{x}^n)}{\partial x_i}\, y_i \tag{3.97}$$

s.t.

$$\sum_i a_{ij}\, y_i \geq b_j \qquad \forall\, j \tag{3.98}$$

The objective function coefficients are as follows:

$$\partial Z(\mathbf{x}^n)/\partial x_1,\ \partial Z(\mathbf{x}^n)/\partial x_2,\ \cdots,\ \partial Z(\mathbf{x}^n)/\partial x_I.$$

These coefficients are the derivatives of the original objective function at $\mathbf{x}^n$, which is known at iteration n. The decision variables of program (3.95)-(3.96) are $\mathbf{y}^n = (y_1^n, y_2^n, \cdots, y_I^n)$ and the decent direction is the vector pointing from $\mathbf{x}^n$ to $\mathbf{y}^n$, i.e., $\mathbf{d}^n = (\mathbf{y}^n - \mathbf{x}^n)$, or in an expanded form, $d_i^n = y_i^n - x_i^n$, $\forall\, i$. Once the decent direction is known, other algorithmic steps involve the determination of the move size and a convergence test.

As in many other descent methods, the move size in the direction of d^n equals the distance to the point along $\mathbf{d}^n$ which minimizes $Z(\mathbf{x})$. The F-W method does not require a special step to bracket the search for an optimal move size in order to maintain feasibility. The new solution, $\mathbf{x}^{n+1}$, must lie between $\mathbf{x}^n$ and $\mathbf{y}^n$. Because $\mathbf{y}^n$ is a solution of the linearized problem, it naturally lies at the boundary of the feasible region. In other words, the search for a descent direction automatically generates a bound for a line search by accounting for all constraints when the descent direction is determined. Since the search interval is bracketed, the bisection or golden section method can be used to find the step size α by solving the minimization of $Z(\mathbf{x})$ along $\mathbf{d}^n = (\mathbf{y}^n - \mathbf{x}^n)$. It follows that

$$\min_{0 \leq \alpha \leq 1} \ Z[\mathbf{x}^n + \alpha(\mathbf{y}^n - \mathbf{x}^n)] \tag{3.99}$$

Once the optimal solution of this line search, α^n, is found, the next point can be generated using the following formula

$$\mathbf{x}^{n+1} = \mathbf{x}^n + \alpha^n(\mathbf{y}^n - \mathbf{x}^n) \tag{3.100}$$

Note that equation (3.100) can be written as

$$\mathbf{x}^n = (1 - \alpha^n)\mathbf{x}^n + \alpha^n \mathbf{y}^n$$

The new solution is thus a convex combination (or a weighted average) of $\mathbf{x}^n$ and $\mathbf{y}^n$. An appropriate convergence criterion is to check the lower bound of the objective function at each iteration. By convexity,

$$Z(\mathbf{x}^*) \geq Z(\mathbf{x}^n) + \nabla Z(\mathbf{x}^n) \cdot (\mathbf{x}^* - \mathbf{x}^n) \tag{3.101}$$

The value of the linearized objective function yields a lower bound at $Z(\mathbf{x}^n)$.

$$LB(\mathbf{x}^n) = Z(\mathbf{x}^n) + \nabla Z(\mathbf{x}^n) \cdot (\mathbf{y} - \mathbf{x}^n) \tag{3.102}$$

An appropriate convergence criterion is

$$\nabla Z(\mathbf{x}^n) \cdot (\mathbf{y} - \mathbf{x}^n)/LB(x^n) < \epsilon \tag{3.103}$$

The numerator of equation (3.103) is sometimes called the *gap*.

The F-W algorithm can be summarized as follows:

Step 0: Initialization.
Find a feasible solution $\mathbf{x}^0$. Set iteration counter $n := 0$.

Step 1: Direction Finding.
Find $\mathbf{y}^n$ that solves the linear program (3.97)-(3.98).

Step 2: Step Size Determination.
Find α^n that solves

$$\min_{0 \leq \alpha \leq 1} Z[\mathbf{x}^n + \alpha(\mathbf{y}^n - \mathbf{x}^n)] \tag{3.104}$$

Step 3: Move.
Set $\mathbf{x}^{n+1} = \mathbf{x}^n + \alpha^n(\mathbf{y}^n - \mathbf{x}^n)$.

Step 4: Convergence Test.
If $\nabla Z(\mathbf{x}^n) \cdot (\mathbf{y} - \mathbf{x}^n)/LB(x^n) < \epsilon$, stop; otherwise, set $n := n+1$ and go to Step 1.

The algorithm converges in a finite number of iterations. Since the F-W algorithm involves a minimization of a linear program as part of the direction-finding step, it is useful only in cases in which this linear program can be solved relatively easily. It is also useful when algorithms which are generally more efficient than the F-W method can not be utilized due to the size of the problem. Minimization problems for dynamic transportation networks possess both properties: they include a large number of variables and the linear program associated with the direction finding step can be efficiently solved.

3.4 Notes

Discrete optimal control problems have been discussed in various optimal control texts. At the elementary level, readers may consult the text by Sage and White (1977) and Bryson and Ho (1975). The necessary conditions for general nonlinear programming problems were provided by Kuhn and Tucker (1951). The material reviewed for mathematical programming can be found in any standard mathematical programming text, such as Luenberger (1984). Further discussion of reducing discrete optimal control problems to mathematical programming problems can be found in texts by Tabak and Kuo (1971) and Canon et al (1970). The computational algorithms for mathematical programming problems are summarized by Bazaraa et al (1993). The F-W algorithm and its applications in static transportation equilibrium problems are described in detail by Sheffi (1985).

Problems

3.1 Solve the discrete optimal control problem

$$\min \qquad J = \sum_{k=1}^{3} [u^2(k) + x^2(k)]$$

s.t.

$$x(k+1) = x(k) + u(k), \qquad x(1) = 1.$$

3.2 Show that the Hessian matrix for function $z(x_1, x_2)$ is positive definite.

$$z(x_1, x_2) = 10x_1^2 - 3x_1x_2 + 5x_2^2 + 10$$

3.3 Does the standard NLP have a unique minimum if $Z(\mathbf{x})$ and the constraint set are convex?

$$\min_{\mathbf{x}} \ Z(\mathbf{x})$$

s.t.

$$\mathbf{g}(\mathbf{x}) \geq \mathbf{b},$$

3.4 Use the KKT conditions to solve the nonlinear programming problem:

$$\min \qquad z = 8x_1^2 - 4x_1x_2 + 6x_2^2$$

s.t.

$$x_1 + x_2 = 10$$

$$x_1, x_2 \geq 0.$$

3.5 Compare the efficiency of the golden section and bisection methods.

3.6 Use the golden section method to find the minimum of $z = x^2 - 4x + 3$ over the interval [0,3]. Determine the optimal value of x within 3% of the initial interval. How many iterations are needed to achieve an estimate of x^* with an error less than 0.01? Solve the same problem using the bisection method.

3.7 Use the Frank-Wolfe method to solve the following problem:

$$\min \qquad z = 5(x_1 - 5)^2 + 2(x_2 - 3)^2$$

s.t.

$$x_1 + x_2 \leq 10$$

$$2x_1 - x_2 \geq 3$$

$$x_1, x_2 \geq 0.$$

Chapter 4

Network Flow Constraints and Definitions of Travel Times

In this chapter, the constraints for dynamic traffic necessary for a urban transportation network are presented. These constraints include flow conservation for links and nodes, flow propagation, first-in-first-out (FIFO) and oversaturation. Associated with these constraints and different needs for dynamic travel time information, two definitions of travel time are considered.

4.1 Flow Conservation Constraints

The multiple origin-destination network flow problem is considered. A traffic network is represented by a directed graph $\mathcal{G} = (\mathcal{N}, \mathcal{A})$, where $\mathcal{N}$ is the set of nodes and $\mathcal{A}$ is the set of directed links. A node can represent an origin and a destination, as well as an intersection. In the following, the index r will denote an origin node and the index s will denote a destination node.

Consider the fixed time period [0,T], which is long enough to allow all travelers departing during the peak period to complete their trips. Let

$x_a(t)$ = number of vehicles traveling on link a at time t;
$x_{ap}^{rs}(t)$ = number of vehicles traveling on link a over route p with origin r and destination s at time t.

In the following, all variables with subscript p and superscripts rs denote the variables with route p, origin r and destination s. It follows that

$$\sum_{rsp} x_{ap}^{rs}(t) = x_a(t) \qquad \forall a. \tag{4.1}$$

Let $u_a(t)$ denote the inflow rate (vehicles/hour or vehicles/minute) into link a at time t and $v_a(t)$ denote the exit flow rate from link a at time t. The inflow $u_a(t)$ and exit flow $v_a(t)$ on link a are both regarded as control variables.

The number of vehicles $x_a(t)$ on link a is defined as the state variable for link a. The state equation for link a can then be written as

$$\frac{dx_{ap}^{rs}(t)}{dt} = u_{ap}^{rs}(t) - v_{ap}^{rs}(t) \qquad \forall a, p, r, s. \tag{4.2}$$

Figure 4.1 illustrates the flow variables for link a.

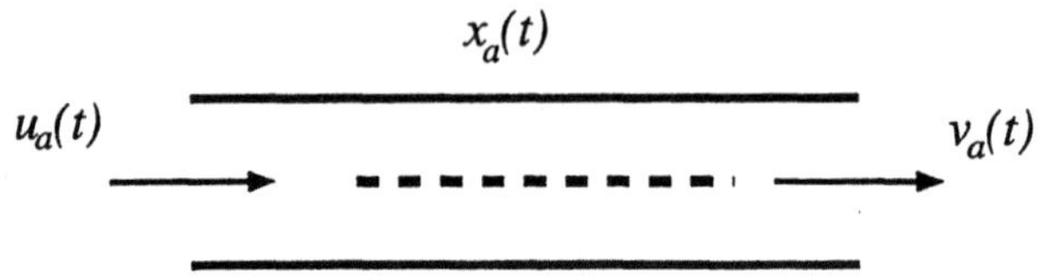

Figure 4.1: Flow Variables for Link a

The number of vehicles on link a at an initial time $t = 0$ is assumed to equal 0:

$$x_{ap}^{rs}(0) = 0, \qquad \forall a, p, r, s. \tag{4.3}$$

Therefore, the number of vehicles on link a at any time t is given by

$$x_{ap}^{rs}(t) = \int_0^t [u_{ap}^{rs}(\omega) - v_{ap}^{rs}(\omega)] d\omega \qquad \forall a, p, r, s. \tag{4.4}$$

If the number of vehicles on link a at an initial time $t = 0$ is not equal to 0,

$$x_{ap}^{rs}(0) > 0, \qquad \forall a, p, r, s, \tag{4.5}$$

then, the number of vehicles on link a at any time t is given by

$$x_{ap}^{rs}(t) = x_{ap}^{rs}(0) + \int_0^t [u_{ap}^{rs}(\omega) - v_{ap}^{rs}(\omega)] d\omega \qquad \forall a, p, r, s. \tag{4.6}$$

In most models in this book, we consider the case when $x_{ap}^{rs}(0) = 0$. For the case when $x_{ap}^{rs}(0) > 0$, the models and corresponding solution algorithms require modification at time 0 accordingly. Finally, all variables must be nonnegative at all times:

$$x_{ap}^{rs}(t) \geq 0, \qquad u_{ap}^{rs}(t) \geq 0, \qquad v_{ap}^{rs}(t) \geq 0, \qquad \forall a, p, r, s. \tag{4.7}$$

Denote the required instantaneous flows from origin node r to destination node s at time t as $f^{rs}(t)$, which is a given function of time in any dynamic route choice problem. Flow conservation at node j ($j \neq r, s$) for route p between O-D pair rs requires that the flow exiting from the link pointing into node j at time t equals the flow entering the link which leave node j at time t. Thus, the flow conservation equations can be expressed as

$$\sum_{a \in B(j)} v_{ap}^{rs}(t) = \sum_{a \in A(j)} u_{ap}^{rs}(t) \qquad \forall j \neq r, s; p, r, s \tag{4.8}$$

where $A(j)$ is the set of links whose tail node is j (after j), and $B(j)$ is the set of links whose head node is j (before j). Figure 4.1 illustrates flow conservation at node j.

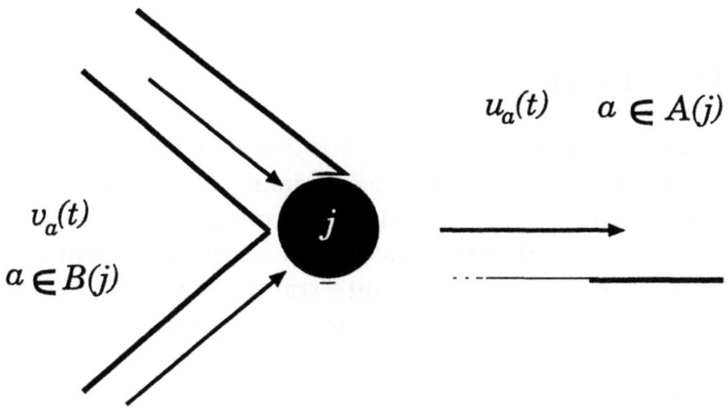

Figure 4.2: Flow Conservation at Node j

Conservation of flow at origin r ($r \neq s$) requires $f^{rs}(t)$, the flow departing from origin r towards destination s at time t, to equal the flow entering links which leave origin r at time t. Thus, the flow conservation equations for the origin nodes can be expressed as

$$\sum_{a \in A(r)} \sum_{p} u_{ap}^{rs}(t) = f^{rs}(t) \qquad \forall r \neq s; s. \tag{4.9}$$

Denote the instantaneous flows arriving at destination s from origin r at time t as the control variable $e^{rs}(t)$, and let $e_p^{rs}(t)$ denote these flows over route p at time t. Conservation of flow at destination s ($s \neq r$) requires the flow exiting at destination s at time t to equal the flow exiting links which lead to destination s at time t. Thus, the flow conservation equations for the destination nodes can be expressed as

$$\sum_{a \in B(s)} \sum_{p} v_{ap}^{rs}(t) = e^{rs}(t) \qquad \forall r; s \neq r. \tag{4.10}$$

Denote the cumulative number of vehicles arriving at destination s from origin r over route p by time t as $E_p^{rs}(t)$. It follows that

$$\frac{dE_p^{rs}(t)}{dt} = e_p^{rs}(t) \qquad \forall p, r, s \neq r. \tag{4.11}$$

At the initial time $t = 0$,

$$E_p^{rs}(0) = 0, \qquad \forall p, r, s. \tag{4.12}$$

Finally, the variables must be nonnegative at all times:

$$e_p^{rs}(t) \geq 0, \qquad E_p^{rs}(t) \geq 0, \qquad \forall p, r, s. \tag{4.13}$$

4.2 Definitions

The *instantaneous* link travel time at any time t is defined as the travel time that would be experienced by vehicles traversing a link when prevailing traffic conditions remain unchanged. The instantaneous route travel time at any time t is the sum of the instantaneous link travel times over all links in this route at time t. Thus, the instantaneous route travel time would be experienced by a vehicle if prevailing traffic conditions do not vary until the vehicle reaches its destination.

The instantaneous travel time $c_a[x_a(t), u_a(t), v_a(t)]$, or simply $c_a(t)$, over link a is assumed to be dependent on the number of vehicles $x_a(t)$, the inflow $u_a(t)$ and the exit flow $v_a(t)$ on link a at time t. We assume the instantaneous travel time $c_a(t)$ on link a is the sum of two components: 1) an instantaneous flow-dependent cruise time $g_{1a}[x_a(t), u_a(t)]$ over link a; 2) an instantaneous queuing delay $g_{2a}[x_a(t), v_a(t)]$. It follows that

$$c_a(t) = g_{1a}[x_a(t), u_a(t)] + g_{2a}[x_a(t), v_a(t)]. \tag{4.14}$$

The two components $g_{1a}[x_a(t), u_a(t)]$ and $g_{2a}[x_a(t), v_a(t)]$ are assumed to be nonnegative and differentiable with respect to $x_a(t), u_a(t)$ and $x_a(t), v_a(t)$, respectively.

Consider the flow which originates at node r at time t and is destined for node s. There is a set of routes $\{p\}$ between O-D pair rs. Define the instantaneous travel time function $\psi_p^{rs}(t)$ for each route p between pair rs as

$$\psi_p^{rs}(t) = \sum_{a \in rsp} c_a[x_a(t), u_a(t), v_a(t)] \qquad \forall r, s, p; \tag{4.15}$$

the summation is over all links a in route p from origin r to destination s.

Define the minimal instantaneous route travel time $\sigma^{rs}(t)$ as the minimal travel time that would be experienced by a vehicle departing from origin r to destination s at time t, if prevailing traffic conditions do not vary until the vehicle reaches its destination. If the instantaneous link travel time $c_a(t)$ is determined, the minimal instantaneous O-D travel time $\sigma^{rs}(t)$ can be computed as $\sigma^{rs}(t) = \min_p \psi_p^{rs}(t)$, where $\sigma^{rs}(t)$ is a functional of all link flow variables at time t, or $\sigma^{rs}(t) = \sigma^{rs}[u(t), v(t), x(t), t]$.

The future link travel time or actual link travel time is the travel time over a link actually experienced by vehicles. This time can also be called the projected time. Define $\tau_a(t)$ as the actual travel time over link a for vehicles entering link a at time t. $\tau_a(t)$ is assumed to be dependent on the number of vehicles $x_a(t)$, the inflow $u_a(t)$ and the exit flow $v_a(t)$ on link a at time t. It follows that

$$\tau_a(t) = \tau_a[x_a(t), u_a(t), v_a(t)] \qquad \forall a \qquad (4.16)$$

Similarly, the future route travel time or actual route travel time is the time actually experienced over a route by vehicles. Define $\eta_p^{rs}(t)$ as the travel time *actually* experienced over route p by vehicles departing origin r toward destination s at time t. We use a recursive formula to compute the route travel time $\eta_p^{rs}(t)$ for all allowable routes. Assume route p consists of nodes $(r, 1, 2, \cdots, i, \cdots, s)$. Denote $\eta_p^{ri}(t)$ as the travel time *actually* experienced over route p from origin r to node i by vehicles departing origin r at time t. Then, a recursive formula for route travel time $\eta_p^{rs}(t)$ is:

$$\eta_p^{ri}(t) = \eta_p^{r(i-1)}(t) + \tau_a[t + \eta_p^{r(i-1)}(t)] \qquad \forall p, r, i; i = 1, 2, \cdots, s;$$

where link $a = (i-1, i)$.

Define $\pi^{rs}(t)$ as the minimal travel time *actually* experienced by motorists departing from origin r to destination s at time t. If the actual link travel time $\tau_a(t)$ is determined, the minimal actual O-D travel time $\tau^{rs}(t)$ can be computed as $\pi^{rs}(t) = \min_p \eta_p^{rs}(t)$. $\pi^{rs}(t)$ is a functional of all link flow variables at time $\omega \geq t$, or $\pi^{rs}(t) = \pi^{rs}[u(\omega), v(\omega), x(\omega)|\omega \geq t]$. This functional is neither a state variable nor a control variable, and it is not fixed. This functional is not available in closed form. Nevertheless, it can be evaluated when $u(\omega)$, $v(\omega)$ and $x(\omega)$ are temporarily fixed.

Various dynamic route choice models are formulated in the following chapters based on actual and instantaneous travel times. More discussion on definitions of travel times and their applications is found in Chapter 13.

4.3 Flow Propagation Constraints

It is necessary to ensure that the entering and exiting flows, as well as the vehicles remaining on links, are consistent with the link travel times. We write these constraints using actual link travel times. In static network models, these flow propagation constraints are not necessary because a flow is assumed to propagate instantaneously over its entire journey from its origin to its destination. Vehicles do not remain on a link for a duration of time in order that a queue can form. Thus, queuing phenomena cannot be captured correctly in any static network model. Flow propagation constraints can be written based on links or nodes. In this book, we suggest using link flow propagation constraints, which can be formulated in three different ways as now described.

4.3.1 Type I

Let $U_{ap}^{rs}(t)$ denote the cumulative number of vehicles entering link a on route p with O-D pair rs by time t, and $V_{ap}^{rs}(t)$ denote the cumulative number of vehicles leaving link a on route p with O-D pair rs by time t. $U_{ap}^{rs}(t)$ and $V_{ap}^{rs}(t)$ are state variables for link a. The state equation for link a can then be written as

$$\frac{dU_{ap}^{rs}(t)}{dt} = u_{ap}^{rs}(t) \qquad \forall a, p, r, s. \tag{4.17}$$

$$\frac{dV_{ap}^{rs}(t)}{dt} = v_{ap}^{rs}(t) \qquad \forall a, p, r, s. \tag{4.18}$$

The cumulative numbers of vehicles entering and exiting link a at an initial time $t = 0$ are assumed to equal 0:

$$U_{ap}^{rs}(0) = 0, \qquad V_{ap}^{rs}(0) = 0, \qquad \forall a, p, r, s. \tag{4.19}$$

When the number of vehicles on link a at an initial time $t = 0$ equals 0, the number of vehicles at any time t is given by

$$x_{ap}^{rs}(t) = U_{ap}^{rs}(t) - V_{ap}^{rs}(t) \qquad \forall a, p, r, s. \tag{4.20}$$

Figure 4.3 illustrates the relationship of cumulative entering and exiting flows on link a. Finally, all variables must be nonnegative at all times:

$$x_{ap}^{rs}(t) \geq 0, \qquad U_{ap}^{rs}(t) \geq 0, \qquad V_{ap}^{rs}(t) \geq 0, \qquad \forall a, p, r, s. \tag{4.21}$$

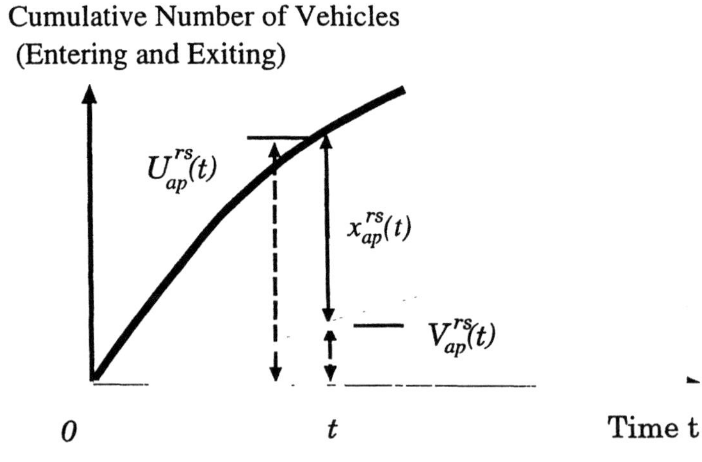

Figure 4.3: Cumulative Entering and Exiting Flows on Link a

The node flow conservation equations defined in the last section do not change. For link a on route p between O-D pair rs, the total number of vehicles entering link a by time t must have exited link a by time $[t + \tau_a(t)]$. It follows that

$$U_{ap}^{rs}(t) = V_{ap}^{rs}[t + \tau_a(t)] \qquad \forall a, p, r, s. \tag{4.22}$$

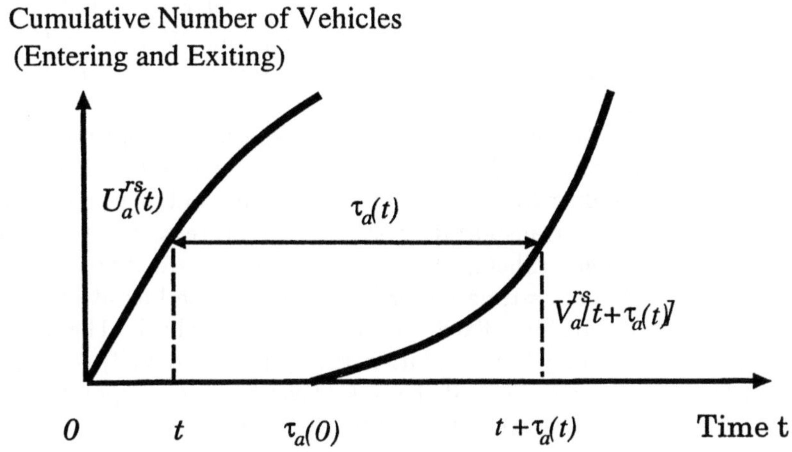

Figure 4.4: Flow Propagation on Link a

Figure 4.4 illustrates the flow propagation on link a. The expression of this flow propagation constraint is very simple and its meaning is also intuitive. However, the following flow propagation constraints are easier to handle in optimal control problems and are used in most of our models.

4.3.2 Type II

We formulate the second type of link flow propagation constraints as follows. Let $x_{ap}^{rs}(t)$ denote the number of vehicles on link a using route p between O-D pair rs at time t. By definition,

$$\sum_{rsp} x_{ap}^{rs}(t) = x_a(t) \qquad \forall a. \tag{4.23}$$

For any intermediate node $j \neq r$ on route p, denote a subroute $\tilde{p}$ as the section of route p from node j to destination s. For any link $a \in B(j)$, vehicles on link a using route p at any time t must result in either:

1. extra vehicles on downstream links on subroute $\tilde{p}$ at time $t + \tau_a(t)$, or
2. increased exiting vehicles at the destination at time $t + \tau_a(t)$.

It follows that

$$x_{ap}^{rs}(t) = \sum_{b \in \tilde{p}} \{x_{bp}^{rs}[t + \tau_a(t)] - x_{bp}^{rs}(t)\} + \{E_p^{rs}[t + \tau_a(t)] - E_p^{rs}(t)\}$$

$$\forall a, j, p, r, s; j \neq r; a \in B(j). \tag{4.24}$$

A detailed discussion of this constraint is given in Chapter 5 following the formulation of an instantaneous DUO route choice model. We note that the flow conservation constraints discussed in the previous section can be used directly when the above link flow propagation constraint is used in the formulation.

In some of our dynamic network models, especially in optimization formulations, we need to write these constraints using *estimates* of actual link travel times. These link travel time estimates must be updated in an iterative procedure known in the transportation science literature as the *relaxation* or *diagonalization technique* (Sheffi, 1985). In this procedure, the travel times over each link a, $\bar{\tau}_a(t)$, are estimated for flows entering the link at each time t. These functions $\bar{\tau}_a(t)$ are held fixed, and the model is solved. Then, the link travel times corresponding to the solution $x_a(t)$, $u_a(t)$ and $v_a(t)$ obtained are compared to the functions $\bar{\tau}_a(t)$. If the link travel times corresponding to the solution are different from $\bar{\tau}_a(t)$, the values of $\bar{\tau}_a(t)$ are reset to these travel times and the process is repeated. Given the robust nature of the relaxation (diagonalization) technique, we expect that the solution will converge to the DUO solution. This procedure is discussed in detail in Chapter 6 as a standard approach for solving a general variational inequality model. The revised flow propagation constraints are as follows:

$$x_{ap}^{rs}(t) = \sum_{b\in\tilde{p}} \{x_{bp}^{rs}[t + \bar{\tau}_a(t)] - x_{bp}^{rs}(t)\} + \{E_p^{rs}[t + \bar{\tau}_a(t)] - E_p^{rs}(t)\}$$

$$\forall a, j, p, r, s; j \neq r; a \in B(j). \qquad (4.25)$$

These constraints are associated with state variables only. This property is fully exploited in the analysis of the optimality conditions for optimal control models in Chapter 5.

In the later chapters, the above two sets of flow propagation constraints are used in appropriate models so to simplify the formulation and to improve the effectiveness of the corresponding computational algorithms. As we noted above, the flow propagation constraints can be formulated using node-based constraints instead of link-based constraints. Thus, other flow propagation constraints can be formulated. However, these constraints should be decided in conjunction with the corresponding formulation of the dynamic model.

4.3.3 Type III

In type I and II flow propagation constraints, it is assumed that some platoon dispersion or concentration is allowed in flow propagation. In other words, the inflow rate $u_{ap}^{rs}(t)$ is not necessarily equivalent to exit flow rate $v_{ap}^{rs}[t + \tau_a(t)]$ after the link travel time $\tau_a(t)$. This type of constraint may incur difficulties in computational solutions for large networks. Therefore, for simplicity of computational efficiency, it can be assumed that the inflow rate $u_{ap}^{rs}(t)$ is equivalent to exit flow rate $v_{ap}^{rs}[t + \tau_a(t)]$ after the link travel time $\tau_a(t)$. It follows that

$$u_{ap}^{rs}(t) = v_{ap}^{rs}[t + \tau_a(t)] \qquad \forall a, p, r, s. \qquad (4.26)$$

Figure 4.5 illustrates the flow propagation constraint on link a. Since the curve representing the cumulative number of exiting vehicles has the same slope at time $[t + \tau_a(t)]$ as that for the cumulative number of entering vehicles at time t, the two curves must have the same shape so that link travel time (horizontal distance between two curves) must be constant at any time t. This conclusion contradicts the concept of time-varying link travel times. Therefore, the continuous link flow propagation constraint (4.26) is invalid, if the curve of the cumulative number of exiting vehicles is continuous.

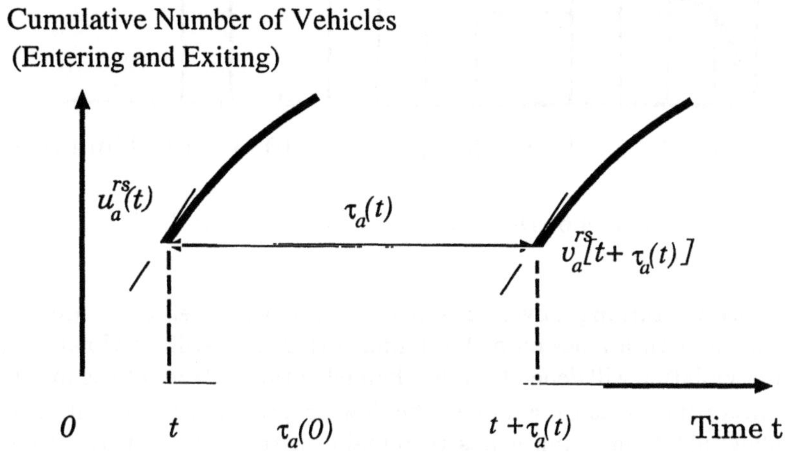

Figure 4.5: No-Dispersion Flow Propagation on Link a

A discrete form of link flow propagation constraint (4.26) is valid and easy to implement computationally, because the discontinuity is handled by the discretization of time. The discrete link flow propagation constraint is:

$$u_{ap}^{rs}(k) = v_{ap}^{rs}[k + \tau_a(k)] \qquad \forall a, p, r, s, k \tag{4.27}$$

where $u_{ap}^{rs}(k)$ is the inflow rate on link a over route p for O-D pair rs during interval k and $v_{ap}^{rs}(k)$ is the exit flow rate on link a over route p for O-D pair rs during interval k. Figure 4.6 illustrates this discrete flow propagation constraint on link a. The link travel times are given as:

$$\tau(1) = 5, \qquad \tau(2) = 6, \qquad \tau(3) = 6, \qquad \tau(4) = 8$$

4.4 First-In-First-Out Constraints

First-In-First-Out (FIFO) conditions may or may not occur in actuality. However, FIFO should be strictly guaranteed when there is only one lane and no

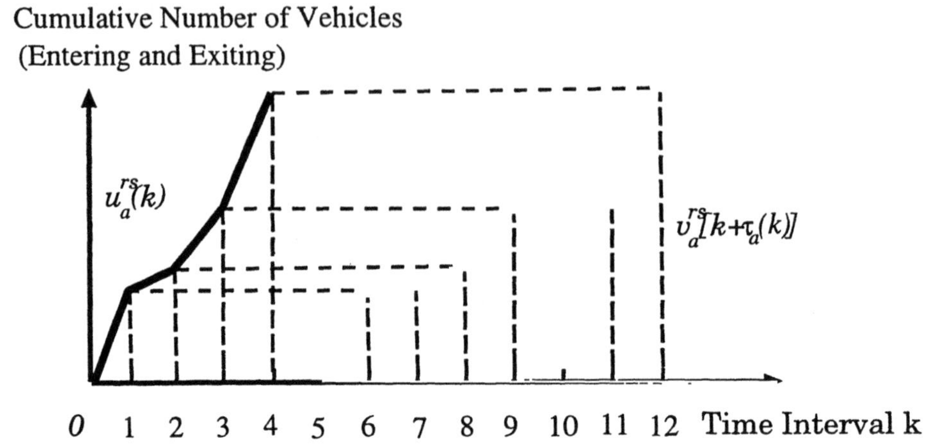

Figure 4.6: Discrete Flow Propagation on Link a

extra space for turning movements at intersections. When there are both left-turn and right-turn lanes on a street link, FIFO may still be violated and the extent of violation will depend on the channelization of lanes at the intersection.

In a continuous time model, the flow propagation constraints state that any inflow $u_a^{rs}(t)$ into link a has to remain on the link for travel time $\tau_a(t)$, regardless of the origin-destination source of this inflow. Thus, the flow propagation constraints imply FIFO constraints in a continuous time model if a rigorous travel time function is used in the flow propagation constraints. In the following, we use a discrete time example to illustrate how the flow propagation constraints should imply FIFO constraints. The example network is shown in Figure 4.7. It is 3-link, 4-node network (links 1-2, 2-3, 3-4) with O-D trips $f^{13}(k) = (0,1,0,1,0,1,0,1,0,1)$ and $f^{14}(k) = (1,0,1,0,1,0,1,0,1,0)$ for time interval $k = 1, \cdots, 10$. Therefore, one vehicle enters link 1-2 in each time interval $k = 1, \cdots, 10$.

Figure 4.7: Example Network

The FIFO condition requires that one vehicle exit link 1-2 and go to node 4 and then another go to node 3, etc. We assume that link 2-3 is blocked due to some incident at $k = 1$. When the first vehicle arrives at node 2 and enters link 2-3, it cannot exit node 3 and enter link 2-4 since the travel time is extremely high and it has to remain link 2-3. Thus, the link travel time on link 2-3 will increase for subsequent vehicles. Our physical flow propagation constraint for link flow and travel time states that the second vehicle entering link 1-2 during period 2 ($k = 2$) cannot arrive at node 3, since link 2-3 is highly congested. This will increase the travel time for vehicles entering link 1-2 during period 3. This process continues and there will be no vehicles exiting node 3. This framework implicitly defines the FIFO constraints.

In the following, we discuss FIFO or its violation, overtaking, in more detail. Overtaking denotes that a late entering vehicle flow propagates faster than an earlier entering vehicle flow and exits earlier than the earlier entering vehicle flow. Overtaking violates the FIFO rule for traffic propagation on links, although it might happen on two-lane links. We consider traffic propagation on link a for two time instants t and $t + \Delta t$ in continuous time problems. Denote the link travel time for flows entering link a at time t as $\tau_a(t)$. The travel time for flows entering link a at time $t + \Delta t$ is $\tau_a(t + \Delta t)$. When the summation of link travel time at time $t + \Delta t$ plus Δt is smaller than the link travel time at time t, i.e., $\tau_a(t) > \tau_a(t + \Delta t) + \Delta t$, overtaking will occur so that FIFO is violated. Because flows entering link a during time $t + \Delta t$ will exit the link after clock time $t+\Delta t+\tau_a(t+\Delta t)$ which is earlier than the clock time $t+\tau_a(t)$, the exiting time for flows entering link a at time t.

Thus, if we require that overtaking should not occur, we must allow the clock time $t+\tau_a(t)$, when flows entering at time t must exit link a, to be smaller than the clock time $t + \Delta t + \tau_a(t + \Delta t)$, the exiting time for flows entering link a at time $t + \Delta t$. It follows that

$$t + \tau_a(t) < t + \Delta t + \tau_a(t + \Delta t) \tag{4.28}$$

Dividing the above equation by Δt, we obtain

$$1 + \frac{\tau_a(t + \Delta t) - \tau_a(t)}{\Delta t} > 0 \tag{4.29}$$

Taking the limit of the above equation ($\Delta t \to 0$),

$$1 + \dot{\tau}_a(t) > 0 \tag{4.30}$$

or

$$\dot{\tau}_a(t) > -1 \tag{4.31}$$

The above condition must be met to avoid overtaking in any dynamic route choice model using link travel time functions in the flow propagation constraints. If the decreasing rate of travel time on any link a exceeds 1, overtaking will occur.

Even with the flow propagation constraints in a general discrete time model, the FIFO constraints may still be violated when the time interval is quite large. For example, assume link travel times at instants t and $t + \Delta t$ are 2 minutes and 40 seconds, respectively. If $\Delta t = 1$ minute, then

$$\frac{\tau_a(t + \Delta t) - \tau_a(t)}{\Delta t} = \frac{40 - 120}{60} = -1.33 < -1 \tag{4.32}$$

Thus, overtaking does occur in this example. However, condition (4.29) can be satisfied by choosing appropriate link length and time interval length in a discrete time environment.

Some nonconvex conditions can be introduced to guarantee FIFO condition (4.29). However, those constraints greatly increase the complexity of the model and its solution. In the continuous time environment, FIFO condition (4.29) is generally satisfied by realistic travel time functions such as those proposed in Chapter 16. As shown there, our proposed link travel time function for a signal-controlled arterial has three parts:

1. an uncongested cruise time over the first part of the link;

2. a queuing delay at the exit part of the link;

3. a uniform delay due to signal setting.

Since the major delay is the queuing delay which assumes FIFO, the condition, $\dot{\tau}_a(t) > -1$, will be satisfied if the link travel time function is validated using real traffic flow data.

Our major concern is the overtaking problem in discrete models. In particular, overtaking may occur in the afternoon peak period when traffic flow is declining rapidly together with travel time. Recall that by equation (4.32), overtaking does not occur when

$$\frac{\tau_a(t+\Delta t) - \tau_a(t)}{\Delta t} > -1 \tag{4.33}$$

Thus, the length of time interval Δt and the link travel time $\tau_a(t)$, or the link length, determine whether overtaking occurs or not. Note that the link length determines the free flow travel time, which is a major factor affecting equation (4.33) because the queuing delay assumes FIFO on the link. If the time interval Δt increases, link travel time $\tau_a(t)$ must be smaller or the link must be shorter so that equation (4.33) holds. If time interval Δt decreases, link travel time $\tau_a(t)$ must be higher or the link must be longer for equation (4.33) to hold.

The detailed values of time interval Δt and link length associated with any specific link travel time function should be determined using numerical experiments in any practical application. In general, overtaking will not occur for most definitions of link lengths (over 100 feet) and time intervals (shorter than 2 minutes) if the free flow speed is assumed to be 50 miles/hour in the experiment. However, as we note in discrete models, no matter how accurate the link traffic dynamics model is, overtaking or a "jump" may still occur when the time interval is too large. On the other hand, for most problems, we are only interested in the aggregate behavior of flows and FIFO is not so important in those situations. Furthermore, we should note that the FIFO assumption itself is also an approximation of reality.

4.5 Link Capacity and Oversaturation

There are two basic constraints for link capacity. The first constraint is the maximal number of vehicles on the link. The second constraint is the maximal exit flow rate from the link. More detailed analyses of their impacts on dynamic network models are discussed in Chapter 13.

4.5.1 Maximal Number of Vehicles on a Link

Let l_a denote the length of link a and e_{am} denote the maximal traffic density (vehicles/mile). The maximal number of vehicles that link a can accommodate is $l_a e_{am}$. The number of vehicles on link a must be less than or equal to the maximal number of vehicles on the link. It follows that

$$x_a(t) \leq l_a e_{am} \qquad \forall a \tag{4.34}$$

This constraint applies for any dynamic network model. Since this constraint involves only state variables, it could be added to the formulation without incurring any analysis problem in our desired optimality conditions for optimal control models. However, the computational algorithm would need to be revised. When solving a route choice problem including the above constraint, we could add a penalty term to the objective function and solve it as an ordinary dynamic route choice problem.

4.5.2 Maximal Exit Flow from a Link

Another constraint concerns the exit flow capacity v_{am} at the exit of a link. It follows that

$$v_a(t) \leq v_{am} \qquad \forall a \tag{4.35}$$

In a network, the exit capacity constraint for an upstream link is also an inflow capacity constraint for downstream links. This constraint can be added directly in the formulation or combined in the link travel time functions. If it is directly added in the optimal control formulation, more analysis of the optimality conditions of dynamic network models is necessary when the exit flow capacity is reached. Moreover, the computational algorithm needs to be revised. A method of combining this constraint with link travel time functions is discussed in Chapter 13.

In most of our dynamic traffic network models, we consider this exit capacity constraint in the travel time functions. Thus, it is not necessary to define an explicit constraint in these network models.

4.5.3 Constraints for Spillback

Oversaturation may occur anywhere and during any time interval when traffic demand exceeds capacity. When queues at critical intersections develop upstream, then they cause the so-called spillback problem.

In an oversaturated situation, continuing excess demand relative to supply could transform local oversaturation to regional oversaturation. Thus, in dynamic network models, corresponding constraints should be formulated to reflect this phenomenon. In an advanced control/assignment framework, those constraints should be consistent with each type of traffic control strategy. The two main types of traffic control strategies are: 1) minimize delay and stops; 2) keep traffic moving or maximize productivity (Lieberman, 1993). We leave the control policy for further study in the context of combined dynamic travel choice/signal control models. Here, we only formulate constraints that describe physical spillback queues.

An extreme case occurs if the queue on link a approaches the maximum, i.e. $x_a(t) = l_a e_{am}$. In this situation, the queue may extend into upstream links, which causes a spillback problem. However, if the above constraint as well as upper bound constraints for cumulative O-D departures and associated access delays at origin nodes are added in the assignment problem, the spillback problem can still be handled in our dynamic traffic network models.

When a spillback queue develops upstream towards origins, such as parking lots, queuing delays at origins will occur. This is an additional constraint for dynamic network models. We construct a dummy link b associated with each origin node r; see Figure 4.8. The state equation for any origin r is:

$$\frac{dx_{bp}^s(t)}{dt} = f_p^{rs}(t) - v_{bp}^s(t) \qquad \forall b, p, r, s; b \in rs. \tag{4.36}$$

$$\sum_p f_p^{rs}(t) = f^{rs}(t) \qquad \forall r, s. \tag{4.37}$$

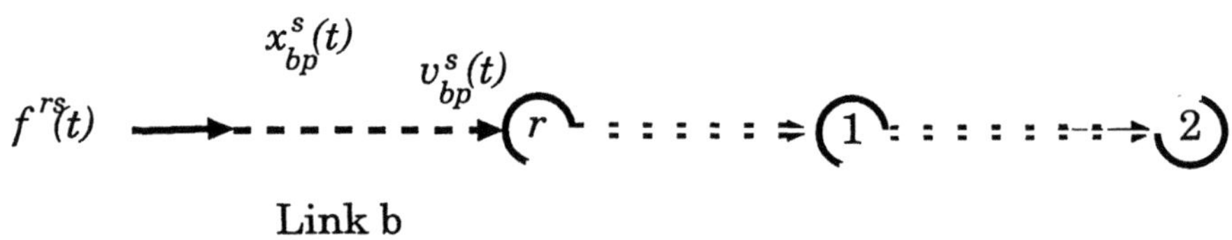

Figure 4.8: Example of Spillback

We assume the number of spillback vehicles at time 0 is 0. It follows that

$$x_{bp}^s(0) = 0 \qquad \forall b, p, r, s; b \in rs. \tag{4.38}$$

Thus, the flow conservation equation for origin r should be revised as

$$\sum_{a \in A(r)} u_{ap}^{rs}(t) = v_{bp}^s(t) \qquad \forall b, p, r, s; b \in rs. \tag{4.39}$$

Of course, we also need the flow propagation constraint for dummy link b. We assume there is no upper bound for the queue length $x_b^s(t)$ since an origin always has enough capacity to accommodate vehicles. The queuing delay at the origin is as follows

$$\tau_b(t) = \tau_b[x_b(t), v_b(t)] \qquad \forall b \in rs \tag{4.40}$$

where

$$x_b(t) = \sum_s \sum_p x^s_{bp}(t) \qquad v_b(t) = \sum_s \sum_p v^s_{bp}(t)$$

The exact form of this formula should be associated with specific queuing patterns at origin nodes. Figure 4.9 illustrates the queuing delay at origin r.

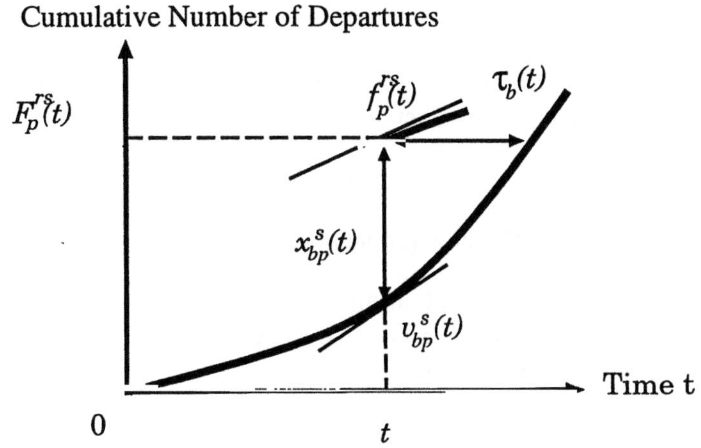

Figure 4.9: Queuing Delay at Origin r

4.6 Notes

Flow conservation constraints are intuitive and essential for any dynamic network model. In contrast with static traffic network models, there are two kinds of travel times in a dynamic traffic network model: instantaneous travel time and actual travel time. These two kinds of travel times reflect the dynamic nature and complexity of dynamic traffic problems.

Flow propagation constraints can also be represented by some alternative constraints which imply flow propagation in flow conservation constraints. Those models include traffic simulation models and the hydrodynamics method. One recent example is the cell transmission model proposed by Daganzo (1993) which assumes traffic flow propagation by cell transmission (flow transfers from cell to cell, where a cell is a small link segment). The advantages of those models are that link flow propagation is directly represented and link travel times are directly provided, while link travel time function is not presented in the model.

For more discussion on FIFO, readers may refer to Carey (1992). Details on oversaturation and spillback can be found in Lieberman (1993).

Problems

4.1 Assume that the link exit flow $g_a(t)$ is dependent on the total number of vehicles $x_a(t)$ on the link. Then, an exit flow function can be defined as:

$$g_a(t) = g_a[x_a(t)].$$

What is the likely shape of this exit flow function? Discuss the advantages and disadvantages of using exit flow functions instead of using exit flow variables.

4.2 Assume that an exit flow function $g_a(t)$ is a continuous, differentiable and nondecreasing function with respect to $x_a(t)$. For example,

$$g_a(t) = \sqrt{x_a(t)}.$$

When does it contradict the flow propagation constraints? Explain.

4.3 Discuss the advantages and disadvantages of using exit flow variables. What kind of conditions are necessary to make them more realistic with regard to the intersection characteristics?

4.4 How many variables are necessary to describe the flow dynamics on a link?

4.5 For the network shown in Figure 4.10, assume that the link travel time functions are

$$c_1(t) = \tau_1(t) = 2 + t$$

$$c_2(t) = \tau_2(t) = 3 + 2t$$

$$c_3(t) = \tau_3(t) = 4 + 1.5t$$

Find the instantaneous route travel time function and actual route travel time function from origin node 1 to destination node 4.

Figure 4.10: Example Network

4.6 All proposed flow propagation constraints in the book are defined as equalities. Is it possible to formulate flow propagation constraints as inequalities? What are the advantages and disadvantages of the inequality formulation?

4.7 For link a, assume there is an inflow pattern $u_a(t) = 2t + 3$ for time period $t \in [0, 3]$ and the link travel time is $\tau_a(t) = 3$ for time period $t \in [0, 10]$. Write the possible link flow propagation constraints and determine the corresponding exit flows. [Hint: assume that there is no flow dispersion on the link.]

4.8 For link a, assume that inflow is $u_a(k) = 3k$ for time intervals $k = 1, 2, 3$ and link travel time is $\tau_a(k) = k+2$ for time intervals $k = 1, 2, 3$. Determine the corresponding exit flows by using the discrete link flow propagation constraints.

4.9 Discuss when and where FIFO should be rigorously satisfied. Provide several examples when FIFO could be violated in the real world.

4.10 Suppose there is a link travel time profile as follows: $\tau_a(k) = (10, 6, 3, 2)$ (minutes) for time intervals $k = 1, 2, 3, 4$. Discuss when FIFO is violated if one interval is 1 minute. Now assume that one time interval is 3 minutes. Will FIFO be violated?

Chapter 5

Ideal Dynamic Route Choice Models

In this chapter, we present both route-based and link-based variational inequality (VI) models for the ideal dynamic user-optimal (DUO) route choice problem. Using a network with two parallel routes, we first present an example to illustrate the ideal dynamic user-optimal concept in Section 5.1. The general definition of ideal dynamic user-optimal state is given in Section 5.2. Then, the route-time-based ideal dynamic user-optimal route choice model is presented in Section 5.3. This model is the most straight-forward formulation of route-time-based, ideal DUO route choice conditions.

Route-time-based VI models have an intuitive interpretation. However, their solution requires explicit route enumeration, a computationally intractable problem for realistic networks. In Section 5.4, therefore, a link-time-based VI model is proposed whose solution does not require explicit route enumeration. We prove that the link-time-based ideal DUO route choice conditions imply the route-time-based ideal DUO route choice conditions. Then, we present a discrete-time numerical example indicating that the ideal dynamic route choice model yields realistic results in Section 5.5.

The above models are also limited in the sense that the travelers are assumed to be homogenous. One way to relax this assumption is to stratify the travelers into classes, in which each class has distinctly different characteristics. In Section 5.6, the the route-based model is extended to multiple classes. In this model, each class of travelers is associated with a disutility or generalized cost function. Thus, the route-based ideal DUO route choice conditions are defined for each class of travelers on the basis of travel disutilities instead of travel time only. Route-cost-based VI models have an intuitive interpretation. However, they are computationally intractable with regard to explicit route enumeration. Therefore, a multi-class link-cost-based VI model is proposed in Section 5.7.

5.1 An Example with Two Parallel Routes

Consider a network with one O-D pair and two parallel routes (see Figure 5.1). Assume that there exists only one bottleneck on each route. For simplicity, we assume that the bottleneck is close to the entry point on each route. Then the route travel times are assumed to depend on the inflow rate and have the following simple form:

$$\tau_1(t) = 10 + 5u_1(t) \tag{5.1}$$

$$\tau_2(t) = 15 + 3u_2(t) \tag{5.2}$$

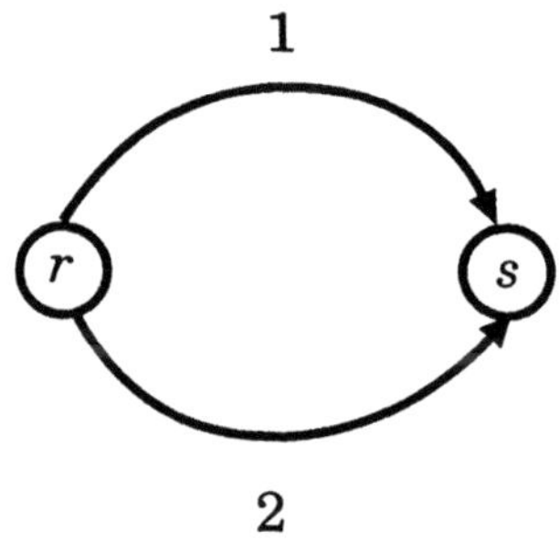

Figure 5.1: Example Network

Suppose there is a departure flow $f(t) = 0.5t$ from origin r to destination s at time $t \in [0,4]$. Flow conservation at origin r requires that

$$u_1(t) + u_2(t) = 0.5t \tag{5.3}$$

In addition, the inflows must be nonnegative, i.e.,

$$u_1(t) \geq 0, \qquad u_2(t) \geq 0 \tag{5.4}$$

The dynamic user-optimal route choice criterion requires that the departing flows use the minimal travel time route. It follows that

$$\tau_1(t) = \tau_2(t) \qquad \text{if } u_1(t) > 0 \text{ and } u_2(t) > 0 \tag{5.5}$$

Solving equations (5.3)-(5.5), we obtain the optimal inflows as follows:

$$u_1^*(t) = \begin{cases} 0.5t & \text{if } t < 2 \\ 0.1875t + 0.625 & \text{if } t \geq 2 \end{cases}$$

$$u_2^*(t) = \begin{cases} 0. & \text{if } t < 2 \\ 0.3125t - 0.625 & \text{if } t \geq 2 \end{cases}$$

The corresponding optimal route travel times are

$$\tau_1^*(t) = \begin{cases} 10. + 2.5t & \text{if } t < 2 \\ 5.125 + 0.9375t & \text{if } t \geq 2 \end{cases}$$

$$\tau_2^*(t) = \begin{cases} 15. & \text{if } t < 2 \\ 5.125 + 0.9375t & \text{if } t \geq 2 \end{cases}$$

The minimal route travel time $\pi^{rs}(t)$ is

$$\pi^{rs}(t) = \begin{cases} \tau_1(t) = 10. + 2.5t & \text{if } t < 2 \\ \tau_1^*(t) = \tau_2^*(t) = 5.125 + 0.9375t & \text{if } t \geq 2 \end{cases}$$

The optimal inflows and optimal route travel times are illustrated in Figure 5.2. Thus, the dynamic user-optimal route choice conditions for this example can be summarized as follows:

$$\tau_a(t) \geq \pi^{rs}(t) \qquad a = 1, 2; \tag{5.6}$$

$$\tau_a(t) = \pi^{rs}(t) \qquad \text{if } u_a > 0 \qquad a = 1, 2; \tag{5.7}$$

$$u_1(t) \geq 0, \qquad u_2(t) \geq 0 \tag{5.8}$$

5.2 Definition of an Ideal State

We now consider a general transportation network. Consider the flow originating at node r at time t and destined for node s. There is a set of routes $\{p\}$ between O-D pair rs. For any link a, link a is defined as being used at time t if $u_a(t) > 0$. Furthermore, a route p between r and s is defined as being used at time t if the departure flow $f_p^{rs}(t) > 0$.

In this chapter, we formulate two alternative variational inequality models, each of which is equivalent to the specific ideal DUO route choice conditions. Our objective here is to explore the complexity of the problem. As shown in Chapter 6, these variational inequality models can be transformed into optimal control models under relaxation. In this chapter, we assume that the time-dependent origin-destination trip pattern is known *a priori*. In other words, the departure times of travelers are given. The ideal dynamic user-optimal (DUO) route choice problem is to determine the dynamic trajectories of link states and inflow and exit flow control variables at each instant of time resulting from drivers using minimal-time routes, given the network, the link travel time functions and the time-dependent O-D departure rate requirements. In this chapter, we consider the following dynamic generalization of the conventional static user-optimal state.

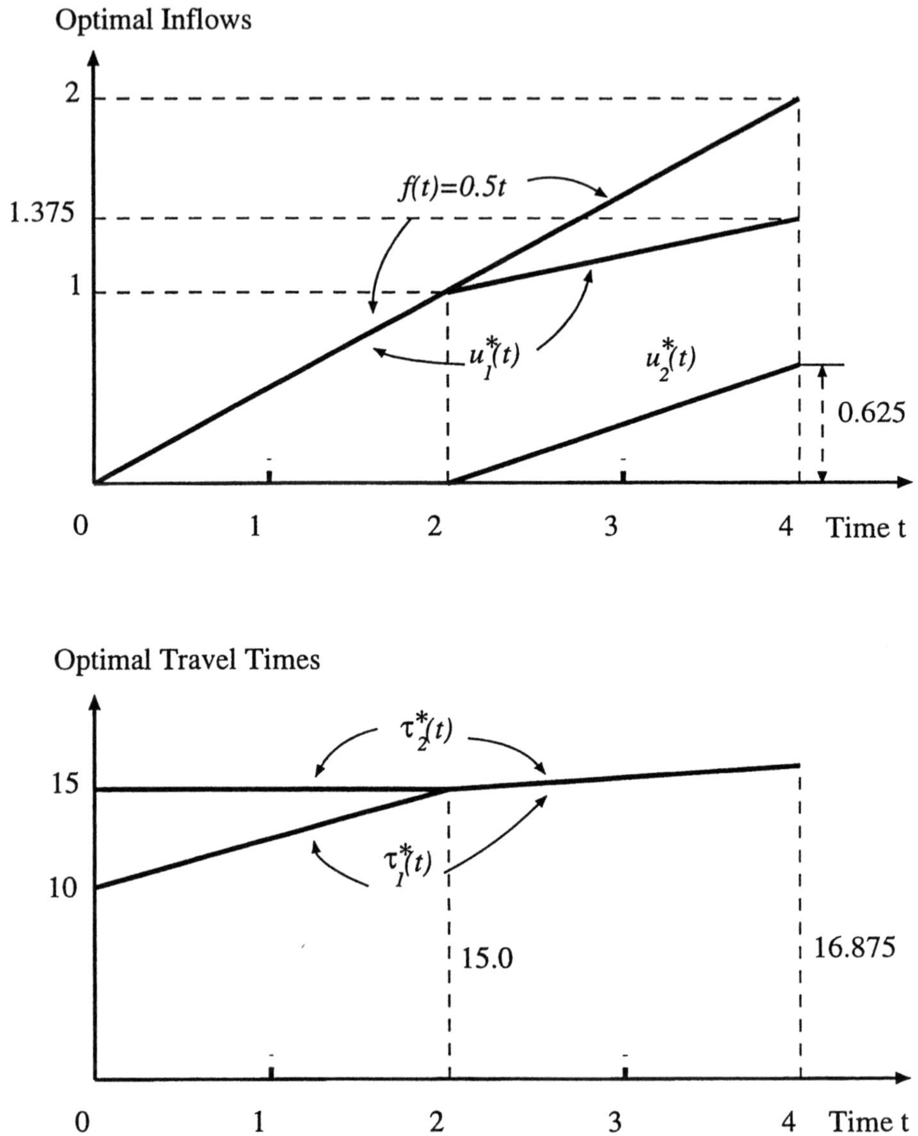

Figure 5.2: Dynamic User-Optimal Inflows and Travel Times

Travel-Time-Based Ideal DUO State: *If, for each O-D pair at each instant of time, the* actual *travel times experienced by travelers departing at the same time are equal and minimal, the dynamic traffic flow over the network is in a travel-time-based* ideal *dynamic user-optimal state.*

The above definition can also be called a predictive or anticipatory DUO model, since the actual route travel time is predicted using the corresponding route choice model. This model assumes each traveler will have perfect information about the future network conditions and will comply with the guidance instructions based on ideal DUO route choice conditions. Travelers will not regret what decisions they have made before their journeys. This model is sometimes described as having "look-ahead" capability. In Section 5.6, we generalize the above definition to the situation of travel disutility and multi-class travelers instead of travel times for a single class of travelers only.

The formulation of the problem in this chapter is based on the underlying choice criterion that each traveler uses the route that minimizes his/her actual travel time when departing from the origin to his/her destination. This route choice rule implies that in a travel-time-based ideal dynamic user-optimal state, the trajectories of the link flow states and inflows and exit flows are such that the actual travel times on all used routes connecting any given O-D pair will be identical and not greater than the actual travel times of routes which are not being used.

5.3 A Route-Time-Based Model

5.3.1 Route-Time-Based Conditions

Recall from Chapter 4 that the actual link travel time is the travel time over a link actually experienced by vehicles. The actual travel time $\tau_a(t)$ over link a for vehicles entering link a at time t is assumed to depend on the number of vehicles $x_a(t)$, the inflow $u_a(t)$ and the exit flow $v_a(t)$ on link a at time t. It follows that

$$\tau_a(t) = \sum_{a \in rsp} \tau_a[x_a(t), u_a(t), v_a(t)] \qquad \forall a \tag{5.9}$$

Suppose we have a set of routes $\{p\}$ and $f_p^{rs}(t)$ is the route inflow from origin r to destination s at time t. Denote $\eta_p^{rs}(t)$ as the actual travel time from origin r to destination s over route p at time t, and $\pi^{rs}(t)$ as the minimal actual route travel time from origin r to destination s at time t. In Chapter 4 we defined a recursive formula to compute the route travel time $\eta_p^{rs}(t)$ for each allowable route as follows. Assume route p consists of nodes $(r, 1, 2, \cdots, i, \cdots, s)$. Denote $\eta_p^{ri}(t)$ as the travel time *actually* experienced over route p from origin r to node i by vehicles departing origin r at time t. Then, a recursive formula for route travel time $\eta_p^{rs}(t)$ is:

$$\eta_p^{ri}(t) = \eta_p^{r(i-1)}(t) + \tau_a[t + \eta_p^{r(i-1)}(t)] \qquad \forall p, r, i; i = 1, 2, \cdots, s;$$

where link $a = (i-1, i)$.

If the actual link travel time $\tau_a(t)$ is determined, the minimal actual O-D travel time $\pi^{rs}(t)$ can be computed as $\pi^{rs}(t) = \min_p \eta_p^{rs}(t)$. $\pi^{rs}(t)$ is a functional of all link flow variables at time ω: $\pi^{rs}(t) = \pi^{rs}[u(\omega), v(\omega), x(\omega)|\omega \geq t]$. This functional is neither a state variable nor a control variable, and it is not fixed. This functional is not available in closed form.

The route-time-based ideal DUO route choice conditions which are equivalent to the ideal DUO definition in the above are now defined as follows.

$$\eta_p^{rs^*}(t) - \pi^{rs^*}(t) \geq 0 \qquad \forall p, r, s; \tag{5.10}$$

$$f_p^{rs^*}(t)\left[\eta_p^{rs^*}(t) - \pi^{rs^*}(t)\right] = 0 \qquad \forall p, r, s; \tag{5.11}$$

$$f_p^{rs}(t) \geq 0 \qquad \forall p, r, s. \tag{5.12}$$

The asterisk in the above equations denotes that the flow variables are the optimal solutions under the travel-time-based ideal DUO state. For any O-D pair rs, if there is a positive inflow over route p, i.e., $f_p^{rs^*}(t) \geq 0$, equation (5.11) requires that

$$\eta_p^{rs^*}(t) = \pi^{rs^*}(t) \qquad \forall p, r, s. \tag{5.13}$$

Thus, route inflow $f_p^{rs^*}(t)$ uses the minimal actual travel time $\pi^{rs^*}(t)$. If the inflow over route p is zero, i.e., $f_p^{rs^*}(t) = 0$, equation (5.11) requires that $[\eta_p^{rs^*}(t) - \pi^{rs^*}(t)]$ be either zero or positive (by equation (5.10)). In other words, route p has either the minimal travel time or higher travel time at time t. On the other hand, if route p has higher travel time at time t, i.e., $\eta_p^{rs^*}(t) > \pi^{rs^*}(t)$, equation (5.11) requires that route p has zero inflow at time t ($f_p^{rs^*}(t) = 0$).

5.3.2 Dynamic Network Constraints

The constraint set for this route-time-based VI model is summarized as follows.

Relationship between state and control variables:

$$\frac{dx_{ap}^{rs}}{dt} = u_{ap}^{rs}(t) - v_{ap}^{rs}(t) \qquad \forall a, p, r, s; \tag{5.14}$$

$$\frac{dE_p^{rs}(t)}{dt} = e_p^{rs}(t) \qquad \forall p, r; s \neq r; \tag{5.15}$$

Flow conservation constraints:

$$f^{rs}(t) = \sum_{a \in A(r)} \sum_p u_{ap}^{rs}(t) \qquad \forall r, s; \tag{5.16}$$

$$\sum_{a \in B(j)} v_{ap}^{rs}(t) = \sum_{a \in A(j)} u_{ap}^{rs}(t) \qquad \forall j, p, r, s; j \neq r, s; \tag{5.17}$$

$$\sum_{a \in B(s)} \sum_{p} v_{ap}^{rs}(t) = e^{rs}(t) \qquad \forall r, s; s \neq r; \tag{5.18}$$

Flow propagation constraints:

$$x_{ap}^{rs}(t) = \sum_{b \in \tilde{p}} \{x_{bp}^{rs}[t + \tau_a(t)] - x_{bp}^{rs}(t)\} + \{E_p^{rs}[t + \tau_a(t)] - E_p^{rs}(t)\}$$

$$\forall a \in B(j); j \neq r; p, r, s; \tag{5.19}$$

Definitional constraints:

$$\sum_{rsp} u_{ap}^{rs}(t) = u_a(t), \qquad \sum_{rsp} v_{ap}^{rs}(t) = v_a(t), \qquad \forall a; \tag{5.20}$$

$$\sum_{rsp} x_{ap}^{rs}(t) = x_a(t), \qquad \forall a; \tag{5.21}$$

Nonnegativity conditions:

$$x_{ap}^{rs}(t) \geq 0, \quad u_{ap}^{rs}(t) \geq 0, \quad v_{ap}^{rs}(t) \geq 0 \quad \forall a, p, r, s; \tag{5.22}$$

$$e_p^{rs}(t) \geq 0, \quad E_p^{rs}(t) \geq 0, \quad \forall p, r, s; \tag{5.23}$$

Boundary conditions:

$$E_p^{rs}(0) = 0, \qquad \forall p, r, s; \tag{5.24}$$

$$x_{ap}^{rs}(0) = 0, \qquad \forall a, p, r, s. \tag{5.25}$$

We illustrate the above network constraints using the example network with three links in Figure 5.3. Assume that O-D flow $f^{14}(t) = 10$ for a short time period $t \in [0, \epsilon]$ ($\epsilon \ll \bar{\tau}_1(t)$) and $f^{14}(t) = 0$ at time $t > \epsilon$. Conservation of flow constraints (5.17) require that $u_1^{14}(t) = 10$ at time $t \in [0, \epsilon]$ (flow enters link 1). Thus, $x_1^{14}(t)$ becomes positive by constraints (5.14) (vehicles are on link 1). When $x_1^{14}(t)$ decreases, constraint (5.14) requires $v_1^{14}(t)$ to become positive (flow exits link 1), and constraint (5.19) for node 2 requires $u_2^{14}(t)$ to equal $v_1^{14}(t)$ (flow enters link 2). Analogously, constraint (5.14) requires $x_2^{14}(t)$ to increase, since $u_2^{14}(t)$ has become positive. Thus, the conservation of flow constraints (together with the flow propagation constraints) require that flow moves from the origin to the destination, successively entering a link, staying on the link, and then exiting the link.

The flow propagation constraints (5.21) for this network are now stated; since there is only one route, we suppress the route subscript p and begin with link 3.

$$x_3^{14}(t) = E^{14}[t + \tau_3(t)] - E^{14}(t) \tag{5.26}$$

Figure 5.3: Example Network with Inflow

$$x_2^{14}(t) = \{x_3^{14}[t+\tau_2(t)] - x_3^{14}(t)\} + \{E^{14}[t+\tau_2(t)] - E^{14}(t)\} \tag{5.27}$$

$$\begin{aligned} x_1^{14}(t) &= \{x_2^{14}[t+\tau_1(t)] - x_2^{14}(t)\} + \{x_3^{14}[t+\tau_1(t)] - x_3^{14}(t)\} \\ &+ \{E^{14}[t+\tau_1(t)] - E^{14}(t)\} \end{aligned} \tag{5.28}$$

These constraints ensure that flows stay on each link for an amount of time consistent with the link's travel time. For example, (5.24) and (5.26) require that $E^{14}[t+\tau_3(t)] = 0$ until $x_3^{14}(t)$ becomes positive. When $x_3^{14}(t)$ does become positive, $E[t+\tau_3(t)]$ must also become positive. Only then can $e[t+\tau_3(t)]$ become positive (see (5.16) and note that this is route $p = 1$), and thus by (5.20), only then can $v_3^{14}[t+\tau_3(t)]$ become positive. This ensures that $v_3^{14}(t)$ cannot become "prematurely" positive; i.e., flows must stay on link 3 for an amount of time consistent with $\tau_3(t)$.

Constraints (5.27) require that flow on link 2 at time t must result in either:

1. *added* flow on link 3 at time $t+\tau_2(t)$ (in case any existing flow on link 3 at time t has not yet cleared by time $t+\tau_2(t)$); or

2. added arrivals at the destination at time $t+\tau_2(t)$ (if link 3 is very short, vehicle flows at the *end of link 2* may have traversed link 3 and exited destination node 4 at time $t+\tau_2(t)$, while vehicle flows at the *beginning of link 2* may still be on link 3 at time $t+\tau_2(t)$).

To illustrate constraint (5.27) further, suppose that vehicle flows first appear on link 2 at time t_o: $x_2^{14}(t_o) > 0$, but $x_2^{14}(t) = 0$ for $t < t_o$. Suppose also that there are no vehicle flows on link 3 and none have exited at time t_o. Thus, $0 = x_2^{14}(t_o - \epsilon)$ (since vehicles first appear on link 2 at time t_o) $= x_3^{14}[t_o - \epsilon + \tau_2(t_o - \epsilon)]$ (by (5.27)) for any $\epsilon \leq t_o$. Thus, $x_3^{14}(\cdot)$ must equal zero at all times prior to time $t_o + \tau_2(t_o)$, so vehicles must stay on link 2 during the entire period $[t_o, \tau_2(t_o)]$. This insures that $x_2^{14}(t)$ propagates consistently with travel time $\tau_2(t)$ over link 2 at each instant t.

Analogously, constraints (5.28) require that flow on link 1 at time t must result in either:

1. added flows on link 2 or 3 at time $t+\tau_1(t)$ (in case flows on links 2 and 3 at time t have not yet cleared by time $t+\tau_1(t)$); or

2. added arrivals at the destination at time $t+\tau_1(t)$.

Thus, this equation insures that $x_1^{14}(t)$ propagates consistently.

5.3.3 The Variational Inequality Problem

The equivalent variational inequality formulation of route-time-based ideal DUO route choice conditions (5.10)-(5.12) may be stated as follows.

Theorem 5.1. *The dynamic traffic flow pattern satisfying network constraint set (5.14)-(5.25) is in a route-time-based ideal DUO route choice state if and only if it satisfies the variational inequality problem:*

$$\int_0^T \sum_{rs}\sum_{p} \eta_p^{rs*}(t)\ \left[f_p^{rs}(t) - f_p^{rs*}(t)\right]\ dt \ \geq\ 0 \tag{5.29}$$

Proof of Necessity. [1]

We need to prove that the ideal DUO route choice conditions (5.10)-(5.12) imply variational inequality (5.29). For any route p, a feasible inflow at time t is

$$f_p^{rs}(t) \geq 0. \tag{5.30}$$

Multiplying ideal DUO route choice condition (5.10) by the above equation, we have

$$f_p^{rs}(t)\ [\eta_p^{rs*}(t) - \pi^{rs*}(t)] \geq 0 \qquad \forall p, r, s. \tag{5.31}$$

We subtract equation (5.11) from equation (5.31) and obtain

$$\left[f_p^{rs}(t) - f_p^{rs*}(t)\right]\ \left[\eta_p^{rs*}(t) - \pi^{rs*}(t)\right] \geq 0 \qquad \forall p, r, s. \tag{5.32}$$

Summing equation (5.32) for all routes p and all O-D pairs rs, it follows that

$$\begin{aligned} &\sum_{rs}\sum_{p} \left[f_p^{rs}(t) - f_p^{rs*}(t)\right]\ \left[\eta_p^{rs*}(t) - \pi^{rs*}(t)\right] \\ =\ &\sum_{rs}\sum_{p} \left[f_p^{rs}(t) - f_p^{rs*}(t)\right]\ \eta_p^{rs*}(t) - \sum_{rs} \pi^{rs*}(t) \sum_{p} \left[f_p^{rs}(t) - f_p^{rs*}(t)\right] \\ =\ &\sum_{rs}\sum_{p} \left[f_p^{rs}(t) - f_p^{rs*}(t)\right]\ \eta_p^{rs*}(t)\ \geq\ 0 \end{aligned} \tag{5.33}$$

where the flow conversation equation

$$\sum_{p} f_p^{rs}(t) = \sum_{p} f_p^{rs*}(t) = f^{rs}(t)$$

holds for each O-D rs at each time t. Integrating the above equation (5.33) from 0 to T, we obtain variational inequality (5.29).

---

[1] The necessity and sufficiency proofs can be skipped without loss of continuity.

Proof of Sufficiency.

We need to prove that any solution $f_p^{rs^*}(t)$ to variational inequality (5.29) satisfies ideal DUO route choice conditions (5.10)-(5.12). We know that the first and third ideal DUO route choice conditions (5.10) and (5.12) hold by definition. Thus, we need to prove that the second ideal DUO route choice condition (5.11) also holds.

Assume that the second ideal DUO route choice condition (5.11) does not hold only for a route q for O-D pair kn during time interval $[t_1-\delta, t_1+\delta] \in [0, T]$, i.e.,

$$f_q^{kn^*}(t) > 0 \quad \text{and} \quad \eta_q^{kn^*}(t) - \pi^{kn^*}(t) > 0 \quad \forall t \in [t_1 - \delta, t_1 + \delta] \tag{5.34}$$

Since the second ideal DUO route choice condition (5.11) holds for all routes other than route q for O-D pair kn at time t, it follows that

$$\int_0^T \sum_{rs} \sum_p f_p^{rs^*}(t) \, [\eta_p^{rs^*}(t) - \pi^{rs^*}(t)] \, dt$$
$$= \int_{t_1-\delta}^{t_1+\delta} f_q^{kn^*}(t) \; [\eta_q^{kn^*}(t) - \pi^{kn^*}(t)] \, dt \; > 0 \tag{5.35}$$

We note that all other terms in the above equation vanish because of ideal DUO route choice condition (5.11).

For each O-D pair rs, we can always find one minimal actual travel time route l for vehicles departing origin r at time t, where route l was evaluated under the optimal flow pattern $\{f_p^{rs^*}(t)\}$. For this route l, the first ideal DUO route choice condition (5.10) becomes an equality by definition. It follows that

$$\eta_l^{rs^*}(t) - \pi^{rs^*}(t) = 0 \qquad \forall l, r, s. \tag{5.36}$$

Next, we need to find a set of feasible route inflows $f_p^{rs}(t)$ so that the following equations

$$f_p^{rs}(t) \; \left[\eta_p^{rs^*}(t) - \pi^{rs^*}(t)\right] = 0 \qquad \forall p, r, s \tag{5.37}$$

always hold. We consider all the departure flows $f^{rs}(t)$ for all O-D pairs at each time t. For each O-D pair rs at each time t, we assign O-D departure flow $f^{rs}(t)$ to the minimal travel time route l, which was evaluated under the optimal flow pattern $\{f_p^{rs^*}(t)\}$. This generates a set of feasible route inflow patterns $\{f_p^{rs}(t)\}$ which always satisfy equation (5.37) because flows are not assigned to routes with non-minimal travel times which were evaluated under the optimal route inflow pattern $\{f_p^{rs^*}(t)\}$. Summing equations (5.37) for all routes p and all O-D pairs rs, it follows that

$$\sum_{rs} \sum_p f_p^{rs}(t) \; \left[\eta_p^{rs^*}(t) - \pi^{rs^*}(t)\right] = 0 \tag{5.38}$$

Integrating the above equation for time period $[0, T]$, we have

$$\int_0^T \sum_{rs} \sum_p f_p^{rs}(t) \left[\eta_p^{rs^*}(t) - \pi^{rs^*}(t)\right] dt = 0 \tag{5.39}$$

We subtract equation (5.35) from equation (5.39) and obtain

$$\begin{aligned}
&\int_0^T \sum_{rs} \sum_p \left[f_p^{rs}(t) - f_p^{rs^*}(t)\right] \left[\eta_p^{rs}(t) - \pi^{rs}(t)\right] dt \\
= &\int_0^T \left\{ \sum_{rs} \sum_p \left[f_p^{rs}(t) - f_p^{rs^*}(t)\right] \eta_p^{rs}(t) \right. \\
- &\left. \sum_{rs} \pi^{rs}(t) \sum_p \left[f_p^{rs}(t) - f_p^{rs^*}(t)\right] \right\} dt \\
= &\int_0^T \sum_{rs} \sum_p \eta_p^{rs^*}(t) \left[f_p^{rs}(t) - f_p^{rs^*}(t)\right] dt < 0
\end{aligned} \tag{5.40}$$

where the flow conservation

$$\sum_p f_p^{rs}(t) = \sum_p f_p^{rs^*}(t) = f^{rs}(t)$$

holds for each O-D pair rs at each time t so that the second term vanishes. The above equation contradicts VI problem (5.29). Therefore, any optimal solution $\{f_p^{rs^*}(t)\}$ to variational inequality (5.29) satisfies the second ideal DUO route choice condition (5.11). Since we proved the necessity and sufficiency of the equivalence of variational inequality (5.29) to route-time-based ideal DUO route choice conditions (5.10)-(5.12), the proof is complete.

5.4 A Link-Time-Based Model

The VI model presented above is route-based. Because the dynamic traffic flow does not have constant flow rate during propagation over links and routes, the route-based VI can not be transformed into a link-based VI. Thus, it is very difficult to develop a solution algorithm for a route-based VI without explicit route enumeration. Although the route-based model is intuitive in terms of understanding, route enumeration is a great burden if the network is large, which is termed the curse of dimensionality in optimal control theory. This issue is the most critical constraint for applying a route-based VI to realistic transportation networks. Therefore, we propose a link-based VI to overcome this problem.

Figure 5.4 shows a 5×5 one-way square grid network with $N = 25$ nodes and $L = 40$ links. The total number of routes from node 1 to node 25

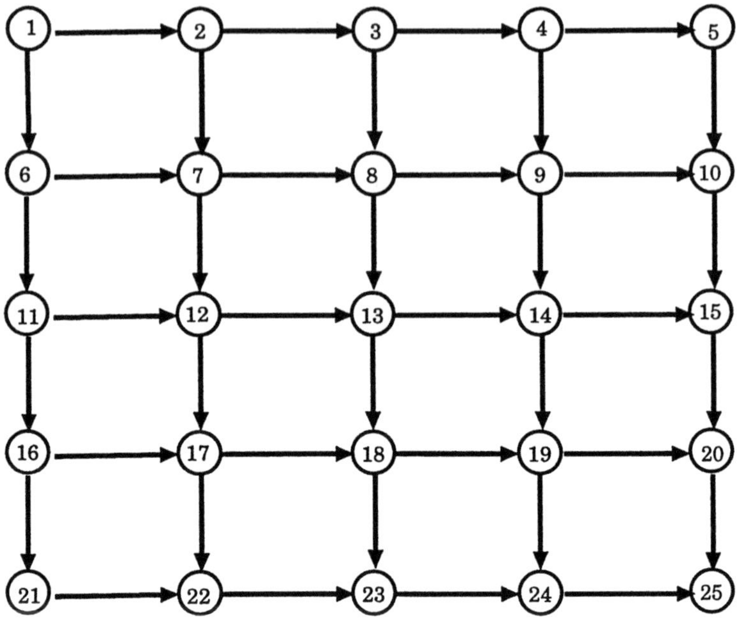

Figure 5.4: Example Grid Network

(both are on the diagonal line) and the total number of routes is 70. Table 5.1 illustrates the increase of links and routes with the increase of nodes in such a grid network. Basically, the number of links increases linearly with the increase of nodes. However, the number of routes increases exponentially with the increase of nodes. For example, when there are only 100 nodes, the number of routes is 48,620. When there are 400 nodes, the number of routes is over 3.5×10^{10}. We note that the routes in these one-way grid networks are efficient routes in the sense of the definition of Dial (1971); i.e., any link on the route takes the vehicles further away from the origin and closer to the destination. From Table 5.1, we can conclude that explicit route enumeration is infeasible for large networks.

Table 5.1: Number of Nodes, Links and Routes

Number of Nodes	4	9	16	25	36	49	64	100	400
Number of Links	4	12	24	40	60	84	112	180	760
Number of Routes	2	6	20	70	252	924	3432	48,620	3.5×10

To distinguish the link-based model from the route-based model, we can focus on the variational inequality instead of the sets of constraints because both constraints are route-based. If the variational inequality can be formulated using link-based variables instead of route-based variables, a carefully designed solution algorithm for such a variational inequality will not require explicit route enumeration. We will demonstrate this point later. This is also true for optimization problems. If the objective function of an optimization program can be formulated using link-based variables instead of route-based variables, a carefully designed solution algorithm will not need explicit route enumeration. Recall that in the mathematical programming formulation for the static user-optimal route choice model, the objective function can be formulated using either link-based variables or route-based variables. However, the link-based objective function is preferred because for such a link-based objective function, the solution algorithm does not need explicit route enumeration. This property is also true for the variational inequality models in this book.

The set of dynamic network constraints for the link-time-based VI model is identical to constraint set (5.14)-(5.25) of the route-time-based VI model in Section 5.3. The basic difference between the two models is that the VI itself of the link-time-based VI model is formulated using link-based flow variables instead of route-based variables as in the route-time-based VI model.

5.4.1 Link-Time-Based Conditions

We introduce a new set of ideal DUO route choice conditions based on link and node variables, instead of route-based variables. As in Section 5.2, the definition of the travel-time-based ideal DUO route choice state is given as follows.

> **Travel-Time-Based Ideal DUO State:** *If, for each O-D pair at each instant of time, the* actual *travel times experienced by travelers departing at the same time are equal and minimal, the dynamic traffic flow over the network is in a travel-time-based* ideal *dynamic user-optimal state.*

We now write the equivalent mathematical inequalities for the above definition using link variables, in contrast to the route-based formulation in Section 5.3. In this case, for any route from origin r to destination s, link a is defined as being used at time t if $u_a^{rs}(t) > 0$. Define $\pi^{ri^*}(t)$ as the minimal travel time *actually* experienced by vehicles departing origin r to node i at time t, the asterisk denoting that the travel time is computed using ideal DUO traffic flows. For link $a = (i, j)$, the minimal travel time $\pi^{rj^*}(t)$ from origin r to j should be equal to or less than the minimal travel time $\pi^{ri^*}(t)$ from origin r to i plus the actual link travel time $\tau_a[t + \pi^{ri}(t)]$ at time instant $[t + \pi^{ri}(t)]$, where this time instant is the earliest clock time when the flow departing origin r at time t can enter link a. It follows that

$$\pi^{ri^*}(t) + \tau_a[t + \pi^{ri^*}(t)] \geq \pi^{rj^*}(t) \qquad \forall a = (i, j), r.$$

If, for each O-D pair rs, any departure flow from origin r at time t enters link a at the earliest clock time $[t+\pi^{ri^*}(t)]$, or $u_a^{rs}[t+\pi^{ri^*}(t)] > 0$, then the ideal DUO route choice conditions require that link a is on the minimal travel time route. In other words, the minimal travel time $\pi^{rj^*}(t)$ from origin r to j should equal the minimal travel time $\pi^{ri^*}(t)$ from origin r to i plus the actual link travel time $\tau_a[t+\pi^{ri^*}(t)]$ at time instant $[t+\pi^{ri^*}(t)]$. It follows that

$$\pi^{rj^*}(t) = \pi^{ri^*}(t) + \tau_a[t+\pi^{ri^*}(t)], \qquad \text{if } u_a^{rs^*}[t+\pi^{ri^*}(t)] > 0 \qquad \forall a=(i,j), r, s.$$

The above equation is also equivalent to the following:

$$\left[\pi^{ri^*}(t) + \tau_a[t+\pi^{ri^*}(t)] - \pi^{rj^*}(t)\right] \; u_a^{rs^*}[t+\pi^{ri^*}(t)] = 0 \qquad \forall a=(i,j), r, s.$$

Thus, the link-time-based ideal DUO route choice conditions can be summarized as follows:

$$\pi^{ri^*}(t) + \tau_a[t+\pi^{ri^*}(t)] - \pi^{rj^*}(t) \;\geq\; 0 \qquad \forall a=(i,j), r; \tag{5.41}$$

$$u_a^{rs^*}[t+\pi^{ri^*}(t)] \; \left[\pi^{ri^*}(t) + \tau_a[t+\pi^{ri^*}(t)] - \pi^{rj^*}(t)\right] \;=\; 0 \qquad \forall a=(i,j), r, s; \tag{5.42}$$

$$u_a^{rs}[t+\pi^{ri^*}(t)] \geq 0 \qquad \forall a=(i,j), r, s. \tag{5.43}$$

Lemma 5.1: *Link-time-based ideal DUO route choice conditions (5.41)-(5.43) imply route-time-based ideal DUO route choice conditions (5.10)-(5.12).*

Proof.

We need to prove for each O-D pair rs that under link-time-based ideal DUO route choice conditions (5.41)-(5.43) any vehicle flows departing from origin r at time t must arrive at destination s at the same time by using the minimal actual travel time routes.

For simplicity, we first consider the case with only two route departure flows $f_1^{rs}(t) > 0$, $f_2^{rs}(t) > 0$ for one O-D pair rs at time t. It follows that

$$f_1^{rs}(t) + f_2^{rs}(t) = f^{rs}(t) \tag{5.44}$$

Suppose $f_1^{rs}(t)$, $f_2^{rs}(t)$ take routes 1 and 2, respectively. Routes 1 and 2 are minimal-travel-time routes generated under the link-time-based ideal DUO route choice conditions. For simplicity, assume that route 1 comprises 4 links: $1=(r,h), 2=(h,i), \cdots, 4=(j,s)$; and route 2 comprises 5 links: $5=(r,k), 6=(k,l), \cdots, 9=(m,s)$. Note that routes 1 and 2 may have overlapping links. Also note that this assumption can be generalized to any route with any number of links. Using link-time-based ideal DUO route choice conditions (5.41)-(5.43), we have

$$u_1^{rs^*}(t) > 0, u_2^{rs}[t+\pi^{rh^*}(t)] > 0, \cdots, u_4^{rs}[t+\pi^{rj^*}(t)] > 0 \tag{5.45}$$

$$u_5^{rs^*}(t) > 0, u_6^{rs}[t+\pi^{rk^*}(t)] > 0, \cdots, u_9^{rs}[t+\pi^{rm^*}(t)] > 0 \tag{5.46}$$

This is because route 1 and route 2 are generated under link-time-based ideal DUO route choice conditions (5.41)-(5.43) so that there are inflows into links $1, 2, \cdots, 9$ over route 1 and route 2. If route 1 and route 2 do not have overlapping links, the inflow on each link over route 1 and 2 is positive at the instant of time when departure flows arrive at the link. It follows that

$$u_{11}^{rs^*}(t) > 0, u_{21}^{rs}[t+\pi^{rh^*}(t)] > 0, \cdots, u_{41}^{rs}[t+\pi^{rj^*}(t)] > 0 \tag{5.47}$$

$$u_{52}^{rs^*}(t) > 0, u_{62}^{rs}[t+\pi^{rk^*}(t)] > 0, \cdots, u_{92}^{rs}[t+\pi^{rm^*}(t)] > 0 \tag{5.48}$$

where the second subscripts 1 and 2 represent routes 1 and 2, respectively. Note that the instants of time when departure flows arrive at the links are ensured by link-time-based ideal DUO route choice conditions (5.41)-(5.43). For example, $[t+\pi^{rh^*}(t)]$ is the instant of time when departure flow $f_1^{rs}(t)$ arrives at link 2. In other words, if departure flows $f_1^{rs}(t) > 0$, $f_2^{rs}(t) > 0$ satisfy link-time-based ideal DUO route choice conditions (5.41)-(5.43), we obtain equations (5.47)-(5.48).

If routes 1 and 2 have overlapping links, conditions (5.47)-(5.48) still hold. For example, if link 2 is identical to link 6 (routes 1 and 2 are overlapping on link 2 or link 6), $[t+\pi^{rh^*}(t)] = [t+\pi^{rk^*}(t)]$ is the instant of time when departure flows arrive at link 2. Both flows must experience the same link travel time $\tau_2[t+\pi^{rh^*}(t)]$ and exit link 2 at the same time $[t+\pi^{ri^*}(t)]$. Then the inflows on subsequent links over routes 1 and 2 still satisfy equations (5.47)-(5.48).

Denote the arrival flows over routes 1 and 2 as $e_1^{rs}[t+\pi^{rs^*}(t)]$ and $e_2^{rs}[t+\pi^{rs^*}(t)]$, which are associated with departure flows $f_1^{rs}(t)$ and $f_2^{rs}(t)$, respectively. Note that routes 1 and 2 are minimal travel time routes. Using route-based flow propagation constraints (5.19) for the last links $4 = (j, s)$ and $9 = (m, s)$ over routes 1 and 2, we obtain that

$$e_1^{rs}[t+\pi^{rs^*}(t)] > 0, \quad e_2^{rs}[t+\pi^{rs^*}(t)] > 0 \tag{5.49}$$

where

$$\pi^{rs^*}(t) = \pi^{rj^*}(t) + \tau_4[t+\pi^{rj^*}(t)] = \pi^{rm^*}(t) + \tau_9[t+\pi^{rm^*}(t)]$$

Note that

$$u_{11}^{rs^*}(t), u_{21}^{rs}[t+\pi^{rh^*}(t)], \cdots, u_{41}^{rs}[t+\pi^{rj^*}(t)], e_1^{rs}[t+\pi^{rs^*}(t)]$$

and

$$u_{52}^{rs^*}(t), u_{62}^{rs}[t+\pi^{rk^*}(t)], \cdots, u_{92}^{rs}[t+\pi^{rm^*}(t)], e_2^{rs}[t+\pi^{rs^*}(t)]$$

are the two sets of inflows over routes 1 and 2, respectively. Since these flows are positive, we conclude that the departure flows $f_1^{rs}(t)$ and $f_2^{rs}(t)$ arrive at destination s at the same time $[t+\pi^{rs^*}(t)]$. Thus, the link-time-based ideal

DUO route choice conditions guarantee for O-D pair rs that flows departing at time t have the same arrival time $[t + \pi^{rs^*}(t)]$.

If we consider a general case having multiple route departure flows $f_p^{rs}(t) > 0$ for any O-D pair rs at time t, the above analysis still applies to any positive departure flow over any route p between O-D pair rs. Therefore, link-time-based ideal DUO route choice conditions (5.41)-(5.43) imply route-time-based ideal DUO route choice conditions (5.10)-(5.12).

5.4.2 The Variational Inequality Problem

Denote $\Omega_a^{rj^*}(t)$ as the difference of the minimal travel time from r to j and the travel time from r to j via minimal travel time route from r to i and link a for vehicles departing from origin at time t. It follows that

$$\Omega_a^{rj^*}(t) = \pi^{ri^*}(t) + \tau_a[t + \pi^{ri^*}(t)] - \pi^{rj^*}(t) \qquad \forall a, r; a = (i, j). \quad (5.50)$$

We then rewrite the link-time-based ideal DUO route choice conditions as:

$$\Omega_a^{rj^*}(t) \geq 0 \qquad \forall a = (i, j), r; \quad (5.51)$$

$$u_a^{rs^*}[t + \pi^{ri^*}(t)]\, \Omega_a^{rj^*}(t) = 0 \qquad \forall a = (i, j), r, s; \quad (5.52)$$

$$u_a^{rs}[t + \pi^{ri^*}(t)] \geq 0 \qquad \forall a = (i, j), r, s. \quad (5.53)$$

Then, the equivalent variational inequality formulation of link-time-based ideal DUO route choice conditions (5.51)-(5.53) may be stated as follows.

> **Theorem 5.2.** *The dynamic traffic flow pattern satisfying constraints (5.14)-(5.25) is in a link-time-based ideal DUO route choice state if and only if it satisfies the variational inequality problem:*

$$\int_0^T \sum_{rs} \sum_{a} \Omega_a^{rj^*}(t) \left\{ u_a^{rs}[t + \pi^{ri^*}(t)] - u_a^{rs^*}[t + \pi^{ri^*}(t)] \right\} dt \geq 0 \quad (5.54)$$

Proof of Necessity.

We need to prove that link-time-based ideal DUO route choice conditions (5.51)-(5.53) imply variational inequality (5.54). For any link a, a feasible inflow at time $[t + \pi^{ri^*}(t)]$ is

$$u_a^{rs}[t + \pi^{ri^*}(t)] \geq 0 \quad (5.55)$$

Multiplying equation (5.55) and DUO route choice condition (5.51), we have

$$u_a^{rs}[t + \pi^{ri^*}(t)]\, \Omega_a^{rj^*}(t) \geq 0 \qquad \forall a, r, s; a = (i, j). \quad (5.56)$$

We subtract the second ideal DUO route choice condition (5.52) from equation (5.56) and obtain

$$\left\{u_a^{rs}[t+\pi^{ri^*}(t)] - u_a^{rs^*}[t+\pi^{ri^*}(t)]\right\} \Omega_a^{rj^*}(t) \geq 0 \quad \forall a, r, s; a=(i,j). \quad (5.57)$$

Summing equation (5.57) for all links a and all O-D pairs rs, it follows that

$$\sum_{rs}\sum_{a}\left\{u_a^{rs}[t+\pi^{ri^*}(t)] - u_a^{rs^*}[t+\pi^{ri^*}(t)]\right\} \Omega_a^{rj^*}(t) \geq 0 \quad (5.58)$$

Integrating the above equation from 0 to T, we obtain variational inequality (5.54).

Proof of Sufficiency.

We need to prove that any solution $u_a^{rs^*}[t+\pi^{ri^*}(t)]$ to variational inequality (5.54) satisfies link-time-based ideal DUO route choice conditions (5.51)-(5.53). We know that the first and third ideal DUO route choice conditions (5.51) and (5.53) hold by definition. Thus, we need to prove that the second ideal DUO route choice condition (5.52) also holds.

Assume that the second ideal DUO route choice condition (5.52) does not hold only for a link $b=(l,m)$ for O-D pair kn during time interval $[t_1-\delta, t_1+\delta] \in [0,T]$, i.e.,

$$u_b^{kn^*}[t+\pi^{kl^*}(t)] > 0 \qquad \text{and} \qquad \Omega_b^{km^*}(t) > 0 \quad (5.59)$$

In other words, we have

$$u_b^{kn^*}[t+\pi^{kl^*}(t)]\, \Omega_b^{km^*}(t) > 0 \quad (5.60)$$

Since the second ideal DUO route choice condition (5.52) holds for all cases other than link $b=(l,m)$ for O-D pair kn during time interval $[t_1-\delta, t_1+\delta]$, it follows that

$$\int_0^T \sum_{rs}\sum_{a} \Omega_a^{rj^*}(t)\, u_a^{rs^*}[t+\pi^{ri^*}(t)]\, dt$$

$$= \int_{t_1-\delta}^{t_1+\delta} u_b^{kn^*}[t+\pi^{kl^*}(t)]\; \Omega_b^{km^*}(t)\, dt \; > 0 \quad (5.61)$$

We note that all other terms in the above equation vanish because of ideal DUO route choice condition (5.52).

For each O-D pair rs, we can always find one minimal travel time route p for vehicles departing origin r at time t, where route p is evaluated under the optimal flow pattern $\{u_a^{rs^*}[t+\pi^{ri^*}(t)]\}$. For this route p, the first ideal DUO route choice condition (5.51) becomes an equality by definition. It follows that

$$\Omega_a^{rj^*}(t) = 0 \qquad \forall a, r, s; a=(i,j); a \in p. \quad (5.62)$$

Next, we need to find a set of feasible inflows $u_a^{rs}[t+\pi^{ri^*}(t)]$ so that the following equations

$$u_a^{rs}[t+\pi^{ri^*}(t)]\,\Omega_a^{rj^*}(t)=0 \qquad \forall a,r,s; a=(i,j) \tag{5.63}$$

always hold. We adjust all the departure flows $f^{rs}(t)$ for all O-D pairs at time t. For each O-D pair rs at each time t, we assign O-D departure flow $f^{rs}(t)$ to the minimal travel time route p, which was evaluated under the optimal flow pattern $\{u_a^{rs^*}[t+\pi^{ri^*}(t)]\}$. This will generate a set of feasible inflow patterns $\{u_a^{rs}[t+\pi^{ri^*}(t)]\}$ which always satisfies equations (5.62) and (5.63) because flows are not assigned to routes with non-minimal actual travel times which were evaluated under the optimal flow pattern $\{u_a^{rs^*}[t+\pi^{ri^*}(t)]\}$. Summing equations (5.63) for all links a and all O-D pairs rs, it follows that

$$\sum_{rs}\sum_{a} u_a^{rs}[t+\pi^{ri^*}(t)]\,\Omega_a^{rj^*}(t)=0 \qquad \text{where } a=(i,j). \tag{5.64}$$

Integrating the above equation from 0 to T, we obtain

$$\int_0^T \sum_{rs}\sum_{a} u_a^{rs}[t+\pi^{ri^*}(t)]\,\Omega_a^{rj^*}(t)\,dt \;=0 \tag{5.65}$$

We subtract equation (5.61) from equation (5.65) and obtain

$$\int_0^T \sum_{rs}\sum_{a} \Omega_a^{rj^*}(t)\,\left\{u_a^{rs}[t+\pi^{ri^*}(t)]-u_a^{rs^*}[t+\pi^{ri^*}(t)]\right\}\,dt\;<0 \tag{5.66}$$

The above equation contradicts variational inequality (5.54). Therefore, any optimal solution $\{u_a^{rs^*}[t+\pi^{ri^*}(t)]\}$ to variational inequality (5.54) satisfies the second ideal DUO route choice condition (5.52). Since we proved the necessity and sufficiency of the equivalence of variational inequality (5.54) to link-time-based ideal DUO route choice conditions (5.51)-(5.53), the proof is complete.

5.5 A Numerical Example

We illustrate the solution of the link-time-based model with the 4-link, 4-node test network shown in Figure 5.5. To simplify the problem, we assume all four links are freeway links each 1 mile long with free flow speed of 60 miles/hour. A symmetrical network is intentionally used to demonstrate that the route travel times in the solution are equal. To convert the continuous variational inequality model into a discrete time model, the assignment time period $[0,T]$ is subdivided into $K=10$ small time intervals, where each time interval is 15 seconds. We then use an algorithm based on the relaxation and Frank-Wolfe techniques to solve this discrete time variational inequality. This algorithm was coded in FORTRAN and solved on a HP715 workstation. The details of the algorithm are presented in Chapter 6; only computational results for a small network are given to illustrate the ideal DUO traffic flows.

The following link travel time functions were used in the computations:

$$\tau_a(k) = g_{1a}(k) + g_{2a}(k)$$

$$g_{1a}(k) = \beta_{1a} + \beta_{2a}[u_a(k)]^2 + \beta_{3a}[x_a(k)]^2$$

$$g_{2a}(k) = \beta_{4a} + \beta_{5a}[v_a(k)]^2 + \beta_{6a}[x_a(k)]^2$$

where time interval $k = 1, 2, \cdots, 10$. Parameter values for each link travel time function are given in Table 5.2, and the trip table is given in Table 5.3. The optimal link flows and corresponding optimal link travel times are given in Table 5.4. The optimal route travel times are given in Table 5.5. In this discrete time example, $x_a(k)$ represents vehicles on the link at the beginning of interval k; $u_a(k)$ and $v_a(k)$ represent inflow and exit flow during interval k.

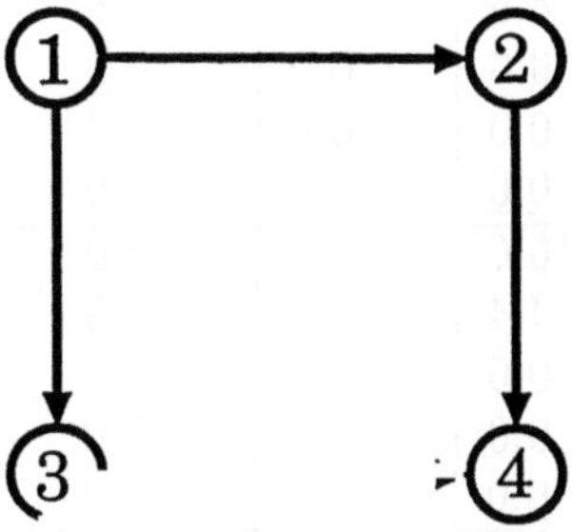

Figure 5.5: Test Network

Table 5.2: Parameters of Link Travel Time Functions

link a	β_{1a}	β_{2a}	β_{3a}	β_{4a}	β_{5a}	β_{6a}
1—2	1.	0.001	0.	0.	0.015	0.003
1—3	1.	0.001	0.	0.	0.015	0.003
2—4	1.	0.001	0.	0.	0.015	0.003
3—4	1.	0.001	0.	0.	0.015	0.003

In this example, note from Table 5.5 that travel times on routes 1–2–4 and 1–3–4 are equal during each interval (except for small differences resulting from incomplete convergence). Thus, the results indicate the existence of dynamic user optimality in the intuitive sense that the two routes have equal travel times in each interval.

Note from Table 5.4 that the *average* travel time that vehicles incur on link 1–2 from the beginning of interval 2 is 1.8 intervals (some vehicles have

Table 5.3: Required Flows from Origin 1 to Destination 4

Time Interval k	1	2	3	4	5
Flow/interval	20.	0.	0.	0.	0.

Table 5.4: Optimal Numbers of Vehicles, Inflows, Exit Flows and Travel Times

Interval k	Link a	Vehicles $x_a(k+1)$	Inflow $u_a(k)$	Exit Flow $v_a(k)$	Vehicles $x_a(k)$	Travel Time $\tau_a(k)$
1	1 — 2	10.0	10.0	0.0	0.0	1.1
2	1 — 2	4.5	0.0	5.5	10.0	1.8
3	1 — 2	0.0	0.0	4.5	4.5	1.4
4	1 — 2	0.0	0.0	0.0	0.0	1.0
5	1 — 2	0.0	0.0	0.0	0.0	1.0
1	1 — 3	10.0	10.0	0.0	0.0	1.1
2	1 — 3	4.4	0.0	5.6	10.0	1.8
3	1 — 3	0.0	0.0	4.4	4.4	1.3
4	1 — 3	0.0	0.0	0.0	0.0	1.0
5	1 — 3	0.0	0.0	0.0	0.0	1.0
1	2 — 4	0.0	0.0	0.0	0.0	1.0
2	2 — 4	5.5	5.5	0.0	0.0	1.0
3	2 — 4	4.5	4.5	5.5	5.5	1.6
4	2 — 4	0.0	0.0	4.5	4.5	1.4
5	2 — 4	0.0	0.0	0.0	0.0	1.0
1	3 — 4	0.0	0.0	0.0	0.0	1.0
2	3 — 4	5.6	5.6	0.0	0.0	1.0
3	3 — 4	4.4	4.4	5.6	5.6	1.6
4	3 — 4	0.0	0.0	4.4	4.4	1.3
5	3 — 4	0.0	0.0	0.0	0.0	1.0

Table 5.5: Dynamic User-Optimal Route Travel Times

Interval	Route Travel Times (minutes)	
k	Route 1-2-4	Route 1-3-4
1	2.1	2.1
2	3.2	3.1
3	2.8	2.6

smaller travel times and some have larger); 10 vehicles are on the link at the beginning of interval 2. Some vehicles exit during interval 2, which reduces the travel time in subsequent intervals. Thus, 5.5 vehicles exit during interval 2 (decreasing the travel time to 1.4 intervals for interval 3) and 4.5 vehicles exit during interval 3. Note also that 5.5 vehicles are on link 2–4 at the beginning of interval 3, and the average travel time is 1.6 from the beginning of interval 3. Also, 4.5 vehicles enter this link during interval 3, 5.5 vehicles exit during interval 3, and 4.5 vehicles exit during interval 4, reducing the travel time to only 1.0 for interval 5. At the beginning of interval 6 (column $x_a(k+1)$ in rows $k = 5$), no vehicle remains on any link.

5.6 A Multi-Class Route-Cost-Based Model

In this section, we define an ideal dynamic user-optimal (DUO) model based on travel costs or disutilities instead of travel times. We stratify travelers into M classes for each O-D pair according to the socio-economic characteristics of each traveler. When $M = 1$, the following definition and VI model reduces to a single class model, but one which is different from the above VI model based on actual travel times.

For multi-class route choice problems, there are several approaches of stratifying travelers into classes. One approach is to classify travelers based on route diversion willingness (see Table 5.6a). There are 3 combinations in this approach. A second approach is to classify travelers based on income and age (see Table 5.6b). There are 9 combinations in this approach.

A third approach is to classify travelers based on driving behavior (see Table 5.6c), which was proposed by Codelli et al (1993). There are 3 combinations in this approach. The above stratifications can also be combined into more detailed classifications. We leave the subsequent analysis of those combinations for empirical studies.

A fourth approach is to stratify travelers into three classes based route choice behavior (see Table 5.6d), as proposed by Lo, Ran and Hongola (1996). Class 1 travelers have predetermined routes; class 2 travelers use Stochastic Dynamic User-Optimal (SDUO) routes based on their perceptions of travel times; and class 3 travelers use DUO routes. The discussion on this type of multi-class model is complicated. Therefore, Sections 5.6 and 5.7 present multi-class models based on various travel disutilities defined by using the previous three stratification methods.

Table 5.6: Stratification of Travelers

a) Route Diversion Willingness

One Route	Few Alternative Routes (2-3)	En Route Diversion

b) Income and Age

	Degree of Change		
Income	High	Middle	Low
Age	Old	Middle	Young

c) Driving Behavior

Cautious Driver	Rushed Driver	Ruthless Driver

d) Route Choice Behavior

Fixed Routes	SDUO Routes	DUO Routes

5.6.1 Multi-Class Route-Cost-Based Conditions

The actual route disutility function depends on the actual route travel time, fuel consumption enroute and automobile operating cost, etc. Denote $\tilde{\tau}_{ma}(t)$ as the actual disutility function for travelers of class m entering link a at time t. It follows that

$$\tilde{\tau}_{ma}(t) = \alpha_{ma} + \beta_{ma}\ \tau_a(t) \qquad \forall m, a \tag{5.67}$$

where α_{ma} is a fixed actual disutility parameter for class m travelers on link a and β_{ma} is a parameter to transform actual link travel time $\tau_a(t)$ into the disutility of class m travelers. Denote $\tilde{\eta}^{rs}_{mp}(t)$ as the actual route travel disutility for class m travelers from origin r to destination s at time t, and $\tilde{\pi}^{rs}_m(t)$ as the minimal actual route travel disutility for class m travelers from origin r to destination s at time t. We also need to use a recursive formula to compute the route travel disutility $\tilde{\eta}^{rs}_{mp}(t)$ for all allowable routes. Assume route p consists of nodes $(r, 1, 2, \cdots, i, \cdots, s)$. Denote $\tilde{\eta}^{rj}_{mp}(t)$ as the travel disutility *actually* experienced over route p from origin r to node j by vehicles departing origin r at time t. Then, a recursive formula for route travel disutility $\tilde{\eta}^{rs}_{mp}(t)$ is:

$$\tilde{\eta}^{rj}_{mp}(t) = \tilde{\eta}^{r(j-1)}_{mp}(t) + \tilde{\tau}_{ma}[t + \eta^{r(j-1)}_p(t)] \qquad \forall m, p, r, j; j = 1, 2, \cdots, s;$$

where link $a = (j-1, j)$ and time $[t + \eta^{r(j-1)}_p(t)]$ is the arrival time at link a for class m travelers. We then define a Multi-Class travel-cost-based ideal DUO state as follows.

> **Multi-Class Route-Cost-Based Ideal DUO State:** *If, for each class m and each O-D pair at each instant of time, the actual travel* disutilities *for all routes that are being used equal the minimal actual route travel* disutility, *the dynamic traffic flow over the network is in a multi-class route-cost-based ideal dynamic user-optimal state.*

Recall that $f_{mp}^{rs}(t)$ is the route inflow of class m from origin r to destination s at time t. The equivalent multi-class ideal DUO route choice conditions can be summarized as follows:

$$\tilde{\eta}_{mp}^{rs^*}(t) - \tilde{\pi}_{m}^{rs^*}(t) \geq 0 \qquad \forall m, p, r, s; \tag{5.68}$$

$$f_{mp}^{rs^*}(t)\left[\tilde{\eta}_{mp}^{rs^*}(t) - \tilde{\pi}_{m}^{rs^*}(t)\right] = 0 \qquad \forall m, p, r, s; \tag{5.69}$$

$$f_{mp}^{rs}(t) \geq 0 \qquad \forall m, p, r, s. \tag{5.70}$$

5.6.2 Dynamic Network Constraints

The revised constraint set for the multi-class route-cost-based model is summarized as follows.

Relationship between state and control variables:

$$\frac{dx_{map}^{rs}}{dt} = u_{map}^{rs}(t) - v_{map}^{rs}(t) \qquad \forall m, a, p, r, s; \tag{5.71}$$

$$\frac{dE_{mp}^{rs}(t)}{dt} = e_{mp}^{rs}(t) \qquad \forall m, p, r; s \neq r; \tag{5.72}$$

Flow conservation constraints:

$$f_{m}^{rs}(t) = \sum_{a \in A(r)} \sum_{p} u_{map}^{rs}(t) \qquad \forall m, r, s; \tag{5.73}$$

$$\sum_{a \in B(j)} v_{map}^{rs}(t) = \sum_{a \in A(j)} u_{map}^{rs}(t) \qquad \forall j, m, p, r, s; j \neq r, s; \tag{5.74}$$

$$\sum_{a \in B(s)} \sum_{p} v_{map}^{rs}(t) = e_{m}^{rs}(t) \qquad \forall m, r, s; s \neq r; \tag{5.75}$$

Flow propagation constraints:

$$x_{ap}^{rs}(t) = \sum_{b \in \tilde{p}} \{x_{bp}^{rs}[t + \tau_a(t)] - x_{bp}^{rs}(t)\} + \{E_{p}^{rs}[t + \tau_a(t)] - E_{p}^{rs}(t)\}$$

$$\forall a \in B(j); j \neq r; p, r, s; \tag{5.76}$$

Definitional constraints:

$$\sum_{mrsp} u_{map}^{rs}(t) = u_a(t), \qquad \sum_{mrsp} v_{map}^{rs}(t) = v_a(t), \qquad \forall a; \tag{5.77}$$

$$\sum_{mrsp} x_{map}^{rs}(t) = x_a(t), \qquad \forall a; \tag{5.78}$$

Nonnegativity conditions:

$$x_{map}^{rs}(t) \geq 0, \quad u_{map}^{rs}(t) \geq 0, \quad v_{map}^{rs}(t) \geq 0 \quad \forall m, a, p, r, s; \tag{5.79}$$

$$e_{mp}^{rs}(t) \geq 0, \qquad E_{mp}^{rs}(t) \geq 0, \qquad \forall m, p, r, s; \tag{5.80}$$

Boundary conditions:

$$E_{mp}^{rs}(0) = 0, \qquad \forall m, p, r, s; \tag{5.81}$$

$$x_{map}^{rs}(0) = 0, \qquad \forall m, a, p, r, s. \tag{5.82}$$

5.6.3 The Variational Inequality Problem

The equivalent variational inequality formulation of multi-class route-cost-based ideal DUO route choice conditions (5.68)-(5.70) may be stated as follows.

Theorem 5.3. *The dynamic traffic flow pattern satisfying network constraint set (5.71)-(5.82) is in a multi-class travel-cost-based ideal DUO route choice state if and only if it satisfies the variational inequality problem:*

$$\int_0^T \sum_{rs} \sum_{mp} \tilde{\eta}_{mp}^{rs^*}(t) \left[f_{mp}^{rs}(t) - f_{mp}^{rs^*}(t) \right] dt \geq 0 \tag{5.83}$$

The proofs of necessity and sufficiency for variational inequality (5.83) follow in the same way as in Section 5.3.3 for the single class route-time-based case.

5.7 A Multi-Class Link-Cost-Based Model

In this section, we consider a multi-class ideal dynamic user-optimal (DUO) model based on travel *costs or disutilities* instead of travel *times*. To be consistent with Section 5.6, we stratify travelers into M classes for each O-D pair. When $M = 1$, the following definition and VI model reduce to a single class model, but it is different from the VI model based on actual travel times. The constraint set for this problem is identical to the constraint set (5.71)-(5.82) for the multi-class route-time-based VI model in Section 5.6.

5.7.1 Multi-Class Link-Cost-Based Conditions

The disutility function depends on travel time, fuel consumption, operating cost, etc. Denote $\tilde{\tau}_{ma}(t)$ as the actual disutility function for class m travelers entering link a at time t. It follows that

$$\tilde{\tau}_{ma}(t) = \alpha_{ma} + \beta_{ma} \, \tau_a(t) \qquad \forall m, a \tag{5.84}$$

where α_{ma} is a fixed actual disutility parameter for class m travelers on link a and β_{ma} is a parameter to transform actual link travel time $\tau_a(t)$ into the disutility of class m travelers.

Denote $\tilde{\eta}^{rs}_{mp}(t)$ as the actual route travel disutility for class m travelers from origin r to destination s at time t. Also denote $\tilde{\pi}^{rs}_m(t)$ as the minimal actual route travel disutility for class m travelers from origin r to destination s at time t, and $\bar{\pi}^{rs}_m(t)$ as the corresponding actual route travel time for class m travelers departing origin r to destination s at time t. We also need to use a recursive formula to compute the route travel disutility $\tilde{\eta}^{rs}_{mp}(t)$ for all allowable routes. Assume route p consists of nodes $(r, 1, 2, \cdots, i, \cdots, s)$. Denote $\tilde{\eta}^{rj}_{mp}(t)$ as the travel disutility *actually* experienced over route p from origin r to node j by vehicles departing origin r at time t. Then, a recursive formula for route travel disutility $\tilde{\eta}^{rs}_{mp}(t)$ is:

$$\tilde{\eta}^{rj}_{mp}(t) = \tilde{\eta}^{r(j-1)}_{mp}(t) + \tilde{\tau}_{ma}[t + \eta^{r(j-1)}_{p}(t)] \qquad \forall m, p, r, j; j = 1, 2, \cdots, s;$$

where link $a = (j-1, j)$ and time $[t + \eta^{r(j-1)}_p(t)]$ is the arrival time instant at link a for class m travelers. Recall the definition of the multi-class travel-cost-based ideal DUO state in Section 7.3 which also applies to the link-cost-based problem.

> **Multi-Class Link-Cost-Based Ideal DUO State:** *If, for each class and each O-D pair at each instant of time, the actual travel* disutilities *for all routes that are being used equal the minimal actual route travel* disutility, *the dynamic traffic flow over the network is in a multi-class link-cost-based ideal dynamic user-optimal state.*

We now write the equivalent mathematical inequalities for the above definition using link variables, in contrast to the route-based formulation in Section 7.3. For class m on link $a = (i, j)$, the minimal travel disutility $\tilde{\pi}^{rj^*}_m(t)$ from origin r to j should be equal or less than the minimal travel disutility $\tilde{\pi}^{ri^*}_m(t)$ from origin r to i plus the actual link travel disutility $\tilde{\tau}_{ma}[t + \bar{\pi}^{ri}_m(t)]$ at time instant $[t + \bar{\pi}^{ri}_m(t)]$, where this time instant is the earliest clock time when the flow departing origin r at time t can enter link a over the minimal travel disutility route. It follows that

$$\tilde{\pi}^{ri^*}_m(t) + \tilde{\tau}_{ma}[t + \bar{\pi}^{ri^*}_m(t)] \geq \tilde{\pi}^{rj^*}_m(t) \qquad \forall m, a = (i, j), r.$$

If, for each class m and each O-D pair rs, any departure flow from origin r at time t enters link a at the earliest clock time $[t + \bar{\pi}^{ri^*}_m(t)]$, or $u^{rs}_{ma}[t + \bar{\pi}^{ri^*}_m(t)] > 0$, then the Multi-Class ideal DUO route choice conditions require that link a is on the minimal travel disutility route. In other words, the minimal travel disutility $\tilde{\pi}^{rj^*}_m(t)$ from origin r to j should equal to the minimal travel disutility $\tilde{\pi}^{ri^*}_m(t)$ from origin r to i plus the actual link travel disutility $\tilde{\tau}_{ma}[t + \bar{\pi}^{ri^*}_m(t)]$ at time instant $[t + \bar{\pi}^{ri^*}_m(t)]$. It follows that

$$\tilde{\pi}^{rj^*}_m(t) = \tilde{\pi}^{ri^*}_m(t) + \tilde{\tau}_{ma}[t + \bar{\pi}^{ri^*}_m(t)], \quad \text{if } u^{rs^*}_{ma}[t + \bar{\pi}^{ri^*}_m(t)] > 0, \quad \forall m, a = (i, j), r, s.$$

The above equation is also equivalent to the following:

$$\left[\bar{\pi}_m^{ri^*}(t) + \tilde{\tau}_{ma}[t + \bar{\pi}_m^{ri^*}(t)] - \bar{\pi}_m^{rj^*}(t)\right] \ u_{ma}^{rs^*}[t + \bar{\pi}_m^{ri^*}(t)] = 0, \quad \forall m, a = (i,j), r, s.$$

Thus, the multi-class link-cost-based ideal DUO route choice conditions can be summarized as follows:

$$\bar{\pi}_m^{ri^*}(t) + \tilde{\tau}_{ma}[t + \bar{\pi}_m^{ri^*}(t)] - \bar{\pi}_m^{rj^*}(t) \ \geq \ 0 \qquad \forall m, a = (i,j), r; \qquad (5.85)$$

$$\left[\bar{\pi}_m^{ri^*}(t) + \tilde{\tau}_{ma}[t + \bar{\pi}_m^{ri^*}(t)] - \bar{\pi}_m^{rj^*}(t)\right] \ u_{ma}^{rs^*}[t + \bar{\pi}_m^{ri^*}(t)] \ = \ 0$$

$$\forall m, a = (i,j), r, s; \qquad (5.86)$$

$$u_{ma}^{rs}[t + \bar{\pi}_m^{ri^*}(t)] \geq 0 \qquad \forall m, a = (i,j), r, s. \qquad (5.87)$$

Similar to Lemma 5.1, we can easily prove that the multi-class link-cost-based ideal DUO route choice conditions imply the multi-class route-cost based ideal DUO route choice conditions. The detailed proof is omitted here.

5.7.2 The Variational Inequality Problem

For class m travelers, denote $\tilde{\Omega}_{ma}^{rj^*}(t)$ as the difference of the minimal travel disutility from r to j and the travel disutility from r to j via minimal travel disutility route from r to i and link a for vehicles departing from origin r at time t. It follows that

$$\tilde{\Omega}_{ma}^{rj^*}(t) = \bar{\pi}_m^{ri^*}(t) + \tilde{\tau}_{ma}[t + \bar{\pi}_m^{ri^*}(t)] - \bar{\pi}_m^{rj^*}(t) \qquad \forall m, a, r; a = (i,j). \quad (5.88)$$

We then rewrite the multi-class link-cost-based ideal DUO route choice conditions as follows:

$$\tilde{\Omega}_{ma}^{rj^*}(t) \ \geq \ 0 \qquad \forall m, a = (i,j), r; \qquad (5.89)$$

$$u_{ma}^{rs^*}[t + \bar{\pi}_m^{ri^*}(t)] \, \tilde{\Omega}_{ma}^{rj^*}(t) = 0 \qquad \forall m, a = (i,j), r, s; \qquad (5.90)$$

$$u_{ma}^{rs}[t + \bar{\pi}_m^{ri^*}(t)] \geq 0 \qquad \forall m, a = (i,j), r, s. \qquad (5.91)$$

The equivalent variational inequality formulation of multi-class link-cost-based ideal DUO route choice conditions (5.89)-(5.91) may then be stated as follows.

Theorem 5.4. *The dynamic traffic flow pattern satisfying network constraint set (5.71)-(5.82) is in a multi-class link-cost-based ideal DUO route choice state if and only if it satisfies the variational inequality problem:*

$$\int_0^T \sum_{rs} \sum_{ma} \tilde{\Omega}_{ma}^{rj^*}(t) \ \left\{u_{ma}^{rs}[t + \bar{\pi}_m^{ri^*}(t)] - u_{ma}^{rs^*}[t + \bar{\pi}_m^{ri^*}(t)]\right\} \ dt \ \geq \ 0 \qquad (5.92)$$

The proof follows in a manner similar to Section 5.4.3 (single class, link-time-based) except that the arrival time at each link $[t + \pi_m^{ri^*}(t)]$ is determined by the travel time $\bar{\pi}_m^{ri^*}(t)$ over the minimal cost route in stead of the minimal travel time $\pi_m^{ri^*}(t)$.

5.8 Notes

There are many possible dynamic generalizations of the static user-optimal route choice model. Some of them may be suitable for providing ideal information and advising vehicles of their best routes in a dynamic route guidance system. Since our ideal DUO route choice models have a direct correspondence to the static UO model, other static UO formulations can be expected to have their DUO counterparts. The ideal DUO route choice models could be extended to more realistic situations when both capacity constraints and oversaturation constraints are taken into account.

In reality, the "travel period" is infinitely long; that is, the origin-destination flows never become permanently zero. Since we want a finite horizon model, it is inherent that there must be a boundary condition to represent the artificial termination period in our models. If the time period is too short, travelers who do not reach the end of their routes at the end of the time horizon are not "using" their route, and thus they can take literally any route. This can be prevented by using a long enough time horizon so that all flow clears the network. The rolling horizon method could be used to represent more realistic traffic flows which never become permanently zero. This approach will be discussed in Chapter 15.

In our model, a continuous flow of traffic is implicitly assumed – when vehicles approach a decision node (intersection), drivers can continue straight ahead or turn left or right without physically blocking the road. In reality, this may not be literally true. For example, suppose two vehicles traveling on a one-lane road arrive at an intersection. The DUO route choice conditions in our model may require the first vehicle to wait and then make a left turn, but allow the second vehicle to go straight ahead without waiting. In this situation, the dynamic physical conditions will be violated since in reality the second vehicle would have to wait for the first one. This implies that our model literally represents only networks with turning lanes of sufficient length to accommodate all turning vehicles.

An advantage of the proposed model is that the flow propagation constraints can serve as a black box. If accurate link travel time functions are not available, the model still works because the flow propagation constraints can be easily replaced by an appropriate traffic flow model. The resulting formulation is a set of nonlinear dynamic equations, or a nonlinear complementarity problem.

Variational inequality approaches originated with static transportation network problems. The static user-optimal route choice problem was formulated as an equivalent set of inequalities by Smith (1979). Later, Dafermos (1980) developed an elastic demand model with disutility functions using the VI approach. An elastic demand model with demand functions was introduced by Dafermos and Nagurney (1984b). Fisk and Boyce (1983) also presented alternative VI formulations for network equilibrium travel choice problems. Nagurney (1993) summarized the modeling and algorithmic aspects of VI models for static traffic assignment problems. Recently, Friesz et al (1993) formulated a VI model for the simultaneous departure time/route choice problem. Smith (1993) also presented a route-based VI formulation using the packet representation of vehicle classes. Both dynamic models are route-based, which require explicit route enumeration for formulation and solution.

Multi-class ideal DUO route choice models are a straightforward extension of the single-class ideal DUO route choice models. Both route-cost-based and link-cost-based models use the notation of disutilities for various classes of travelers. Of course, there are many other types of multi-class DUO route choice models. The multi-class route choice model of Lo, Ran and Hongola (1996) is one example. In this model, travelers with predetermined routes and minimal time routes are stratified into two classes. Other travelers are stratified into various classes based on the modeling needs. Most of these multi-class route choice models are micro-simulation models, such as INTEGRATION (Van Aerde, 1992), DYNASMART (Mahmassani et al, 1993) and MITSIM (Ben-Akiva et al, 1994).

Problems

5.1 Discuss the assumptions of ideal DUO route choice. Give three possible scenarios where ideal DUO could be achieved.

5.2 For the network shown in Figure 5.6, determine inflows on links 1 and 2 for time period $[0, 10]$, by using an ideal DUO route choice model. Also find the optimal route travel times. The travel time functions are

$$\tau_1(t) = 2u_1(t) + 5, \qquad \tau_2(t) = 5,$$

$$\tau_3(t) = 10, \qquad \tau_4(t) = u_4(t) + 10$$

The O-D flow is $f^{14}(t) = 20t$ units. [Hint: assume no flow dispersion in the link flow propagation.]

5.3 Discuss the major differences of route-time-based and link-time-based ideal DUO route choice models. Are they equivalent? Explain your answer.

5.4 Discuss when the route-time-based model is superior to the link-time-based model. Also discuss when the link-time-based model is superior to the route-time-based model. When applying both models to realistic transportation networks, what are the appropriate networks for each of them?

5.5 A two-link network is shown in Figure 5.7. Assume the link travel time functions are

$$\tau_1(t) = 5 + u_1(t), \qquad \tau_2(t) = 2 + 2u_2(t)$$

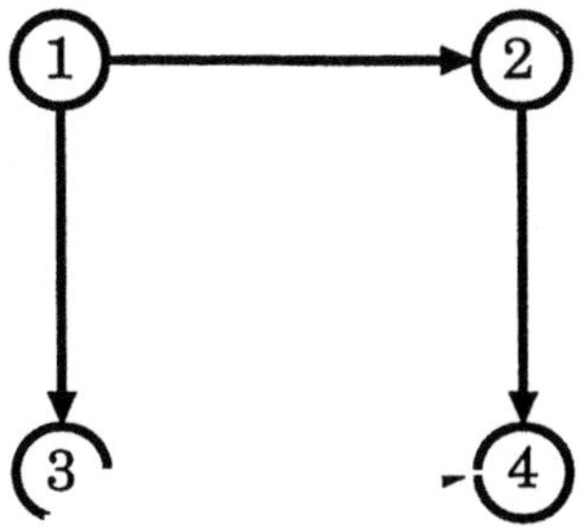

Figure 5.6: Network for Problem 5.2

The actual disutility functions $\tilde{\tau}_a(t)$ are

$$\tilde{\tau}_1(t) = 5 + \tau_1(t), \qquad \tilde{\tau}_2(t) = 3 + 1.5\tau_2(t)$$

The O-D flow is 10 units for time period $[0, 5]$. Determine route departure flows and route travel times for time period $[0, 5]$ based on the link-time-based and the link-cost-based ideal DUO route choice criteria, respectively. Compare the difference between the two cases.

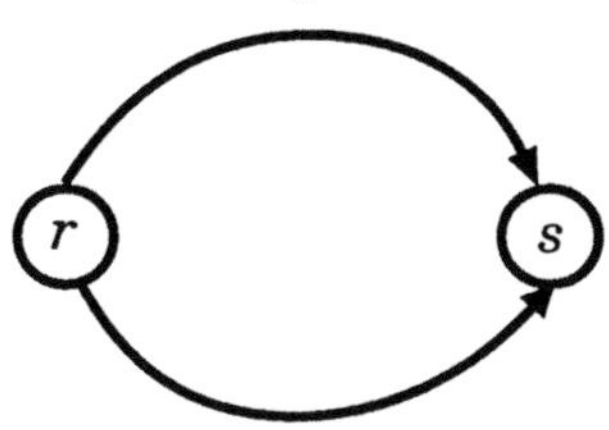

Figure 5.7: Network for Problems 5.5 and 5.6

5.6 For the two-link network shown in Figure 5.7, there are two classes of travelers. The O-D departure flows for class 1 and 2 are 5 units and 10 units for time period $[0, 5]$, respectively. For both classes ($m = 1, 2$), the actual disutility functions $\tilde{\tau}_{ma}(t)$ are

$$\tilde{\tau}_{11}(t) = 4 + u_1(t), \qquad \tilde{\tau}_{21}(t) = 6 + u_1(t)$$

$$\tilde{\tau}_{12}(t) = 2 + 2u_2(t), \qquad \tilde{\tau}_{22}(t) = 3 + 2.5u_2(t)$$

Determine route departure flows and route travel times for time period $[0, 5]$ based on the multi-class link-cost-based ideal DUO route choice criterion. [Hint:

the solution may not be unique.]

5.7 Discuss the increase of complexity of the multi-class route choice problem with the increase of the number of traveler classes. Is this increase a linear relationship or a nonlinear relationship?

Chapter 6

A Solution Algorithm for an Ideal Route Choice Model

In this chapter, solution algorithms are considered for solving the ideal DUC route choice models presented in Chapter 5. A capability to solve the DUC route choice problem is needed for several reasons. First, it appears that prop· erties of alternative models can only be fully understood by computing solu tions to hypothetical and real test problems. Unlike their static counterparts dynamic models are sufficiently opaque that they are difficult to understanc analytically. Second, computational solutions for standard test problems basec on actual networks are needed to evaluate how well alternative models describ‹ reality. Third, solutions for large networks are required to evaluate the poten tial effectiveness of proposed in-vehicle navigation and route guidance systems Ultimately, such models might be used to guide the operation of such systems however, the requirements of such systems are still not sufficiently defined, s‹ any discussion of algorithmic requirements remains rather speculative.

The objective of this chapter is to describe in detail an algorithm fo solving one of the ideal DUO route choice models and to illustrate its perfor mance with a toy network. Through the development and implementation o the algorithm, additional insights into the model's properties have been gained These properties are also discussed. In this chapter, the link-time-based idea DUO route choice model is selected as the sample model to be solved by th‹ solution algorithm. This algorithm is a combination of relaxation and Frank Wolfe techniques.

In Section 6.1, the relaxation method is first investigated. Under re laxation, the link-time-based VI model is reformulated as an optimal contrc problem. In Section 6.1, we reformulate the ideal DUO route choice model as discrete-time nonlinear program (NLP). Then the relaxation and Frank-Wolf techniques are employed to solve the NLP. In the relaxation procedure, the es timated link travel time is updated iteratively. Then we apply the Frank-Wolf technique to solve the NLP. An expanded time-space network is constructed i

Section 6.2 so that each LP subproblem can be decomposed according to O-D pairs and can be viewed as a set of minimal-cost route problems. The flow propagation constraints representing the relationship of link flows and travel times are satisfied in modified minimal-cost route searches in Section 6.3 so that only flow conservation constraints for links and nodes remain. Since the model is convex, the discrete version should be efficient to solve for large networks. A numerical example is given in Section 6.4.

6.1 Statement of the Algorithm

6.1.1 Discrete VI Model for the Link-Time-Based Case

To convert our continuous VI model into a discrete VI model, the time period $[0, T]$ is subdivided into N small time increments. Each time increment is a unit of time. Then, $u_a(n)$ represents the inflow into link a during interval n and $v_a(n)$ represents the exit flow from link a during interval n. To simplify the formulation, we modify the estimated actual travel time on each link in the following way so that each estimated travel time is equal to a multiple of the time increment.

$$\bar{\tau}_a(n) = i \quad \text{if} \quad i - 0.5 \leq \bar{\tau}_a(n) < i + 0.5,$$

where i is an integer and $0 \leq i \leq K$. We note that this round-off method is used only in the flow propagation constraints. More accurate flow propagation constraints can be obtained by making the time intervals smaller. We also note that evaluation of the link travel time function and objective function does not have this round-off error so that the subsequent minimal-cost route search does not have this round-off error.

We now consider the relationship between VI models and optimization models. We do not discuss each VI model in this chapter. Instead, we focus our analysis on the link-time-based VI model for the ideal DUO route choice problem. We show in the following that the VI model can be reformulated as a discrete optimization problem under relaxation. Similar analyses can be performed for the other VI models for the various ideal DUO route choice problems.

In this section, we discuss the following VI problem:

$$\sum_{n=1}^{N} \sum_{rs} \sum_{a} \Omega_a^{rj^*}(n) \left\{ u_a^{rs}[n + \pi^{ri^*}(n)] - u_a^{rs^*}[n + \pi^{ri^*}(n)] \right\} \geq 0 \qquad (6.1)$$

To simplify our analysis, we assume the time period $[0, T]$ is long enough so that all departure flows can be cleared at the final time T. In other words, any positive departure from origin r at time t will arrive at destination s at time $t + \pi^{rs^*}(t) \leq T$.

6.1.2 Relaxation Procedure and Optimization Problem

For each relaxation (or relaxation) iteration, we temporarily fix:

1. actual travel time $\tau_a(n)$ in the link flow propogation constraints as $\bar{\tau}_a(n)$;
2. minimal travel times $\pi^{ri^*}(n)$ as $\bar{\pi}^{ri^*}(n)$ and $\pi^{rj^*}(n)$ as $\bar{\pi}^{rj^*}(n)$ for each link and each origin node.

Then, the objective function of an optimization problem (NLP) can be formulated, which is equivalent to the discrete VI under relaxation, as follows

$$\begin{aligned}\min_{u,v,x,E} \quad Z &= \sum_{k=1}^{K}\sum_{a}\left\{\int_0^{u_a(k)} \tau_a[x_a(k),\omega,v_a(k)]d\omega\right. \\ &+ \left.\sum_r u_a^r(k)[\bar{\pi}^{ri}(\xi_a^r) - \bar{\pi}^{rj}(\xi_a^r)]\right\} \qquad (6.2)\end{aligned}$$

where the new time interval index k is defined as

$$k = \xi_a^r + \bar{\pi}^{ri}(\xi_a^r) \qquad (6.3)$$

with departure time interval ξ_a^r calculated by using the above equation once interval k, origin r and arrival node i are given. Note that both minimal travel times $\bar{\pi}^{ri}(\xi_a^r)$ and $\bar{\pi}^{rj}(\xi_a^r)$ are fixed temporarily at each relaxation iteration.

To ensure that the optimization program with the above objective function is indeed equivalent to the discrete VI, we need to demonstrate that the gradient of the above discrete objective function (6.2) is equivalent to the cost term of discrete variational inequality (6.1) under relaxation, i.e.,

$$\begin{aligned}\frac{\partial Z}{\partial u_a^{rs}[n+\bar{\pi}^{ri}(n)]} &= \tau_a[n+\bar{\pi}^{ri}(n)] + \bar{\pi}^{ri}(n) - \bar{\pi}^{rj}(n) \\ &= \bar{\Omega}_a^{rj}(n) \qquad \forall a,r,s;n. \qquad (6.4)\end{aligned}$$

In the following, we prove that equation (6.4) holds. Taking partial derivative of objective function (6.2) with respect to each link inflow, it follows that

$$\frac{\partial Z}{\partial u_a^{rs}(k)} = \tau_a(k) + \bar{\pi}^{ri}(\xi_a^r) - \bar{\pi}^{rj}(\xi_a^r) \qquad \forall a,r,s;k. \qquad (6.5)$$

Since the above equation applies for each time interval k, it applies for each time interval ξ_a^r. Replacing the departure time interval ξ_a^r by n in the above equation, we have

$$\frac{\partial Z}{\partial u_a^{rs}(k)} = \tau_a(k) + \bar{\pi}^{ri}(n) - \bar{\pi}^{rj}(n) \qquad \forall a,r,s;k. \qquad (6.6)$$

Substituting equation (6.3) (replacing departure time interval ξ_a^r by n) into the above equation, we obtain equation (6.4). Therefore, we have demonstrated that the gradient of discrete objective function (6.2) is equivalent to the cost term of discrete variational inequality (6.1) under relaxation. In other words, objective function (6.2) can be used in the relaxation optimization problem.

Note that the network constraints apply for each time interval, regardless whether the index of time interval is n or k. Thus, for simplicity, we use the time interval index k in the network constraint set. The discrete network flow constraints are now summarized as follows:

$$x_{ap}^{rs}(k+1) = x_{ap}^{rs}(k) + u_{ap}^{rs}(k) - v_{ap}^{rs}(k) \qquad \forall a,p,r,s; k=1,\cdots,K; \quad (6.7)$$

$$E^{rs}(k+1) = E^{rs}(k) + \sum_{a\in B(s)} \sum_{p} v_{ap}^{rs}(k) \qquad \forall r; s\neq r; k=1,\cdots,K; \quad (6.8)$$

$$\sum_{a\in A(r)} \sum_{p} u_{ap}^{rs}(k) = f^{rs}(k) \qquad \forall r\neq s; k=1,\cdots,K; \quad (6.9)$$

$$\sum_{a\in B(j)} v_{ap}^{rs}(k) - \sum_{a\in A(j)} u_{ap}^{rs}(k) = 0 \qquad \forall j,p,r,s; j\neq r,s; k=1,\cdots,K; \quad (6.10)$$

$$x_{ap}^{rs}(k) = \sum_{b\in\tilde{p}} \{x_{bp}^{rs}[k+\bar{\tau}_a(k)] - x_{bp}^{rs}(k)\} + \{E_p^{rs}[k+\bar{\tau}_a(k)] - E_p^{rs}(k)\}$$

$$\forall a\in B(j); j\neq r; p,r,s; k=1,\cdots,K+1; \quad (6.11)$$

$$u_{ap}^{rs}(k)\geq 0, \quad v_{ap}^{rs}(k)\geq 0, \quad x_{ap}^{rs}(k+1)\geq 0, \quad \forall a,p,r,s; k=1,\cdots,K; \quad (6.12)$$

$$E_p^{rs}(k+1)\geq 0, \qquad \forall p,r,s; k=1,\cdots,K; \quad (6.13)$$

$$E_p^{rs}(1) = 0 \qquad \forall p,r,s; \quad (6.14)$$

$$x_{ap}^{rs}(1) = 0, \qquad \forall a,p,r,s. \quad (6.15)$$

In summary, in each relaxation step, we need to solve a discrete NLP consisting of objective function (6.2) and constraints (6.7)-(6.15).

6.1.3 The Frank-Wolfe Method

The Frank-Wolfe algorithm (Frank and Wolfe, 1956) is reasonably efficient for solving nonlinear programming problems (NLP) with network constraints, and has been widely used for solving static UO models on urban networks (LeBlanc et al, 1975). They showed that solving the static UO model simplifies to solving a sequence of minimal-cost route (shortest path) problems and line searches. Thus, the computing times can be reduced by orders of magnitude for large-scale networks, as compared with solving a sequence of linear programming problems.

Denote the subproblem variables as $p, q, y, \bar{E}$, corresponding to the main problem variables u, v, x, E. Applying the Frank-Wolfe algorithm to the minimization of the discretized DUO route choice program requires, at each iteration, a solution of the following linear program (LP):

$$\min_{p,q,y,\bar{E}} \hat{Z} = \nabla_u Z(u,v,x,E)\, p^T + \nabla_v Z(u,v,x,E)\, q^T + \nabla_x Z(u,v,x,E)\, y^T + \nabla_E Z(u,v,x,E)\, \bar{E}^T \qquad (6.16)$$

s.t.

$$y_{ap}^{rs}(k+1) = y_{ap}^{rs}(k) + p_{ap}^{rs}(k) - q_{ap}^{rs}(k) \qquad \forall a,p,r,s; k=1,\cdots,K; \quad (6.17)$$

$$\bar{E}^{rs}(k+1) = \bar{E}^{rs}(k) + \sum_{a\in B(s)} \sum_{p} q_{ap}^{rs}(k) \qquad \forall r; s\neq r; k=1,\cdots,K; \quad (6.18)$$

$$\sum_{a\in A(r)} \sum_{p} p_{ap}^{rs}(k) = f^{rs}(k) \qquad \forall r\neq s; k=1,\cdots,K; \quad (6.19)$$

$$\sum_{a\in B(j)} q_{ap}^{rs}(k) - \sum_{a\in A(j)} p_{ap}^{rs}(k) = 0 \qquad j,p,r,s; \forall j\neq r,s; k=1,\cdots,K; \quad (6.20)$$

$$y_{ap}^{rs}(k) = \sum_{b\in\tilde{p}} \{y_{bp}^{rs}[k+\bar{\tau}_a(k)] - y_{bp}^{rs}(k)\} + \{\bar{E}_p^{rs}[k+\bar{\tau}_a(k)] - \bar{E}_p^{rs}(k)\}$$

$$\forall a\in B(j); j\neq r; p,r,s; k=1,\cdots,K+1; \quad (6.21)$$

$$p_{ap}^{rs}(k)\geq 0,\ \ q_{ap}^{rs}(k)\geq 0,\ \ y_{ap}^{rs}(k+1)\geq 0, \ \ \forall a,p,r,s; k=1,\cdots,K; \quad (6.22)$$

$$\bar{E}_p^{rs}(k+1)\geq 0, \qquad \forall p,r,s; k=1,\cdots,K; \quad (6.23)$$

$$\bar{E}_p^{rs}(1) = 0, \qquad \forall p,r,s; \quad (6.24)$$

$$y_{ap}^{rs}(1) = 0, \qquad \forall a,p,r,s. \quad (6.25)$$

Objective function (6.16) is equivalent to:

$$\hat{Z} = \sum_{k=1}^{K}\sum_{rs}\sum_{ap}\left[\frac{\partial Z}{\partial u_{ap}^{rs}(k)} p_{ap}^{rs}(k) + \frac{\partial Z}{\partial v_{ap}^{rs}(k)} q_{ap}^{rs}(k) + \frac{\partial Z}{\partial x_{ap}^{rs}(k+1)} y_{ap}^{rs}(k+1)\right] + \sum_{k=1}^{K}\sum_{rsp} \frac{\partial Z}{\partial E_p^{rs}(k+1)} \bar{E}_p^{rs}(k+1) \qquad (6.26)$$

The components of the gradient of $Z(u,v,x,E)$ with respect to control and state variables u, v, x, E are

$$t_{1a}^{r}(k) = \frac{\partial Z(u,v,x,E)}{\partial u_a^r(k)} = \tau_a[x_a(k), u_a(k), v_a(k)] + \bar{\pi}^{ri}(n) - \bar{\pi}^{rj}(n)$$

$$\forall a,r; k=1,\cdots,K; \quad (6.27)$$

$$t_{2a}(k) = \frac{\partial Z(u,v,x,E)}{\partial v_a(k)} = \int_0^{u_a(k)} \frac{\partial \tau_a[x_a(k),\omega,v_a(k)]}{\partial v_a(k)} d\omega \qquad \forall a; k=1,\cdots,K; \tag{6.28}$$

$$t_{3a}(k) = \frac{\partial Z(u,v,x,E)}{\partial x_a(k)} = \int_0^{u_a(k)} \frac{\partial \tau_a[x_a(k),\omega,v_a(k)]}{\partial x_a(k)} d\omega \qquad \forall a; k=2,\cdots,K; \tag{6.29}$$

$$t_{3a}(K+1) = \frac{\partial Z(u,v,x,E)}{\partial x_a(K+1)} = 0 \qquad \forall a; \tag{6.30}$$

$$t_4^{rs}(k) = \frac{\partial \hat{Z}(u,v,x,E)}{\partial E^{rs}(k)} = 0 \qquad \forall r,s; k=2,\cdots,K+1. \tag{6.31}$$

The objective function can be rewritten as

$$\hat{Z} = \sum_{k=1}^{K} \sum_{rsap} [t_{1a}^r(k) p_{ap}^{rs}(k) + t_{2a}(k) q_{ap}^{rs}(k) + t_{3a}(k+1) y_{ap}^{rs}(k+1)] \tag{6.32}$$

Since τ_a is a nonnegative and increasing function, it follows that

$$t_{1a}^r(k) \geq 0, \quad t_{2a}(k) \geq 0, \quad t_{3a}(k+1) \geq 0, \qquad \forall a,r; k=1,\cdots,K. \tag{6.33}$$

This property has a significant impact on the solvability of this model for large networks, since these components will be link cost coefficients in a minimal-cost network subproblem.

Note that there are no capacity constraints on the links. In addition to the flow propagation constraints (6.21) and link definitional constraints (6.17), the only constraints are nonnegativity and conservation of flow. Furthermore, the constraints apply to each origin-destination pair independently, so linear program (6.16)-(6.25) can be decomposed by origin-destination pair. The resulting subproblem for each O-D pair rs is given by

$$\min_{p,q,y,\bar{E}} \sum_{k=1}^{K} \sum_{ap} [t_{1a}^r \, p_{ap}^{rs}(k) + t_{2a} \, q_{ap}^{rs}(k) + t_{3a}^{rs}(k+1) \; y_{ap}^{rs}(k+1)] \tag{6.34}$$

s.t.

$$y_{ap}^{rs}(k+1) = y_{ap}^{rs}(k) + p_{ap}^{rs}(k) - q_{ap}^{rs}(k) \qquad \forall a,p; k=1,\cdots,K; \tag{6.35}$$

$$\bar{E}^{rs}(k+1) = \bar{E}^{rs}(k) + \sum_{a\in B(s)} \sum_p q_{ap}^{rs}(k) \qquad \forall k=1,\cdots,K; \tag{6.36}$$

$$\sum_{a\in A(r)} \sum_p p_{ap}^{rs}(k) = f^{rs}(k) \qquad \forall k=1,\cdots,K; \tag{6.37}$$

$$\sum_{a \in B(j)} q_{ap}^{rs}(k) - \sum_{a \in A(j)} p_{ap}^{rs}(k) = 0 \qquad \forall j, p; j \neq r, s; k = 1, \cdots, K; \quad (6.38)$$

$$y_{ap}^{rs}(k) = \sum_{b \in \bar{p}} \{y_{bp}^{rs}[k + \bar{\tau}_a(k)] - y_{bp}^{rs}(k)\} + \{\bar{E}_p^{rs}[k + \bar{\tau}_a(k)] - \bar{E}_p^{rs}(k)\}$$

$$\forall a \in B(j); j \neq r; p, r, s; k = 1, \cdots, K+1; \quad (6.39)$$

$$p_{ap}^{rs}(k) \geq 0, \quad q_{ap}^{rs}(k) \geq 0, \quad y_{ap}^{rs}(k+1) \geq 0, \quad \forall a; p; k = 1, \cdots, K; \quad (6.40)$$

$$\bar{E}_p^{rs}(k+1) \geq 0, \qquad \forall p; k = 1, \cdots, K; \quad (6.41)$$

$$E_p^{rs}(1) = 0; \qquad \forall p; \quad (6.42)$$

$$y_{ap}^{rs}(1) = 0 \qquad \forall a, p. \quad (6.43)$$

For each O-D flow $f^{rs}(i)$ for each time interval $i = 1, \cdots, K$, the above subproblem can be further decomposed as follows, where each variable with index i denotes the value caused by O-D flow $f^{rs}(i)$ for each interval $i = 1, \cdots, K$.

$$\min_{p,q,y,\bar{E}} \quad \sum_{k=1}^{K} \sum_{ap} [t_{1a}^{r}\, p_{ap}^{rs}(k,i) + t_{2a}\, q_{ap}^{rs}(k,i)$$
$$+ \quad t_{3a}^{rs}\,(k+1) y_{ap}^{rs}(k+1,i)] \quad (6.44)$$

s.t.

$$y_{ap}^{rs}(k+1,i) = y_{ap}^{rs}(k,i) + p_{ap}^{rs}(k,i) - q_{ap}^{rs}(k,i) \qquad \forall a, p; k = 1, \cdots, K; \quad (6.45)$$

$$\bar{E}^{rs}(k+1,i) = \bar{E}^{rs}(k,i) + \sum_{a \in B(s)} \sum_{p} q_{ap}^{rs}(k,i) \qquad \forall k = 1, \cdots, K; \quad (6.46)$$

$$\sum_{a \in A(r)} \sum_{p} p_{ap}^{rs}(i,i) = f^{rs}(i) \qquad \forall k = i; \quad (6.47)$$

$$\sum_{a \in B(j)} q_{ap}^{rs}(k,i) - \sum_{a \in A(j)} p_{ap}^{rs}(k,i) = 0 \qquad \forall j, p; j \neq r, s; k = 1, \cdots, K; \quad (6.48)$$

$$y_{ap}^{rs}(k,i) = \sum_{b \in \bar{p}} \{y_{bp}^{rs}[k + \bar{\tau}_a(k), i] - y_{bp}^{rs}(k,i)\}$$
$$+ \{\bar{E}_p^{rs}[k + \bar{\tau}_a(k), i] - \bar{E}_p^{rs}(k,i)\}$$

$$\forall a \in B(j); j \neq r; p, r, s; k = 1, \cdots, K+1; \quad (6.49)$$

$$p_{ap}^{rs}(k,i) \geq 0, \quad q_{ap}^{rs}(k,i) \geq 0, \quad y_{ap}^{rs}(k+1,i) \geq 0, \quad \forall a, p; k = 1, \cdots, K; \quad (6.50)$$

$$\bar{E}_p^{rs}(k+1,i) \geq 0, \qquad \forall p; k = 1, \cdots, K; \quad (6.51)$$

$$E_p^{rs}(1,i) = 0; \qquad \forall p; \quad (6.52)$$

$$y_{ap}^{rs}(1,i) = 0 \qquad \forall a, p. \tag{6.53}$$

The above LP subproblem for each O-D flow $f^{rs}(i)$ for each time interval $i = 1, \cdots, K$ between each O-D pair rs can be viewed as a one-to-one minimal-cost route problem over an expanded time-space network using an artificial origin (see the next subsection and Section 6.2). Note that the minimal-cost route searching is based on actual travel time. It can be solved by determining the minimal-cost routes from the artificial origin to a super destination and completing an all-or-nothing assignment. By revising the costs for some artificial links, the minimal cost route is searched while the flow propagation constraints are automatically satisfied by construction of the links of the expanded time-space networks (see Section 6.2). The flow variables $p_{ap}^{rs}(k)$, $q_{ap}^{rs}(k)$, $y_{ap}^{rs}(k+1)$, and $\bar{E}_p^{rs}(k+1)$ are determined by solving the minimal-cost route problems for all artificial origin-destination pairs between each original O-D pair rs for each time interval and assigning the O-D flows to the links.

In this combined algorithm, we define the relaxation procedure as the outer iteration and the F-W procedure as the inner iteration. Denote the new solution at inner F-W iteration $l+1$ as

$$u_a^{l+1}(k) = u_a^l(k) + \alpha^l[u_a^l(k) - p_a^l(k)] \qquad \forall a; k = 1, \cdots, K; \tag{6.54}$$

$$v_a^{l+1}(k) = v_a^l(k) + \alpha^l[v_a^l(k) - q_a^l(k)] \qquad \forall a; k = 1, \cdots, K; \tag{6.55}$$

$$x_a^{l+1}(k) = x_a^l(k) + \alpha^l[x_a^l(k) - y_a^l(k)] \quad \forall a; k = 1, \cdots, K+1; \tag{6.56}$$

where α^l is the optimal step size of the one-dimensional search problem in the F-W algorithm. The one-dimensional search problem is to find step size α^l that

$$\min_{0 \le \alpha^l \le 1} \quad \sum_{k=1}^{K} \sum_{a} \Bigg\{ \int_0^{u_a^{l+1}(k)} \tau_a[x_a^{l+1}(k), \omega, v_a^{l+1}(k)] d\omega$$
$$+ \quad \sum_{r} [u_a^r(k)]^{l+1} [\bar{\pi}^{ri}(\xi_a^r) - \bar{\pi}^{rj}(\xi_a^r)] \Bigg\} \tag{6.57}$$

where $u_a^{l+1}(k)$, $v_a^{l+1}(k)$ and $x_a^{l+1}(k)$ must be substituted using the above definitional equations.

The algorithm for solving our ideal DUO route choice model is illustrated in the flowchart in Figure 6.1 and is summarized as follows:

Step 0: Initialization.
Find an initial feasible solution $\{x_a^1(k)\}$, $\{u_a^1(k)\}$, $\{v_a^1(k)\}$, $\{E^1(k)\}$. Set the outer iteration counter $m = 1$.

Step 1: Relaxation.
Find a new estimate of the actual link travel time $\bar{\tau}_a^l(k)$ and solve the ideal DUO program. Set the inner iteration counter $n = 1$.

[Step 1.1]: *Update.* Calculate $t_{1a}^r(k)$, $t_{2a}(k)$ and $t_{3a}(k)$ using equations (6.27)-(6.30).

[Step 1.2]: *Direction Finding.* Based on $\{t_{1a}^r(k)\}$, $\{t_{2a}(k)\}$ and $\{t_{3a}(k)\}$ and satisfying the flow propagation constraints (6.49), search the minimal-cost route forward from each artificial origin to the super destination over an expanded time-space network for each O-D pair rs. Perform an all-or-nothing assignment, yielding subproblem solution $\{p_a(k)\}, \{q_a(k)\}, \{y_a(k)\}, \{\bar{E}^{rs}(k)\}$.

[Step 1.3]: *Line Search.* Find the optimal step size α^l that solves the one-dimensional search problem.

[Step 1.4]: *Move.* Find a new solution by combining $\{u_a(k)\}$, $\{v_a(k)\}$, $\{x_a(k)\}$, $\{E^{rs}(k)\}$ and $\{p_a(k)\}$, $\{q_a(k)\}$, $\{y_a(k)\}$, $\{\bar{E}^{rs}(k)\}$.

[Step 1.5]: *Convergence Test for Inner Iterations.* If n equals a prespecified number, go to step 2; otherwise, set $n = n+1$ and go to step 1.1.

Step 2: Convergence Test for Outer Iterations.
If $\bar{\tau}_a^{(m)}(k) \simeq \bar{\tau}_a^{(m+1)}(k)$, stop. The current solution, $\{u_a(k)\}$, $\{v_a(k)\}$, $\{x_a(k)\}$, $\{E^{rs}(k)\}$, is in a near ideal DUO state; otherwise, set $m = m+1$ and go to step 1.

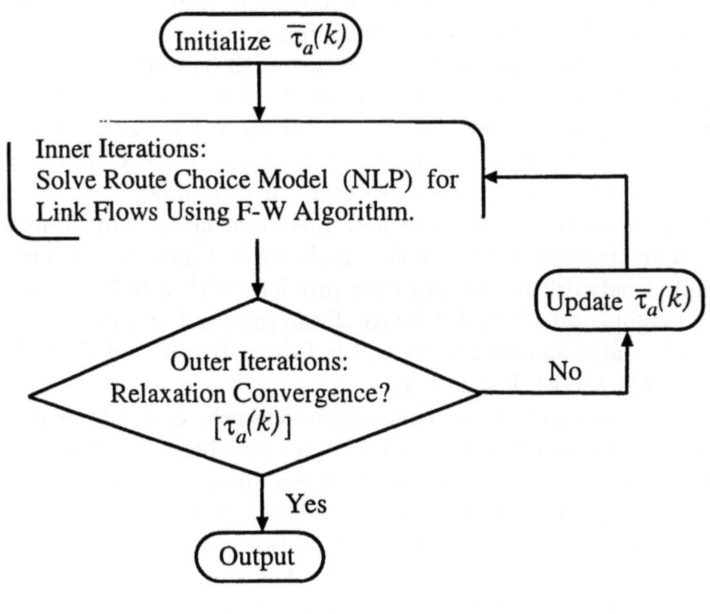

Figure 6.1: Flowchart of the Solution Algorithm

The number of inner iterations l and the number of outer iterations m are interrelated. If we set m larger, then l should be set smaller accordingly, and vice versa. In general, $l = 1$ to 3 is sufficient to speed up convergence. Similar experience with a relaxation and F-W algorithm for solving static UO traffic equilibrium problems was reported by Mahmassani and Mouskos (1988). In order to speed up convergence, an incremental assignment technique is suggested for finding a good starting solution before the relaxation procedure. Since the linear subproblem can be decomposed by each artificial origin-destination pair, this problem is a good candidate for solution with parallel computing techniques.

To speed up the serial computing speed, we can construct a super origin which connects all artificial origins. Thus, we can search the minimal-cost route forward from the super origin to the super destination over an expanded time-space network for each physical destination s. Therefore, the number of iterations within Step 1.2 can be reduced by the order of the number of origins. A significant saving of total CPU time can be achieved for a large network with many O-D pairs, although a marginal increase of CPU time is incurred due to the slight enlargement of the expanded time-space network. (A super origin is created and is connected with all artificial origins.)

6.2 Solving the LP Subproblem

In addition to flow propagation constraints, there are 3 types of constraints in the LP subproblem: link flow state equations; node flow conservation equations; and nonnegativity/boundary constraints. Since there are 3 variables associated with each physical link, we replace each link with 3 separate artificial links for each time period, by adding artificial nodes to define the new links. Figure 6.2 shows the expansion for link a which points from l to m for 3 time periods. The initial state for link a is assumed to be $x_a 1 = 0$. Note that a total of 9 nodes is required for the expansion of each physical link.

We now describe how these new links and nodes can be numbered. If node m is a destination, the expansion is shown in Figure 6.3. Figure 6.4 shows the expanded network for an example problem with 3 links, 3 nodes, 3 time periods, 1 origin (node 1) and 2 destinations (nodes 2 and 3). The initial state for each link is also assumed to be $x_a 1 = 0$ $(a = 1, 2, 3)$ and the O-D flows are given as $f^{12}(k)$, $f^{13}(k)$, $k = 1, 2, 3$.

In the expansion of the time-space network, each physical node is expanded into additional nodes only (i.e., no extra links are introduced). However, since there are 3 separate variables per link, each link is expanded into 3 links, which requires additional nodes as well. Since the LP problem can be decomposed by each artificial origin-destination pair for all O-D pairs, we initially only consider a one-to-one network. The subproblem variables $p_{ap}^{rs}(k,i)$, $q_{ap}^{rs}(k,i)$, $y_{ap}^{rs}(k+1,i)$, $\bar{E}_p^{rs}(k+1,i)$ are viewed as flows over the artificial links of the expanded time-space network, which is constructed as follows.

1. **Nodes**: Each node j is expanded to K nodes, one for each time period. We denote these as $j(k)$, $k = 1, \cdots, K$, with new node numbers $j(k) =$

$(N+A)(k-1)+j$. For example, we consider the time-space expansion of a 3-node, 2-link network for 3 time periods (Figure 6.5). The physical nodes in the original network are numbered as 1, 2, 3 and physical links are numbered as 1 = (1,2) and 2 = (2,3). Node 1 is an origin and node 3 is a destination. Thus, $N = 3$ and $A = 2$. For the expansion of physical nodes 1, 2, 3, the new node numbers are 1, 2, 3 for period 1; 6, 7, 8 for period 2; and 11, 12, 13 for period 3.

2. **Links**: Since there are 3 variables for each link, each link $a=(l,m)$ is expanded to K nodes and $3K$ links, as follows:
   a) K transshipment nodes $n_a(k)$, $k = 1, \cdots, K$, with new node numbers $n_a(k) = (N+A)(k-1)+N+a$. For the expansion of physical link 1 and 2 in Figure 6.5, the new transshipment node numbers are 4 and 5 for period 1, 9 and 10 for period 2, and 14 and 15 for period 3.
   b) K links $(l(k), n_a(k))$, $k = 1, \cdots, K$, where each new link $\hat{a}$ has link number $[3A(k-1)+a]$, flow $p_a(k,i)$ and cost $t_{1a}(k)$. In Figure 6.5, these new links are 1 = (1,4) and 2 = (2,5) for period 1; 8 = (6,9) and 9 = (7,10) for period 2; and 15 = (11,14) and 16 = (12,15) for period 3.
   c) K links $(n_a(k), m(k))$, $k = 1, \cdots, K$, where each new link $\hat{a}$ has link number $[3A(k-1)+A+a]$, flow $q_a(k,i)$ and cost $t_{2a}(k)$. In Figure 6.5, these new links are 3 = (4,2) and 4 = (5,3) for period 1; 10 = (9,7) and 11 = (10,8) for period 2; and 17 = (14,12) and 18 = (15,13) for period 3.
   d) $(K-1)$ links $(n_a(k), n_a(k+1))$, $k = 1, \cdots, K-1$, where each new link $\hat{a}$ has link number $[3A(k-1)+2A+a]$, flow $y_a^{rs}(k+1,i)$ and cost $t_{3a}(k+1)$ for corresponding O-D pair (r,s). In Figure 6.5, these new links are 5 = (4,9) and 6 = (5,10) for period 1; 12 = (9,14) and 13 = (10,15) for period 2;
   e) one link $\hat{a} = (n_a(K), S)$ with new link number $[3A(K-1)+2A+a]$, flow $y_a^{rs}(K+1,i)$ and cost $\hat{t}_{\hat{a}}^{rs}$ for corresponding O-D pair (r,s); in Figure 6.5, these new links are 19 = (14,16) and 20 = (15,16) for period 3.

3. **Origin Nodes**: Each origin node r is expanded to K origin nodes $r(k)$, $k = 1, \cdots, K$. In Figure 6.5, origin 1 is expanded to nodes 1, 6 and 11.

4. **Destination Nodes**: The destination node s is expanded to:
   a) one super destination S; in Figure 6.5, the physical destination node 3 is expanded to super destination 16;
   b) K nodes $s(k)$, $k = 1, \cdots, K$; (nodes 3, 8 and 13 in Figure 6.5);
   c) K links $\hat{a} = (s(k), s(k+1))$, $k = 1, \cdots, K-1$ and $\hat{a} = (s(K), S)$, with new link number $[3AK+k]$, flow $\bar{E}^{rs}(k+1,i)$ and cost $t_4^{rs}(k+1) = 0$ for corresponding artificial origin-destination pair for all O-D pair (r,s); (links 7 = (3,8), 14 = (8,13), 21 = (13,16) for periods 1,2,3).

In summary, the expanded time-space network has $[(3A+1)K]$ links and $[(N+A)K+1]$ nodes. The expanded time-space network of the example (Figure 6.5) has 21 links and 16 nodes.

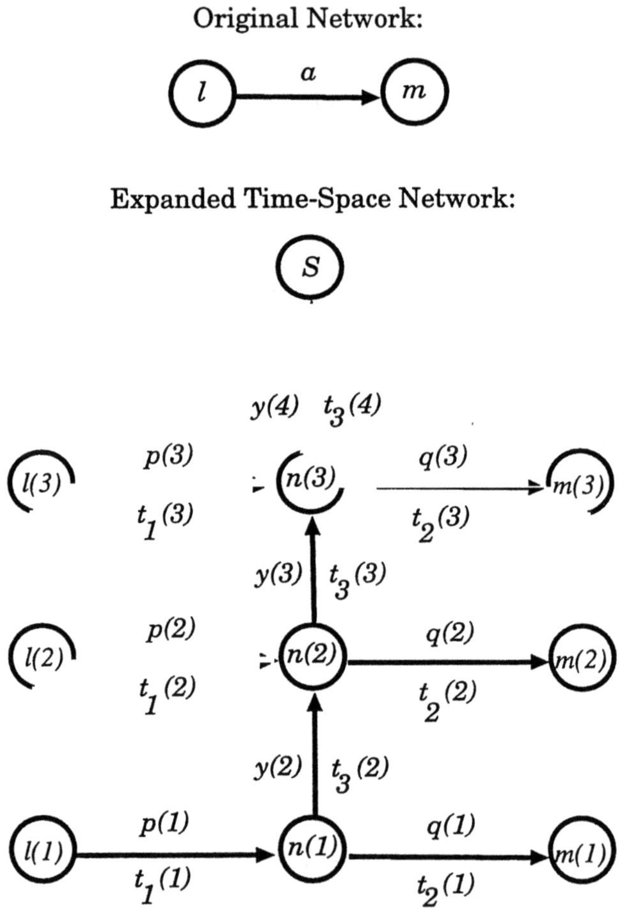

Figure 6.2: Expansion for Link a for 3 Time Periods

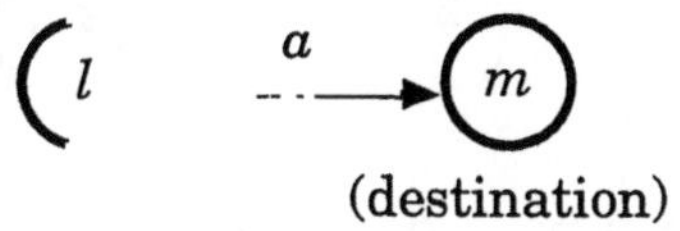

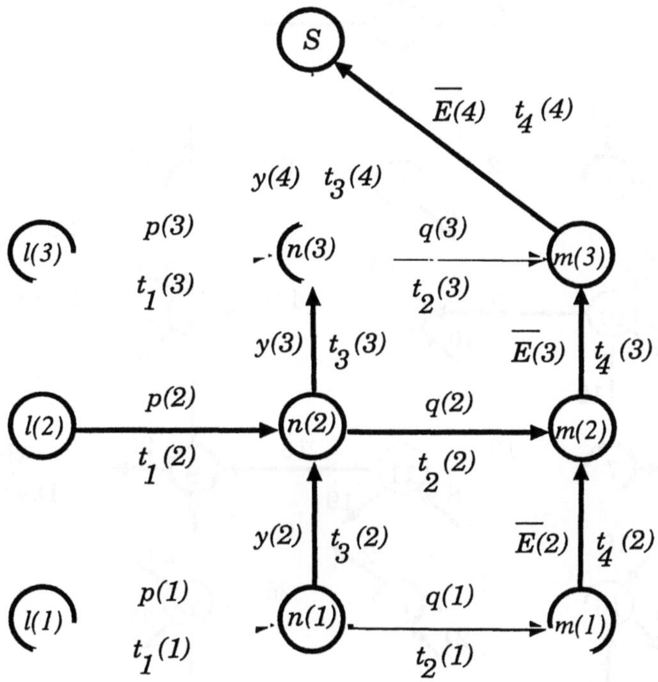

Figure 6.3: Expansion for Link a for 3 Time Periods (m is a destination)

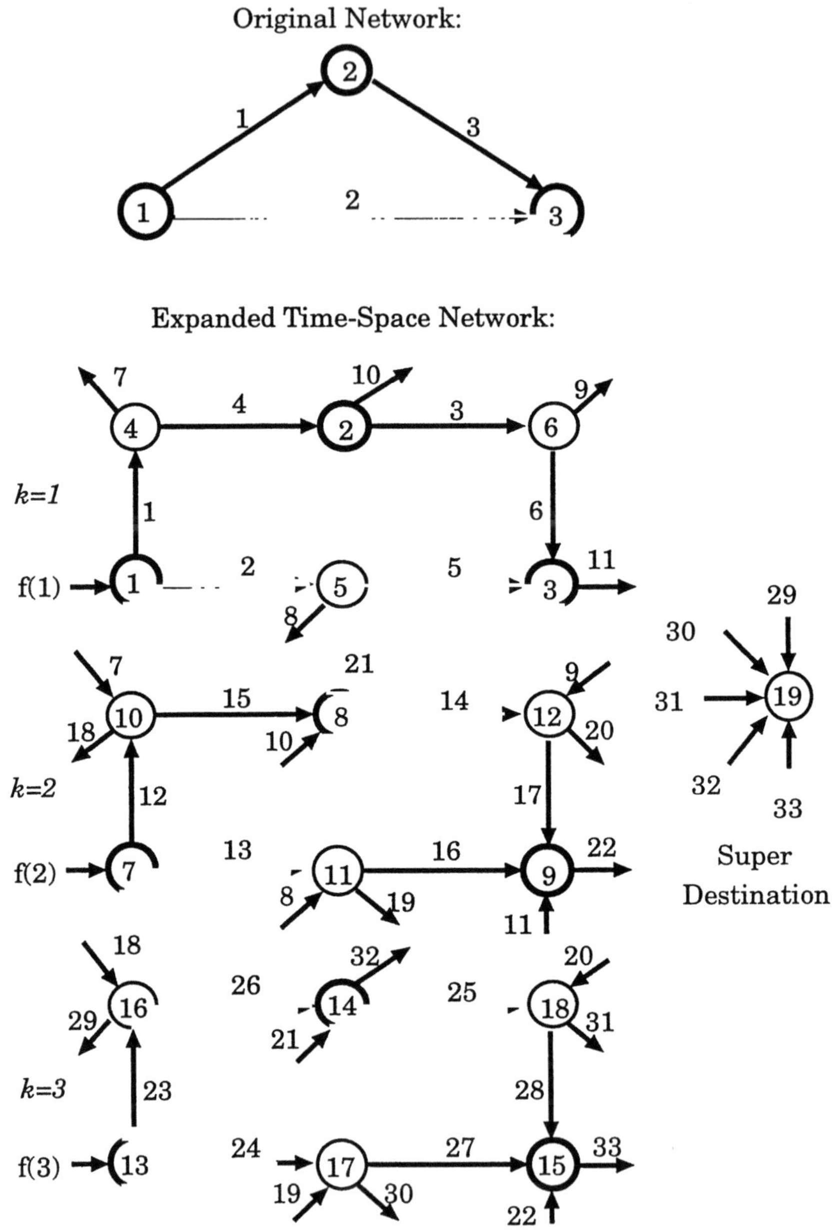

Figure 6.4: Expansion for 3-Link Network for 3 Time Periods

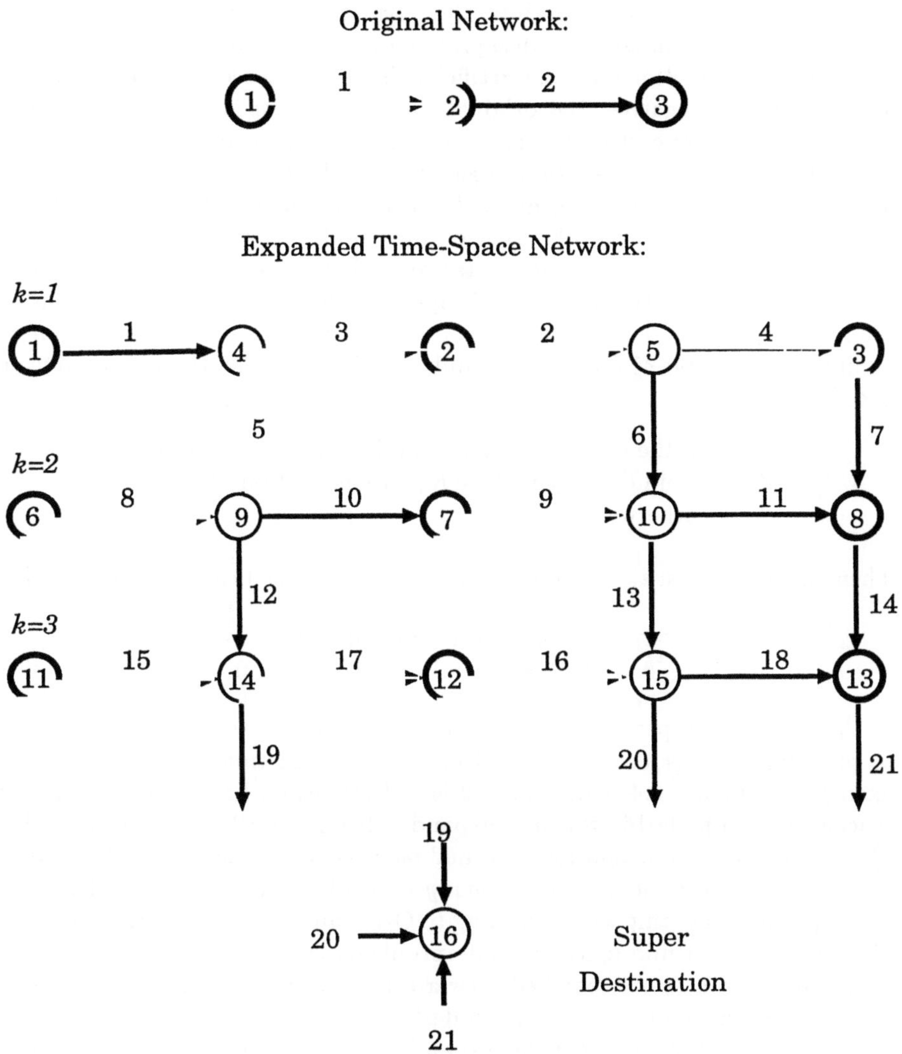

Figure 6.5: Expansion of Example Network (2 Links)

Link state equation (6.45) for each physical link a can be viewed as a set of flow conservation equations for artificial nodes $n_a(k)$, $k = 1, \cdots, K$ in the expanded time-space network. For example, in Figure 6.2, the conservation of flow constraint for node $n(3)$ is $p(3,i)+y(3,i) = q(3,i)+y(4,i)$. This is equivalent to link state equation (6.45). Together with node flow conservation equations and nonnegativity constraints, these equations constitute the constraints for a one-to-one minimal-cost route problem with flow propagation constraints. Since the cost functions for the artificial links over the expanded time-space network are nonnegative, the original LP subproblem is transformed into a one-to-one minimal-cost route problem with flow propagation constraints.

The following explanation describes how the travel cost $\hat{t}_{\hat{a}}^{rs}$ for artificial link $\hat{a} = (n_a(K), \mathcal{S})$ is determined for corresponding O-D pair rs. In the expanded time-space network, each dummy node $n_a(K)$, which is expanded from each link a for the last time interval K, has a link pointing to the super destination $\mathcal{S}$. However, it is not always true that every physical link in the original network will lead to destination s. In order to represent this situation, an indicator variable $\delta_{\hat{a}}^s$ for each new link $\hat{a}$ in the expanded time-space network is defined as follows:

$$\delta_{\hat{a}}^s = \begin{cases} 1 & \text{if the expanded link } \hat{a} \text{ can be reached backwards} \\ & \text{from dummy node } s(K) \text{ expanded from destination } s \\ 0 & \text{otherwise} \end{cases}$$

Then, the travel cost $\hat{t}_{\hat{a}}^{rs}$ for dummy link $\hat{a} = (n_a(K), \mathcal{S})$ can be determined as:

$$\hat{t}_{\hat{a}}^{rs} = \begin{cases} t_{3a}(K+1) = 0 & \text{if} \quad \delta_{\hat{a}}^s = 1 \\ +\infty & \text{if} \quad \delta_{\hat{a}}^s = 0 \end{cases}$$

In this way, the property that flows cannot move backwards from destination s to some links $\{a\}$ is also guaranteed in the expanded time-space network. For example, in Figure 6.4, when node 2 is a destination, link 27 and 28 are not reachable from node 14 (which is expanded from node 2 for period 3). Thus, $\delta_{27}^2 = 0$ and $\delta_{28}^2 = 0$ so that $\hat{t}_{27} = \infty$ and $\hat{t}_{28} = \infty$ when node 2 is a destination.

We now assume a *many-to-many* network. Since the original LP subproblem can be decomposed according to O-D pairs, the LP subproblem can be viewed as a set of minimal-cost route problems with propagation constraints. Minimal-cost routes are searched forward from each artificial origin to the super destination. Suppose there are S destinations in a many-to-many network. Each destination node s can be expanded to: one super destination $\mathcal{S}$; K nodes $s(k)$, $k = 1, \cdots, K$; and K links $\hat{a} = (s(k), s(k+1))$, $k = 1, \cdots, K-1$ and $\hat{a} = (s(K), \mathcal{S})$, with new link number $[3AK+(s-1)K+k]$, flow $\bar{E}^{rs}(k+1, i)$ and cost $\hat{t}_{\hat{a}}^{rs}(k+1)$.

For the many-to-many case, we use the same network notation for the expanded time-space network as for the one-to-one network so that the computational code can be simplified. Therefore, we use one super destination $\mathcal{S}$ to represent the super destinations expanded from all destinations $\{s\}$ and define the cost $\hat{t}^{rs}_{\hat{a}}(K+1)$ for each dummy link $\hat{a} = (s(K), s(k+1))$ $k = 1, \cdots, K-1$ and $\hat{a} = (s(K), \mathcal{S})$ as follows:

$$\hat{t}^{rs}_{\hat{a}}(k+1) = \begin{cases} t^{rs}_{4}(k+1) = 0 & \text{if minimal cost route between any artificial} \\ & \text{O-D pair } (r(k), \mathcal{S}) \text{ is searched} \\ +\infty & \text{otherwise} \end{cases}$$

where artificial O-D pair $(r(k), \mathcal{S})$ is expanded from original O-D pair rs for time interval k. The above cost setting reflects that when considering artificial O-D pair $(r(k), \mathcal{S})$ for any O-D pair rs, other destinations $\hat{s}$ become intermediate nodes. There is no cumulative effect for these nodes $\hat{s}$. In summary, the expanded time-space network has $[(3A+S)K]$ links and $[(N+A)K+1]$ nodes for a many-to-many network.

In searching for the minimal-cost route between each artificial origin-destination pair $(r(k), \mathcal{S})$, flow propagation constraint (6.49) is automatically satisfied by temporally adjusting the costs for $2K$ artificial links $\hat{a} = (n_a(k), n_a(k+1))$, $k = 1, \cdots, K-1$, $\hat{a} = (n_a(K), \mathcal{S})$ and $\hat{a} = (n_a(k), m(k))$, $k = 1, \cdots, K$. These artificial links are expanded from original link a=(l, m).

The temporal cost adjustment procedure begins when an artificial node $l(k)$ is being searched in the minimal-cost route search. This procedure is as follows:

if $i - 0.5 \leq \bar{\tau}_a(k) < i + 0.5$ $(i = 0, 1, \cdots, K)$,
then the feasible subroute will be $[l(k), n(k), \cdots, n(k+i), m(k+i)]$
for the time-space subnetwork expanded from link a=(l, m).

This subroute is guaranteed to have the minimal cost by setting the costs of other artificial links in the time-space subnetwork (except links on the above subroute) temporally equal to infinity.

In Figure 6.3, we assume the minimal-cost route is searched from $l1$ to $\mathcal{S}$. If $\bar{\tau}_a 1 = 2$, the feasible subroute is $[l1, n1, n(2), n(3), m(3)]$. This cost adjustment procedure is performed for each artificial origin-destination pair.

6.3 Computational Experience

The algorithm was coded in FORTRAN and solved on a HP-715 workstation. For the purpose of illustration, we give computational results for a small test problem with 12 arcs and 9 nodes. The network is shown in Figure 6.6. This problem required 20 incremental iterations for the initial solution, 3 inner F-W iterations per outer iteration and 40 outer relaxation iterations to converge. The following link travel cost functions were used in the computations.

$$\tau_a(k) = g_{1a}(k) + g_{2a}(k)$$

$$g_{1a}(k) = \beta_{1a} + \beta_{2a}[u_a(k)]^2 + \beta_{3a}[x_a(k)]^2$$

$$g_{2a}(k) = \beta_{4a} + \beta_{5a}[v_a(k)]^2 + \beta_{6a}[x_a(k)]^2$$

The parameters for each link travel cost function are given in Table 6.1. The number of vehicles x_a1 on each link a at initial time $k = 1$ is assumed to equal 0. Two O-D pairs and 7 time intervals are considered, and the corresponding trip table for each time interval is given in Table 6.2. The optimal link flow trajectories and corresponding optimal link travel costs are given in Table 6.3.

Table 6.3 shows that $10.0+10.0 = 20.0$ vehicles enter the network during interval 1 and $5.0+5.0 = 10.0$ enter during interval 2 (on links 1-2 and 1-4). For node 5, 18.1 vehicles enter links 2-5 and 4-5 during intervals 2-4. No vehicles remain on these 2 links at the end of interval 4, and 3.1 vehicles enter links 5-6 and 5-8 pointing out of node 5 continuing to node 9 during intervals 2-5. Thus, $18.1 - 3.1 = 15.0$ vehicles exit the network at node 5 during intervals 1-5 so that the requirements for O-D pair 1-5 for intervals 1-5 are met. At node 9, $3.5 + 3.1 + 0.7 + 3.6 + 3.1 + 1.0 = 15.0$ vehicles exit links 6-9 and 8-9 into node 9 during intervals 5-7. Note that no vehicles remain on links 6-9 and 8-9 at the end of interval 7 and no vehicles remain on other links of the network at the end of interval 6.

Inspection of Table 6.3 reveals that several links are being *used* in the sense defined in Chapter 5; also a few routes are being used. It is interesting to ask whether the actual route travel times are equal. Table 6.4 provides a comparison of actual route travel times for the 2 O-D flows. The travel times for the 2 routes used from node 1 to node 5 are equal in each time interval. The travel times for the 6 routes from node 1 to node 9 are equal in each interval. This simple example illustrates the inherent complexity of the dynamic route choice model, as compared with its static counterpart.

6.4 Notes

An algorithm for solving the ideal dynamic user-optimal route choice model was presented in this chapter. One significant aspect of our algorithm is that by using a time-space expanded network, the Frank-Wolfe LP subproblem requires only the solution of minimal-cost route problems for each O-D pair. This expansion technique allows standard algorithms for *static* traffic assignment to solve dynamic route choice models. Thus, the DUO route choice model and solution algorithm have an elegant correspondence with the static UO route choice model and its Frank-Wolfe algorithm. It may be possible to extend other static UO formulations and solution algorithms to dynamic versions. This algorithm should be tested on a large-scale transportation network; other efficient algorithms also need to be investigated.

Problems

6.1 Provide a physical interpretation for the cost terms $t^r_{1a}(k)$, $t_{2a}(k)$, $t_{3a}(k)$ in the LP subproblem objective function (6.32).

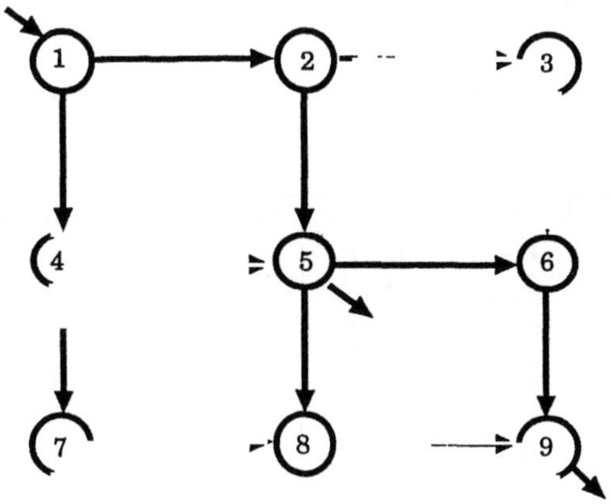

Figure 6.6: Test Network

Table 6.1: Parameters of Link Cost Functions

link a	β_{1a}	β_{2a}	β_{3a}	β_{4a}	β_{5a}	β_{6a}
1—2	1.	0.001	0.	0.	0.015	0.002
2—3	1.	0.001	0.	0.	0.015	0.002
1—4	1.	0.001	0.	0.	0.015	0.002
2—5	1.	0.001	0.	0.	0.015	0.002
3—6	1.	0.001	0.	0.	0.015	0.002
4—5	1.	0.001	0.	0.	0.015	0.002
5—6	1.	0.001	0.	0.	0.015	0.002
4—7	1.	0.001	0.	0.	0.015	0.002
5—8	1.	0.001	0.	0.	0.015	0.002
6—9	1.	0.001	0.	0.	0.015	0.002
7—8	1.	0.001	0.	0.	0.015	0.002
8—9	1.	0.001	0.	0.	0.015	0.002

Table 6.2: O-D Trip Table for Each Time Interval k

Time Interval k	1	2	3	4	5	6	7
$f^{15}(k)$	10.	5.	0.	0.	0.	0.	0.
$f^{19}(k)$	10.	5.	0.	0.	0.	0.	0.

Table 6.3: Optimal Trajectories of Link Flows and Travel Times

Interval k	Link a	Vehicles $x_a(k+1)$	Inflow $u_a(k)$	Exit Flow $v_a(k)$	Vehicles $x_a(k)$	Travel Time $c_a(k)$
1	1—2	10.0	10.0	0.0	0.0	1.10
2	1—2	7.9	5.0	7.1	10.0	1.99
3	1—2	1.7	0.0	6.2	7.9	1.70
4	1—2	0.0	0.0	1.7	1.7	1.05
5	1—2	0.0	0.0	0.0	0.0	1.00
6	1—2	0.0	0.0	0.0	0.0	1.00
1	2—3	0.0	0.0	0.0	0.0	1.00
2	2—3	3.0	3.0	0.0	0.0	1.01
3	2—3	2.6	2.6	3.0	3.0	1.16
4	2—3	0.4	0.4	2.6	2.6	1.12
5	2—3	0.0	0.0	0.4	0.4	1.00
6	2—3	0.0	0.0	0.0	0.0	1.00
1	1—4	10.0	10.0	0.0	0.0	1.10
2	1—4	7.9	5.0	7.1	10.0	1.99
3	1—4	1.6	0.0	6.2	7.9	1.71
4	1—4	0.0	0.0	1.6	1.6	1.04
5	1—4	0.0	0.0	0.0	0.0	1.00
6	1—4	0.0	0.0	0.0	0.0	1.00
1	2—5	0.0	0.0	0.0	0.0	1.00
2	2—5	4.2	4.2	0.0	0.0	1.02
3	2—5	3.6	3.6	4.2	4.2	1.31
4	2—5	1.2	1.2	3.6	3.6	1.22
5	2—5	0.0	0.0	1.2	1.2	1.03
6	2—5	0.0	0.0	0.0	0.0	1.00
1	3—6	0.0	0.0	0.0	0.0	1.00
2	3—6	0.0	0.0	0.0	0.0	1.00
3	3—6	3.0	3.0	0.0	0.0	1.01
4	3—6	2.6	2.6	3.0	3.0	1.16
5	3—6	0.4	0.4	2.6	2.6	1.12
6	3—6	0.0	0.0	0.4	0.4	1.00
1	4—5	0.0	0.0	0.0	0.0	1.00
2	4—5	4.2	4.2	0.0	0.0	1.02
3	4—5	3.6	3.6	4.2	4.2	1.32
4	4—5	1.3	1.3	3.6	3.6	1.23
5	4—5	0.0	0.0	1.3	1.3	1.03
6	4—5	0.0	0.0	0.0	0.0	1.00

Table 6.3: Optimal Trajectories of Link Flows and Travel Times (continued)

Interval k	Link a	Vehicles $x_a(k+1)$	Inflow $u_a(k)$	Exit Flow $v_a(k)$	Vehicles $x_a(k)$	Travel Time $c_a(k)$
1	5—6	0.0	0.0	0.0	0.0	1.00
2	5—6	0.0	0.0	0.0	0.0	1.00
3	5—6	0.5	0.5	0.0	0.0	1.00
4	5—6	0.5	0.5	0.5	0.5	1.01
5	5—6	0.2	0.2	0.5	0.5	1.00
6	5—6	0.0	0.0	0.2	0.2	1.00
1	4—7	0.0	0.0	0.0	0.0	1.00
2	4—7	2.9	2.9	0.0	0.0	1.01
3	4—7	2.6	2.6	2.9	2.9	1.15
4	4—7	0.3	0.3	2.6	2.6	1.11
5	4—7	0.0	0.0	0.3	0.3	1.00
6	4—7	0.0	0.0	0.0	0.0	1.00
1	5—8	0.0	0.0	0.0	0.0	1.00
2	5—8	0.0	0.0	0.0	0.0	1.00
3	5—8	0.7	0.7	0.0	0.0	1.00
4	5—8	0.5	0.5	0.7	0.7	1.01
5	5—8	0.7	0.7	0.5	0.5	1.01
6	5—8	0.0	0.0	0.7	0.7	1.01
1	6—9	0.0	0.0	0.0	0.0	1.00
2	6—9	0.0	0.0	0.0	0.0	1.00
3	6—9	0.0	0.0	0.0	0.0	1.00
4	6—9	3.5	3.5	0.0	0.0	1.01
5	6—9	3.1	3.1	3.5	3.5	1.22
6	6—9	0.7	0.7	3.1	3.1	1.16
7	6—9	0.0	0.0	0.7	0.7	1.01
1	7—8	0.0	0.0	0.0	0.0	1.00
2	7—8	0.0	0.0	0.0	0.0	1.00
3	7—8	2.9	2.9	0.0	0.0	1.01
4	7—8	2.6	2.6	2.9	2.9	1.15
5	7—8	0.3	0.3	2.6	2.6	1.11
6	7—8	0.0	0.0	0.3	0.3	1.00
1	8—9	0.0	0.0	0.0	0.0	1.00
2	8—9	0.0	0.0	0.0	0.0	1.00
3	8—9	0.0	0.0	0.0	0.0	1.00
4	8—9	3.6	3.6	0.0	0.0	1.01
5	8—9	3.1	3.1	3.6	3.6	1.23
6	8—9	1.0	1.0	3.1	3.1	1.17
7	8—9	0.0	0.0	1.0	1.0	1.02

Table 6.4: Comparison of Actual Route Travel Times

Interval	Routes from 1 to 5	
k	1-2-5	1-4-5
1	2.12	2.12
2	3.21	3.22
3	2.73	2.74
4	2.08	2.07
5	2.03	2.03
6	2.00	2.00

Interval	Routes from 1 to 9					
k	1-4-7-8-9	1-4-5-8-9	1-4-5-6-9	1-2-3-6-9	1-2-5-6-9	1-2-5-8-9
1	4.13	4.12	4.13	4.13	4.13	4.13
2	5.38	5.40	5.23	5.39	5.37	5.39
3	4.73	4.77	4.75	4.71	4.74	4.76

6.2 Suppose there are additional constraints in the constraint set, including explicit capacity and oversaturation constraints. Will the new constraints affect the relaxation procedure? Explain your answer. How will they affect the Frank-Wolfe procedure?

6.3 For the relaxation procedure for solving the VI problem, formulate a different optimization subproblem objective function than equation (6.2). What variables should be temporarily fixed in the relaxation?

6.4 For the one-dimentional minimization problem (6.57), discuss which method should be used, the golden section method or the bisection method? Is it possible to use a more efficient method?

6.5 For the network shown in Figure 6.7, construct an expanded time-space network for 6 time intervals. Assume the link travel times are

Figure 6.7: Network for Problem 6.5

$$\tau_1(k) = 2, \qquad \tau_2(k) = 1, \qquad \tau_3(k) = 2, \qquad \forall k = 1, 2, \cdots, 6.$$

(a) Determine the feasible route and corresponding route flows on the expanded time-space network for inflow $u_1(1) = 10$.
(b) If node 4 is a destination, construct a revised time-space network. Assume that the travel time on link 2 is changed to

$$\tau_2(k) = 3 \qquad \forall k = 1, 2, \cdots, 6.$$

Determine the feasible route and corresponding route flows on the expanded time-space network for inflow $u_1(1) = 10$.

6.6 The round-off of link travel times is given by the following criterion:

$$\bar{\tau}_a(n) = i \quad \text{if} \quad i - 0.5 \leq \bar{\tau}_a(n) < i + 0.5,$$

where i is an integer and $0 \leq i \leq K$. Discuss the impact of the round-off errors on the convergence of the relaxation algorithm.

6.7 Without using the expanded time-space network, redesign the solution algorithm for the ideal DUO route choice problem.

Chapter 7

Instantaneous Dynamic Route Choice Models

In this chapter, we present several variational inequality (VI) models for instantaneous dynamic user-optimal route choice problems for a network with multiple origin-destination pairs. In Section 7.1, the definition of instantaneous dynamic user-optimal (DUO) state is presented. Then, a route-time-based VI model is proposed in Section 7.2. The equivalence of the VI model with the route-time-based instantaneous DUO route choice conditions is demonstrated.

Since explicit route enumeration is needed to solve the route-based VI model, we also formulate a link-based VI model. In Section 7.3, a link-time-based VI model is proposed and the equivalence of the VI model with the travel-time-based instantaneous DUO route choice conditions is demonstrated. In Section 7.4, we discuss the relationships between VI models and optimization models. As an example, we demonstrate that the link-time-based VI model presented in Section 7.3 can be reduced to an optimal control model under relaxation. Thus, the relaxation algorithm and the F-W technique presented in Chapter 6 is used to solve this link-time-based VI model. In the solution, we note that explicit route enumeration is unnecessary. Finally, a numerical example is given to demonstrate the solution properties.

7.1 Definition of an Instantaneous State

We now consider a general transportation network. We define a decision node for each route p of each O-D pair as any node on the route including the origin. The instantaneous route travel time between a decision node and the destination node is calculated using the currently prevailing link travel times. Significantly, with many current traveler information systems, such as radio broadcasts of traffic conditions, the information provided to travelers on freeways is the estimated *instantaneous* route travel time. Thus, at present many travelers do choose routes based on current or instantaneous travel times.

Consider the flow originating at node r at time t and destined for node s. There is a set of routes $\{p\}$ between O-D pair rs. Denote $f_p^{rs}(t)$ as the route inflow from origin r to destination s over route p at time t. For any link a, link a is defined as being used at time t if $u_a(t) > 0$. Furthermore, a route p between r and s is defined as being used at time t if its departure flow $f_p^{rs}(t) > 0$.

In this chapter, we formulate two alternative variational inequality models, each of which is equivalent to the instantaneous DUO route choice conditions. We assume that the time-dependent origin-destination trip pattern is known *a priori*. In other words, the departure times of travelers are given. The instantaneous dynamic user-optimal (DUO) route choice problem is to determine vehicle flows at each instant of time on each link resulting from drivers using minimal-time routes under the currently prevailing travel times. In this chapter, we consider the following dynamic generalization of the conventional static user-optimal state.

Travel-Time-Based Instantaneous DUO State: *If, for each O-D pair at each decision node at each instant of time, the instantaneous travel times for all routes that are being used equal the minimal instantaneous route travel time, the dynamic traffic flow over the network is in a travel-time-based* instantaneous *dynamic user-optimal state.*

In Chapter 9, we generalize the above definition to the situation of travel disutility and multi-class travelers instead of travel times for a single class of travelers.

Although the instantaneous user-optimal travel times for all routes that are being used are equal at each decision node at each instant of time, route flows with the same departure time and the same origin-destination may actually experience somewhat different route travel times. This is because the route time may subsequently change due to changing network traffic conditions, even though at each decision node the flows select the route that is currently the best. The difference between instantaneous DUO and ideal DUO is discussed further in Chapter 8.

The dynamic user-optimal route choice problem is to find the dynamic trajectories of link states and inflow and exit flow control variables, given the time-dependent O-D flow requirements, the network and the link travel time functions. The formulation of the problem in this chapter is based on the underlying choice criterion that each traveler uses the route that minimizes his/her instantaneous travel time when departing from the origin or any intermediate node to his/her destination. This route choice rule implies that in a travel-time-based instantaneous dynamic user-optimal state, the trajectories of the link flow states and inflows and exit flows are such that the instantaneous travel times at each decision node of all used routes connecting any given O-D pair are identical and not greater than the instantaneous travel times of routes which are not being used.

7.2 A Route-Time-Based Model

7.2.1 Route-Time-Based Conditions

Recall from Chapter 4 that the instantaneous travel time $c_a(t)$ over link a is assumed to be dependent on the number of vehicles $x_a(t)$, the inflow $u_a(t)$ and the exit flow $v_a(t)$ on link a at time t. This instantaneous link time is the travel time that would be incurred if traffic conditions on the link remain unchanged while traversing the link. Note that for most cases, we don't distinguish the instantaneous link travel time and actual link travel time. We assume the instantaneous travel time $c_a(t)$ on link a is the sum of two components: 1) an instantaneous flow-dependent running time $g_{1a}[x_a(t), u_a(t)]$ over link a; and 2) an instantaneous queuing delay $g_{2a}[x_a(t), v_a(t)]$. It follows that

$$c_a(t) = g_{1a}[x_a(t), u_a(t)] + g_{2a}[x_a(t), v_a(t)]. \tag{7.1}$$

The two components $g_{1a}[x_a(t), u_a(t)]$ and $g_{2a}[x_a(t), v_a(t)]$ of the time-dependent link travel time function $c_a[x_a(t), u_a(t), v_a(t)]$ are assumed to be nonnegative and differentiable with respect to $x_a(t), u_a(t)$ and $x_a(t), v_a(t)$, respectively.

Consider the flow which originates at node r at time t and is destined for node s. There is a set of routes $\{p\}$ between O-D pair rs. Define the instantaneous travel time function $\psi_p^{rs}(t)$ for each route p between rs as

$$\psi_p^{rs}(t) = \sum_{a \in rsp} c_a[x_a(t), u_a(t), v_a(t)] \qquad \forall r, s, p; \tag{7.2}$$

where the summation is over all links in route p from origin r to destination s.

We now write the route-time-based instantaneous DUO route choice conditions which are equivalent to the definition of travel-time-based instantaneous DUO. Denote $\sigma^{rs}(t)$ as the minimal instantaneous route travel time from origin r to destination s at time t. The equivalent instantaneous DUO route choice conditions can be stated as follows:

$$\psi_p^{rs^*}(t) - \sigma^{rs^*}(t) \geq 0 \qquad \forall p, r, s; \tag{7.3}$$

$$f_p^{rs^*}(t) \left[\psi_p^{rs^*}(t) - \sigma^{rs^*}(t)\right] = 0 \qquad \forall p, r, s; \tag{7.4}$$

$$f_p^{rs}(t) \geq 0 \qquad \forall p, r, s. \tag{7.5}$$

Equation (7.4) states that if there is a positive route inflow $f_p^{rs^*}(t) > 0$, the instantaneous route travel time $\psi_p^{rs^*}(t)$ must equal the minimal instantaneous route travel time $\sigma^{rs^*}(t)$. Otherwise, the instantaneous route travel time $\psi_p^{rs^*}(t)$ may be greater than or equal to the minimal instantaneous route travel time $\sigma^{rs^*}(t)$. The above route-time-based definition of an instantaneous DUO state and its corresponding route choice conditions are defined for each O-D pair.

7.2.2 Dynamic Network Constraints

The constraint set for the route-time-based VI model is summarized as follows.

Relationship between state and control variables:

$$\frac{dx_{ap}^{rs}}{dt} = u_{ap}^{rs}(t) - v_{ap}^{rs}(t) \qquad \forall a, p, r, s; \tag{7.6}$$

$$\frac{dE_p^{rs}(t)}{dt} = e_p^{rs}(t) \qquad \forall p, r; s \neq r; \tag{7.7}$$

Flow conservation constraints:

$$f^{rs}(t) = \sum_{a \in A(r)} \sum_p u_{ap}^{rs}(t) \qquad \forall r, s; \tag{7.8}$$

$$\sum_{a \in B(j)} v_{ap}^{rs}(t) = \sum_{a \in A(j)} u_{ap}^{rs}(t) \qquad \forall j, p, r, s; j \neq r, s; \tag{7.9}$$

$$\sum_{a \in B(s)} \sum_p v_{ap}^{rs}(t) = e^{rs}(t) \qquad \forall r, s; s \neq r; \tag{7.10}$$

Flow propagation constraints:

$$x_{ap}^{rs}(t) = \sum_{b \in \bar{p}} \{x_{bp}^{rs}[t + \tau_a(t)] - x_{bp}^{rs}(t)\} + \{E_p^{rs}[t + \tau_a(t)] - E_p^{rs}(t)\}$$

$$\forall a \in B(j); j \neq r; p, r, s; \tag{7.11}$$

Definitional constraints:

$$\sum_{rsp} u_{ap}^{rs}(t) = u_a(t), \qquad \sum_{rsp} v_{ap}^{rs}(t) = v_a(t), \qquad \forall a; \tag{7.12}$$

$$\sum_{rsp} x_{ap}^{rs}(t) = x_a(t), \qquad \forall a; \tag{7.13}$$

Nonnegativity conditions:

$$x_{ap}^{rs}(t) \geq 0, \quad u_{ap}^{rs}(t) \geq 0, \quad v_{ap}^{rs}(t) \geq 0 \quad \forall a, p, r, s; \tag{7.14}$$

$$e_p^{rs}(t) \geq 0, \qquad E_p^{rs}(t) \geq 0, \qquad \forall p, r, s; \tag{7.15}$$

Boundary conditions:

$$E_p^{rs}(0) = 0, \qquad \forall p, r, s; \tag{7.16}$$

$$x_a^{rs}(0) = 0, \qquad \forall a, r, s. \tag{7.17}$$

We note that other constraints, such as FIFO constraints, capacity constraints and oversaturation constraints, can be added to this VI model. To simplify the analysis, we ignore them here.

7.2.3 The Variational Inequality Problem

The equivalent variational inequality formulation of the route-time-based instantaneous DUO route choice conditions (7.3)-(7.5) may be stated as follows.

Theorem 7.1. *The dynamic traffic flow pattern satisfying the network constraint set (7.6)-(7.17) is in a route-time-based instantaneous DUO route choice state if and only if it satisfies the variational inequality problem:*

$$\int_0^T \sum_{rs} \sum_p \psi_p^{rs^*}(t) \left[f_p^{rs}(t) - f_p^{rs^*}(t) \right] dt \geq 0 \tag{7.18}$$

Proof of Necessity.

We need to prove that the route-time-based instantaneous DUO route choice conditions (7.3)-(7.5) imply variational inequality (7.18). For any route p, a feasible inflow at time t is

$$f_p^{rs}(t) \geq 0. \tag{7.19}$$

Multiplying instantaneous DUO route choice condition (7.3) by the above equation, we have

$$f_p^{rs}(t) \left[\psi_p^{rs^*}(t) - \sigma^{rs^*}(t)\right] \geq 0 \qquad \forall p, r, s. \tag{7.20}$$

We subtract equation (7.4) from equation (7.20) and obtain

$$\left[f_p^{rs}(t) - f_p^{rs^*}(t) \right] \left[\psi_p^{rs^*}(t) - \sigma^{rs^*}(t) \right] \geq 0 \qquad \forall p, r, s. \tag{7.21}$$

Summing equation (7.21) for all routes p and all O-D pairs rs, it follows that

$$\begin{aligned}
& \sum_{rs} \sum_p \left[f_p^{rs}(t) - f_p^{rs^*}(t) \right] \left[\psi_p^{rs^*}(t) - \sigma^{rs^*}(t) \right] \\
= & \sum_{rs} \sum_p \left[f_p^{rs}(t) - f_p^{rs^*}(t) \right] \psi_p^{rs^*}(t) - \sum_{rs} \sigma^{rs^*}(t) \sum_p \left[f_p^{rs}(t) - f_p^{rs^*}(t) \right] \\
= & \sum_{rs} \sum_p \left[f_p^{rs}(t) - f_p^{rs^*}(t) \right] \psi_p^{rs^*}(t) \geq 0
\end{aligned} \tag{7.22}$$

where the flow conversation equation

$$\sum_p f_p^{rs}(t) = \sum_p f_p^{rs^*}(t) = f^{rs}(t)$$

holds for each O-D rs at each time t. Integrating the above equation from 0 to T, we obtain variational inequality (7.18).

Proof of Sufficiency.

We need to prove that any solution $f_p^{rs*}(t)$ to variational inequality (7.18) satisfies the route-time-based instantaneous DUO route choice conditions (7.3)-(7.5). We know that the first and third instantaneous DUO route choice conditions (7.3) and (7.5) hold by definition. Thus, we need to prove that the second instantaneous DUO route choice condition (7.4) also holds.

Assume that the second instantaneous DUO route choice condition (7.4) does not hold only for a route q for O-D pair kn during time interval $[t_1 - \delta, t_1 + \delta] \in [0, T]$, i.e.,

$$f_q^{kn*}(t) > 0 \quad \text{and} \quad \psi_q^{kn*}(t) - \sigma^{kn*}(t) > 0 \quad \forall\, t \in [t_1 - \delta, t_1 + \delta] \tag{7.23}$$

Since the second instantaneous DUO route choice condition (7.4) holds for all other routes other than route q for O-D pair kn at any time t and for O-D pair kn at any time $t \notin [t_1 - \delta, t_1 + \delta]$, it follows that

$$\begin{aligned} &\int_0^T \sum_{rs} \sum_p f_p^{rs*}(t)\, [\psi_p^{rs*}(t) - \sigma^{rs*}(t)]\, dt \\ &= \int_{t_1-\delta}^{t_1+\delta} f_q^{kn*}(t)\ [\psi_q^{kn*}(t) - \sigma^{kn*}(t)]\, dt\ > 0 \end{aligned} \tag{7.24}$$

We note that all other terms in the above equation vanish because of instantaneous DUO route choice condition (7.4).

For each O-D pair rs, we can always find one minimal instantaneous travel time route l for vehicles departing origin r at time t, where route l was evaluated under the optimal route inflow pattern $\{f_p^{rs*}(t)\}$. For this route l, the first instantaneous DUO route choice condition (7.3) becomes an equality by definition. It follows that

$$\psi_l^{rs*}(t) - \sigma^{rs*}(t) = 0 \qquad \forall l, r, s. \tag{7.25}$$

Next, we need to find a set of feasible route inflows $f_p^{rs}(t)$ so that the following equations

$$f_p^{rs}(t)\ \left[\psi_p^{rs*}(t) - \sigma^{rs*}(t)\right] = 0 \qquad \forall p, r, s \tag{7.26}$$

always hold. We consider departure flows $f^{rs}(t)$ for all O-D pairs at each time t. For each O-D pair rs at each time t, we assign O-D departure flow $f^{rs}(t)$ to the minimal travel time route l, which was evaluated under the optimal route inflow pattern $\{f_p^{rs*}(t)\}$. This generates a set of feasible route inflow patterns $\{f_p^{rs}(t)\}$ which always satisfy equation (7.26) because flows are not assigned to routes with non-minimal travel times which were evaluated under the optimal route inflow pattern $\{f_p^{rs*}(t)\}$. Summing equations (7.26) for all routes p and all O-D pairs rs, it follows that

$$\sum_{rs} \sum_p f_p^{rs}(t)\ \left[\psi_p^{rs*}(t) - \sigma^{rs*}(t)\right] = 0 \tag{7.27}$$

Integrating the above equation for time period $[0, T]$, we have

$$\int_0^T \sum_{rs} \sum_p f_p^{rs}(t) \left[\psi_p^{rs^*}(t) - \sigma^{rs^*}(t)\right] dt = 0 \tag{7.28}$$

We subtract equation (7.24) from equation (7.28) and obtain

$$\begin{aligned} & \int_0^T \sum_{rs} \sum_p \left[f_p^{rs}(t) - f_p^{rs^*}(t)\right] \left[\psi_p^{rs^*}(t) - \sigma^{rs^*}(t)\right] dt \\ = & \int_0^T \left\{ \sum_{rs} \sum_p \left[f_p^{rs}(t) - f_p^{rs^*}(t)\right] \psi_p^{rs^*}(t) \right. \\ - & \left. \sum_{rs} \sigma^{rs^*}(t) \sum_p \left[f_p^{rs}(t) - f_p^{rs^*}(t)\right] \right\} dt \\ = & \int_0^T \sum_{rs} \sum_p \psi_p^{rs^*}(t) \left[f_p^{rs}(t) - f_p^{rs^*}(t)\right] dt < 0 \end{aligned} \tag{7.29}$$

where the flow conversation equation

$$\sum_p f_p^{rs}(t) = \sum_p f_p^{rs^*}(t) = f^{rs}(t)$$

holds for each O-D rs at each time t so that the second term of the equation vanishes. The above equation contradicts variational inequality (7.18). Therefore, any optimal solution $\{f_p^{rs^*}(t)\}$ to variational inequality (7.18) satisfies the second instantaneous DUO route choice condition (7.4). Since we proved the necessity and sufficiency of the equivalence of variational inequality (7.18) to route-time-based instantaneous DUO route choice conditions (7.3)-(7.5), the proof is complete.

7.3 A Link-Time-Based Model

The VI model presented above is route-based; its solution requires explicit route enumeration. Although the route-based model is intuitive in terms of understanding, route enumeration is intractable if the network is large, which is termed the curse of dimensionality.

To distinguish the link-based model from the route-based model, we can focus on the variational inequality instead of their constraints because both constraints are route-based. If the variational inequality can be formulated using link-based variables instead of route-based variables, a carefully designed solution algorithm for such a variational inequality will not require explicit route enumeration. We will demonstrate this point later. This is also true for optimization problems. If the objective function of an optimization program can be formulated using link-based variables instead of route-based variables, a carefully designed solution algorithm will not need explicit route enumeration. Recall that in the nonlinear programming formulation of the static user-optimal route choice model, the objective function can be formulated using either link-based variables or route-based variables. However, the link-based objective function is preferred because for such a link-based objective function, the Frank-Wolfe algorithm does not need explicit route enumeration. This property is also true for the variational inequality models in this book.

Because dynamic traffic flow does not have a constant flow rate during propagation over links and routes, route-based VI models can not be readily transformed into link-based VI models. This issue is the most critical constraint for applying route-based VI to realistic transportation networks. Therefore, we propose a link-based VI which overcomes this problem. In addition to this contribution, we note that our formulation approach is different from others (Smith, 1993; Friesz et al, 1993).

The set of dynamic network constraints for the link-time-based VI model is identical to constraint set (7.6)-(7.17) of the route-time-based VI model in Section 7.2. The basic difference between the two models is that the variational inequality of the link-time-based VI model is formulated using link-based flow variables instead of route-based variables in the route-time-based VI model.

7.3.1 Link-Time-Based Conditions

In contrast to the previous sections, we now introduce a new set of instantaneous DUO route choice conditions based on link and node variables, instead of route-based variables. In this problem, link a is defined as *being used* at time t if $u_a(t) > 0$. A route p from decision node i to destination s is defined as *being used* at time t if $u^s_{ap}(t) > 0$ for the first link $a \in p$. In the following, we define an instantaneous DUO route choice state based on link and node variables.

> **Link-Time-Based Instantaneous DUO State:** *If, for any departure flow from each decision node to each destination node at each instant of time, the instantaneous travel times equal the minimal instantaneous route travel time, the dynamic traffic flow over the network is in a link-time-based* instantaneous *dynamic user-optimal state.*

Note that the route-time-based instantaneous DUO state defined in Section 7.1 is a subset of the link-time-based instantaneous state because the decision node includes the origin so that each O-D pair is one decision node-destination pair.

Define $\sigma^{ri^*}(t)$ as the minimal instantaneous travel time for vehicles departing from origin r to node i at time t, the asterisk denoting that the travel time is computed using link-time-based instantaneous DUO traffic flows. For link $a = (i,j)$, the minimal instantaneous travel time $\sigma^{ri^*}(t)$ from origin r to node i should be equal to or less than the minimal instantaneous travel time

$\sigma^{rj^*}(t)$ from origin r to node j plus the instantaneous link travel time $c_a^*(t)$ at time instant t. It follows that

$$\sigma^{ri^*}(t) + c_a^*(t) \geq \sigma^{rj^*}(t) \qquad \forall a = (i,j), s. \tag{7.30}$$

If any departure flow from origin r to node i enters link a at time t, or $u_a^r(t) > 0$, then the link-time-based instantaneous DUO route choice conditions require that link a is on the minimal instantaneous travel time route. In other words, the instantaneous minimal travel time $\sigma^{rj^*}(t)$ for vehicles departing origin r toward node j at time t should equal the minimal instantaneous travel time $\sigma^{ri^*}(t)$ from origin r to node i plus the instantaneous link travel time $c_a^*(t)$ at time t. It follows that

$$\sigma^{ri^*}(t) + c_a^*(t) = \sigma^{rj^*}(t), \qquad \text{if } u_a^{r^*}(t) > 0 \qquad \forall a = (i,j), r. \tag{7.31}$$

The above equations are also equivalent to the following:

$$\left[\sigma^{ri^*}(t) + c_a^*(t) - \sigma^{rj^*}(t)\right] u_a^{r^*}(t) = 0 \qquad \forall a = (i,j), r. \tag{7.32}$$

Let $\Theta_a^{r^*}(t)$ be the difference between the minimal instantaneous travel time from origin r to node j and the instantaneous travel time from origin r to node i plus the instantaneous travel time on link a at time t. It follows that

$$\Theta_a^{r^*}(t) = \sigma^{ri^*}(t) + c_a^*(t) - \sigma^{rj^*}(t) \qquad \forall a, r; a = (i,j). \tag{7.33}$$

Thus, the link-time-based instantaneous DUO route choice conditions can be summarized as follows:

$$\Theta_a^{r^*}(t) \geq 0 \qquad \forall a = (i,j), r; \tag{7.34}$$

$$u_a^{r^*}(t)\, \Theta_a^{r^*}(t) = 0 \qquad \forall a = (i,j), r; \tag{7.35}$$

$$u_a^r(t) \geq 0 \qquad \forall a = (i,j), r. \tag{7.36}$$

Now, we state a lemma concerning the relationship between the link-time-based and the route-time-based instantaneous DUO route choice conditions.

Lemma 7.1. *Link-time-based instantaneous DUO route choice conditions (7.34)-(7.36) imply route-time-based instantaneous DUO route choice conditions (7.3)-(7.5).*

Proof.

If we consider node j as an destination s, then link-time-based instantaneous DUO route choice conditions (7.34)-(7.36) apply to O-D pair rs. Thus, equations (7.34)-(7.36) can be written as

$$\sigma^{ri^*}(t) + c_a^*(t) - \sigma^{rs^*}(t) \geq 0 \qquad \forall a = (i, s), r, s; \tag{7.37}$$

$$u_a^{rs^*}(t)\left[\sigma^{ri^*}(t) + c_a^*(t) - \sigma^{rs^*}(t)\right] = 0 \qquad \forall a = (i, s), r, s; \tag{7.38}$$

$$u_a^{rs}(t) \geq 0 \qquad \forall a = (i, s), r, s \tag{7.39}$$

Suppose there are P routes from origin r to destination s via link a. Since equation (7.55) applies to all links a exiting origin r, we can define an instantaneous travel time on route p as $\psi_p^{rs^*}(t)$. If link a is on route p, this instantaneous route travel time $\psi_p^{rs^*}(t)$ must be greater than or equal to the minimal instantaneous travel time from origin r to node i plus the instantaneous travel time on link a at time t. It follows that

$$\psi_p^{rs^*}(t) \geq \sigma^{ri^*}(t) + c_a^*(t) \tag{7.40}$$

By equation (7.37), we have

$$\psi_p^{rs^*}(t) \geq \sigma^{rs^*}(t) \tag{7.41}$$

which is identical to equation (7.3). Furthermore, if there is a flow on link a, i.e., $u_a^{rs^*}(t) > 0$, then by equations (7.37)-(7.38), we have

$$\sigma^{ri^*}(t) + c_a^*(t) - \sigma^{rs^*}(t) = 0 \tag{7.42}$$

Specify p as the route via link a and the minimal instantaneous travel time subroute from origin r to node i. Thus, the flow $f_p^{rs^*}(t)$ on route p is positive at time t and we have

$$\psi_p^{rs^*}(t) = \sigma^{ri^*}(t) + c_a^*(t) \tag{7.43}$$

In other words, we have

$$f_p^{rs^*}(t)\left[\psi_p^{rs^*}(t) - \sigma^{rs^*}(t)\right] = 0 \tag{7.44}$$

For any positive link inflow $u_a^{rs^*}(t) > 0$, we can generate a corresponding positive route inflow $f_p^{rs^*}(t)$ and the corresponding equation (7.44). Thus, equation (7.44) applies to any positive route inflow $f_p^{rs^*}(t)$. We conclude that it applies to any route between any O-D pair rs. Note that the above equation applies to any zero route inflow $f_p^{rs^*}(t) = 0$ as well. Therefore, equations (7.41) and (7.44) imply the route-time-based instantaneous DUO route choice conditions (7.3)-(7.5). The proof is complete.

7.3.2 The Variational Inequality Problem

The equivalent variational inequality formulation of link-time-based instantaneous DUO route choice conditions (7.34)-(7.36) may be stated as follows.

Theorem 7.3. *The dynamic traffic flow pattern satisfying network constraint set (7.6)-(7.17) is in a link-time-based instantaneous DUO route choice state if and only if it satisfies the variational inequality problem:*

$$\int_0^T \sum_r \sum_a \Theta_a^{r^*}(t) \left[u_a^r(t) - u_a^{r^*}(t)\right] dt \geq 0 \tag{7.45}$$

Proof of Necessity.

We need to prove that link-time-based instantaneous DUO route choice conditions (7.34)-(7.36) imply variational inequality (7.45). For any link a, a feasible inflow at time t is

$$u_a^r(t) \geq 0. \qquad \forall a = (i,j), r. \tag{7.46}$$

Multiplying equation (7.46) and equation (7.34) we have

$$u_a^r(t)\, \Theta_a^{r^*}(t) \geq 0 \qquad \forall a, r; a = (i,j). \tag{7.47}$$

We subtract equation (7.35) from equation (7.47) and obtain

$$\left[u_a^r(t) - u_a^{r^*}(t)\right] \Theta_a^{r^*}(t) \geq 0 \qquad \forall a, r; a = (i,j). \tag{7.48}$$

Summing equation (7.48) for all links a and all origins r, it follows that

$$\sum_r \sum_a \left[u_a^r(t) - u_a^{r^*}(t)\right] \Theta_a^{r^*}(t) \geq 0 \qquad \text{where } a = (i,j). \tag{7.49}$$

Integrating the above equation from 0 to T, we obtain variational inequality (7.45).

Proof of Sufficiency.

We need to prove that any solution $u_a^{r^*}(t)$ to variational inequality (7.45) satisfies link-time-based instantaneous DUO route choice conditions (7.34)-(7.36). We know that the first and third instantaneous DUO route choice conditions (7.34) and (7.36) hold by definition. Thus, we need to prove that the second instantaneous DUO route choice condition (7.35) also holds.

Assume that the second instantaneous DUO route choice condition (7.35) does not hold only for a link $b = (l,m)$ for a origin n during a time interval $[d-\delta, d+\delta] \in [0,T]$, i.e.,

$$u_b^{n^*}(t) > 0 \quad \text{and} \quad \Theta_b^{n^*}(t) > 0 \quad t \in [d-\delta, d+\delta] \tag{7.50}$$

Thus, we have

$$u_b^{n^*}(t)\ \Theta_b^{n^*}(t) > 0 \tag{7.51}$$

where

$$\Theta_b^{n^*}(t) = \sigma^{nl^*}(t) + c_b^*(t) - \sigma^{nm^*}(t) > 0 \qquad \text{where } b = (l, m). \tag{7.52}$$

Note that the second instantaneous DUO route choice condition (7.35) holds for all links other than link $b = (l, m)$ for origin n at time t. Equation (7.35) also holds for link $b = (l, m)$ for origins $r \neq n$ at time t and for link $b = (l, m)$ for origin n at time $t \notin [d - \delta, d + \delta]$. It follows that

$$\int_0^T \sum_r \sum_a \Theta_a^{r^*}(t)\ u_a^{r^*}(t)\ dt \ = \int_{d-\delta}^{d+\delta} u_b^{n^*}(t)\ \ \Theta_b^{n^*}(t)\ dt\ > 0 \tag{7.53}$$

We note that all other terms in the above equation vanish because of instantaneous DUO route choice condition (7.35).

For each O-D pair rs, we can always find one minimal travel time route p for vehicles departing origin r at time t, where route p was evaluated under optimal flow pattern $\{u_a^{r^*}(t)\}$. For each link a on this route p, the first instantaneous DUO route choice condition (7.34) becomes an equality by definition. It follows that

$$\Theta_a^{r^*}(t) = \sigma^{ri^*}(t) + c_a^*(t) - \sigma^{rj^*}(t) = 0 \qquad \forall a, r; a = (i, j); a \in p. \tag{7.54}$$

Next, we need to find a set of feasible inflows $u_a^r(t)$ so that equations

$$u_a^r(t)\ \Theta_a^{r^*}(t) = 0 \qquad\qquad \forall a, r; a = (i, j) \tag{7.55}$$

always hold. We adjust all the departure flows $f^{rs}(t)$ for all O-D pairs at time t. For each O-D pair rs at each time t, we assign O-D departure flow $f^{rs}(t)$ to the minimal travel time route p, which was evaluated under the optimal flow pattern $\{u_a^{r^*}(t)\}$. This will generate a set of feasible inflow patterns $\{u_a^r(t)\} = \sum_r u_a^{rs}(t)$ which always satisfies equations (7.54) and (7.55) because flows are not assigned to routes with non-minimal instantaneous travel times which were evaluated under the optimal flow pattern $\{u_a^{r^*}(t)\}$. Summing equations (7.55) for all links a and all origins r, it follows that

$$\sum_r \sum_a u_a^r(t)\ \Theta_a^{r^*}(t) = 0 \qquad\qquad \text{where } a = (i, j). \tag{7.56}$$

Integrating the above equation (7.56) for interval $[0, T]$, we have

$$\int_0^T \sum_r \sum_a u_a^r(t)\ \Theta_a^{r^*}(t)\ dt\ = 0 \tag{7.57}$$

We subtract equation (7.53) from equation (7.57) and obtain

$$\int_0^T \sum_r \sum_a \Theta_a^{r^*}(t)\ \left[u_a^r(t) - u_a^{r^*}(t)\right]\ dt\ < 0 \tag{7.58}$$

The above equation contradicts variational inequality (7.45). Therefore, any optimal solution $\{u_a^{r^*}(t)\}$ to variational inequality (7.45) satisfies the second instantaneous DUO route choice condition (7.35). Since we proved the necessity and sufficiency of the equivalence of variational inequality (7.45) to link-time-based instantaneous DUO route choice conditions (7.34)-(7.36), the proof is complete.

7.4 Solution Algorithm

7.4.1 Discrete VI Model for the Link-Time-Based Case

To convert our instantaneous DUO route choice model into an NLP, the time period $[0, T]$ is subdivided into K small time increments. Each time increment is a unit of time. Then, $u_a(k)$ represents the inflow into link a during interval k and $v_a(k)$ represents the exit flow from link a during interval k. To simplify the formulation, we modify the estimated actual travel time on each link in the following way so that each estimated travel time is equal to a multiple of the time increment.

$$\bar{\tau}_a(k) = i \quad \text{if} \quad i - 0.5 \leq \bar{\tau}_a(k) < i + 0.5,$$

where i is an integer and $0 \leq i \leq K$. We note that this round-off method is used only in the flow propagation constraints. More accurate flow propagation constraints can be obtained by making the time intervals smaller. We also note that evaluation of the instantaneous link travel time function and objective function does not have this round-off error so that the subsequent minimal-cost route search does not have this round-off error.

In this section, we discuss the following discrete VI problem:

$$\sum_{k=1}^{K} \sum_{r} \sum_{a} \Theta_a^{r^*}(k) \left[u_a^r(k) - u_a^{r^*}(k) \right] \geq 0 \tag{7.59}$$

where

$$\Theta_a^{r^*}(k) = \sigma^{ri^*}(k) + c_a^*(k) - \sigma^{rj^*}(k) \qquad \forall a, r; a = (i, j). \tag{7.60}$$

To simplify our analysis, we assume the time period $[0, T]$ is long enough so that all departure flows can be cleared at final time T.

7.4.2 Relaxation Procedure and Optimization Program

For each relaxation iteration, we temporarily fix:

1. the actual travel time $\tau_a(k)$ in link flow propogation constraints as $\bar{\tau}_a(k)$;

2. minimal instantaneous travel times $\sigma^{ri^*}(k)$ as $\bar{\sigma}^{ri^*}(k)$ and $\sigma^{rj^*}(k)$ as $\bar{\sigma}^{rj^*}(k)$ for each link and each origin node.

Then, the objective function of an optimization program (NLP) can be formulated, which is equivalent to the discrete VI under relaxation as follows

$$\min_{u,v,x,E} \quad Z \quad = \quad \sum_{k=1}^{K}\sum_{a}\left\{\int_{0}^{u_a(k)} c_a[x_a(k),\omega,v_a(k)]d\omega \right.$$
$$\left. + \quad \sum_{r} u_a^r(k)[\bar{\sigma}^{ri}(k) - \bar{\sigma}^{rj}(k)]\right\} \qquad (7.61)$$

Note that both instantaneous minimal travel times $\bar{\sigma}^{ri}(k)$ and $\bar{\sigma}^{rj}(k)$ are fixed temporarily at each relaxation iteration.

To ensure that the optimization program with the above objective function is indeed equivalent to the discrete VI, we should demonstrate that the gradient of the above discrete objective function (7.61) is equivalent to the cost term of discrete variational inequality (7.59) under relaxation, i.e.,

$$\frac{\partial Z}{\partial u_a^{rs}(k)} \quad = \quad c_a(k) + \bar{\sigma}^{ri}(k) - \bar{\sigma}^{rj}(k) = \bar{\Theta}_a^{r^*}(k) \qquad \forall a, r, s, k. \quad (7.62)$$

It is obvious that the above equation holds. In other words, this objective function (7.61) can be used in the relaxation optimization problem.

Note that the network constraints apply for each time interval k. The discrete network flow constraints are now summarized as follows:

$$x_{ap}^{rs}(k+1) = x_{ap}^{rs}(k) + u_{ap}^{rs}(k) - v_{ap}^{rs}(k) \qquad \forall a,p,r,s; k = 1,\cdots,K; \quad (7.63)$$

$$E^{rs}(k+1) = E^{rs}(k) + \sum_{a\in B(s)}\sum_{p} v_{ap}^{rs}(k) \qquad \forall r; s \neq r; k = 1,\cdots,K; \quad (7.64)$$

$$\sum_{a\in A(r)}\sum_{p} u_{ap}^{rs}(k) = f^{rs}(k) \qquad \forall r \neq s; k = 1,\cdots,K; \qquad (7.65)$$

$$\sum_{a\in B(j)} v_{ap}^{rs}(k) - \sum_{a\in A(j)} u_{ap}^{rs}(k) = 0 \qquad \forall j,p,r,s; j \neq r,s; k = 1,\cdots,K; \quad (7.66)$$

$$x_{ap}^{rs}(k) = \sum_{b\in\bar{p}}\{x_{bp}^{rs}[k+\bar{\tau}_a(k)] - x_{bp}^{rs}(k)\} + \{E_p^{rs}[k+\bar{\tau}_a(k)] - E_p^{rs}(k)\}$$
$$\forall a \in B(j); j \neq r; p,r,s; k = 1,\cdots,K+1; \qquad (7.67)$$

$$u_{ap}^{rs}(k) \geq 0, \quad v_{ap}^{rs}(k) \geq 0, \quad x_{ap}^{rs}(k+1) \geq 0, \quad \forall a,p,r,s; k = 1,\cdots,K; \qquad (7.68)$$

$$E_p^{rs}(k+1) \geq 0, \qquad \forall p,r,s; k = 1,\cdots,K; \qquad (7.69)$$

$$E_p^{rs}(1) = 0 \qquad \forall p,r,s; \qquad (7.70)$$

$$x_{ap}^{rs}(1) = 0, \qquad \forall a, p, r, s. \tag{7.71}$$

In summary, in each relaxation step, we need to solve a discrete NLP consisting of objective function (7.61) and constraints (7.63)-(7.71). The solution procedure in Chapter 6 solves a VI model for the ideal DUO route choice problem. Similar solution procedures can be developed for solving the link-based VI model for the instantaneous DUO route choice problem.

7.4.3 The Frank-Wolfe Method

Denote the subproblem variables as $p, q, y, \bar{E}$, corresponding to the main problem variables u, v, x, E. Applying the Frank-Wolfe algorithm to the minimization of the discretized DUO route choice program requires, at each iteration, a solution of the following linear program (LP) :

$$\begin{aligned} \min_{p,q,y,\bar{E}} \hat{Z} &= \nabla_u Z(u,v,x,E)\, p^T + \nabla_v Z(u,v,x,E)\, q^T \\ &+ \nabla_x Z(u,v,x,E)\, y^T + \nabla_E Z(u,v,x,E)\, \bar{E}^T \end{aligned} \tag{7.72}$$

s.t.

$$y_{ap}^{rs}(k+1) = y_{ap}^{rs}(k) + p_{ap}^{rs}(k) - q_{ap}^{rs}(k) \qquad \forall a, p, r, s; k = 1, \cdots, K; \tag{7.73}$$

$$\bar{E}^{rs}(k+1) = \bar{E}^{rs}(k) + \sum_{a \in B(s)} \sum_{p} q_{ap}^{rs}(k) \qquad \forall r; s \neq r; k = 1, \cdots, K; \tag{7.74}$$

$$\sum_{a \in A(r)} \sum_{p} p_{ap}^{rs}(k) = f^{rs}(k) \qquad \forall r \neq s; k = 1, \cdots, K; \tag{7.75}$$

$$\sum_{a \in B(j)} q_{ap}^{rs}(k) - \sum_{a \in A(j)} p_{ap}^{rs}(k) = 0 \qquad j, p, r, s; \forall j \neq r, s; k = 1, \cdots, K; \tag{7.76}$$

$$y_{ap}^{rs}(k) = \sum_{b \in \tilde{p}} \{y_{bp}^{rs}[k + \bar{\tau}_a(k)] - y_{bp}^{rs}(k)\} + \{\bar{E}_p^{rs}[k + \bar{\tau}_a(k)] - \bar{E}_p^{rs}(k)\}$$

$$\forall a \in B(j); j \neq r; p, r, s; k = 1, \cdots, K+1; \tag{7.77}$$

$$p_{ap}^{rs}(k) \geq 0, \quad q_{ap}^{rs}(k) \geq 0, \quad y_{ap}^{rs}(k+1) \geq 0, \quad \forall a, p, r, s; k = 1, \cdots, K; \tag{7.78}$$

$$\bar{E}_p^{rs}(k+1) \geq 0, \qquad \forall p, r, s; k = 1, \cdots, K; \tag{7.79}$$

$$\bar{E}_p^{rs}(1) = 0, \qquad \forall p, r, s; \tag{7.80}$$

$$y_{ap}^{rs}(1) = 0, \qquad \forall a, p, r, s. \tag{7.81}$$

Objective function (7.72) is equivalent to:

$$\begin{aligned} \hat{Z} &= \sum_{k=1}^{K} \sum_{r,s} \sum_{ap} \left[\frac{\partial Z}{\partial u_{ap}^{rs}(k)} p_{ap}^{rs}(k) + \frac{\partial Z}{\partial v_{ap}^{rs}(k)} q_{ap}^{rs}(k) + \frac{\partial Z}{\partial x_{ap}^{rs}(k+1)} y_{ap}^{rs}(k+1) \right] \\ &+ \sum_{k=1}^{K} \sum_{r,s,p} \frac{\partial Z}{\partial E_p^{rs}(k+1)} \bar{E}_p^{rs}(k+1) \end{aligned} \tag{7.82}$$

The components of the gradient of $Z(u, v, x, E)$ with respect to control and state variables u, v, x, E are

$$t_{1a}^{r}(k) = \frac{\partial Z(u,v,x,E)}{\partial u_a^r(k)} = c_a[x_a(k), u_a(k), v_a(k)] + \bar{\sigma}^{ri}(k) - \bar{\sigma}^{rj}(k) \quad \forall a, r; k = 1, \cdots, K; \tag{7.83}$$

$$t_{2a}(k) = \frac{\partial Z(u,v,x,E)}{\partial v_a(k)} = \int_0^{u_a(k)} \frac{\partial c_a[x_a(k), \omega, v_a(k)]}{\partial v_a(k)} d\omega \quad \forall a; k = 1, \cdots, K; \tag{7.84}$$

$$t_{3a}(k) = \frac{\partial Z(u,v,x,E)}{\partial x_a(k)} = \int_0^{u_a(k)} \frac{\partial c_a[x_a(k), \omega, v_a(k)]}{\partial x_a(k)} d\omega \quad \forall a; k = 2, \cdots, K; \tag{7.85}$$

$$t_{3a}(K+1) = \frac{\partial Z(u,v,x,E)}{\partial x_a(K+1)} = 0 \qquad \forall a; \tag{7.86}$$

$$t_4^{rs}(k) = \frac{\partial \hat{Z}(u,v,x,E)}{\partial E^{rs}(k)} = 0 \qquad \forall r, s; k = 2, \cdots, K+1. \tag{7.87}$$

The objective function can be rewritten as

$$\hat{Z} = \sum_{k=1}^{K} \sum_{rsap} \left[t_{1a}^{r}(k) p_{ap}^{rs}(k) + t_{2a}(k) q_{ap}^{rs}(k) + t_{3a}(k+1) y_{ap}^{rs}(k+1) \right] \tag{7.88}$$

Since c_a is a nonnegative and increasing function, it follows that

$$t_{1a}^{r}(k) \geq 0, \quad t_{2a}(k) \geq 0, \quad t_{3a}(k+1) \geq 0, \qquad \forall a, r; k = 1, \cdots, K. \tag{7.89}$$

This property has a significant impact on the solvability of this model for large networks, since these components will be link cost coefficients in a minimal-cost network subproblem.

The above LP subproblem for each O-D flow $f^{rs}(i)$ for each time interval $i = 1, \cdots, K$ between each O-D pair rs can be viewed as a one-to-one minimal-cost route problem over an expanded time-space network using an artificial origin. (The detailed discussion is presented in Section 6.2.) It can be solved by determining the minimal-cost routes from the artificial origin to a super destination and completing an all-or-nothing assignment. Note that the minimal-cost routes are determined by using instantaneous travel times instead of actual travel times as in Chapter 6. By revising the costs for some artificial links, the minimal cost route is searched while the flow propagation constraints are automatically satisfied by construction of the links of the expanded time-space networks (see Section 6.2). The flow variables $p_{ap}^{rs}(k)$, $q_{ap}^{rs}(k)$, $y_{ap}^{rs}(k+1)$,

$\bar{E}_p^{rs}(k+1)$ are determined by solving the minimal-cost route problems for all artificial origin-destination pairs between each original O-D pair rs for each time interval and assigning the O-D flows to the links.

In this combined algorithm, we define the relaxation procedure as the outer iteration and the F-W procedure as the inner iteration. Denote the new solution at inner F-W iteration $n+1$ as

$$u_a^{n+1}(k) = u_a^n(k) + \alpha^n[u_a^n(k) - p_a^n(k)] \qquad \forall a; k = 1, \cdots, K; \tag{7.90}$$

$$v_a^{n+1}(k) = v_a^n(k) + \alpha^n[v_a^n(k) - q_a^n(k)] \qquad \forall a; k = 1, \cdots, K; \tag{7.91}$$

$$x_a^{n+1}(k) = x_a^n(k) + \alpha^n[x_a^n(k) - y_a^n(k)] \quad \forall a; k = 1, \cdots, K+1; \tag{7.92}$$

where α^n is the optimal step size of the one-dimensional search problem in the F-W algorithm. The one-dimensional search problem is to find step size α^n that

$$\min_{0 \le \alpha^n \le 1} \quad \sum_{k=1}^{K} \sum_a \left\{ \int_0^{u_a^{n+1}(k)} c_a[x_a^{n+1}(k), \omega, v_a^{n+1}(k)] d\omega \right.$$
$$\left. + \quad \sum_r [u_a^r(k)]^{n+1} [\bar{\sigma}^{ri}(k) - \bar{\sigma}^{rj}(k)] \right\} \tag{7.93}$$

where $u_a^{n+1}(k)$, $v_a^{n+1}(k)$, and $x_a^{n+1}(k)$ must be substituted using the above definitional equations.

The algorithm for solving our instantaneous DUO route choice model is illustrated in the flowchart in Figure 7.1 and is summarized as follows:

Step 0: Initialization.
Find an initial feasible solution $\{x_a^1(k)\}$, $\{u_a^1(k)\}$, $\{v_a^1(k)\}$, $\{E^1(k)\}$. Set the outer iteration counter $m = 1$.

Step 1: Relaxation.
Find a new estimate of the actual link travel time $\bar{\tau}_a^n(k)$ and solve the instantaneous DUO program. Set the inner iteration counter $n = 1$.

[Step 1.1]: *Update.* Calculate $t_{1a}^r(k)$, $t_{2a}(k)$ and $t_{3a}(k)$ using equations (7.83)-(7.86).

[Step 1.2]: *Direction Finding.* Based on $\{t_{1a}^r(k)\}$, $\{t_{2a}(k)\}$ and $\{t_{3a}(k)\}$ and satisfying the flow propagation constraints (7.77), search the minimal-cost route forward from each artificial origin to the super destination over an expanded time-space network for each O-D pair (r, s). Perform an all-or-nothing assignment, yielding subproblem solution $\{p_a(k)\}, \{q_a(k)\}, \{y_a(k)\}, \{\bar{E}^{rs}(k)\}$.

[Step 1.3]: *Line Search.* Find the optimal step size α^n that solves the one-dimensional search problem.

[Step 1.4]: *Move.* Find a new solution by combining $\{u_a(k)\}$, $\{v_a(k)\}$, $\{x_a(k)\}$, $\{E^{rs}(k)\}$ and $\{p_a(k)\}$, $\{q_a(k)\}$, $\{y_a(k)\}$, $\{\bar{E}^{rs}(k)\}$.

[Step 1.5]: *Convergence Test for Inner Iterations.* If n equals a pre-specified number, go to step 2; otherwise, set $n = n + 1$ and go to step 1.1.

Step 2: Convergence Test for Outer Iterations.
If $\bar{\tau}_a^m(k) \simeq \bar{\tau}_a^{m+1}(k)$, stop. The current solution, $\{u_a(k)\}$, $\{v_a(k)\}$, $\{x_a(k)\}$, $\{E^{rs}(k)\}$, is in a near instantaneous DUO state; otherwise, set $m = m+1$ and go to step 1.

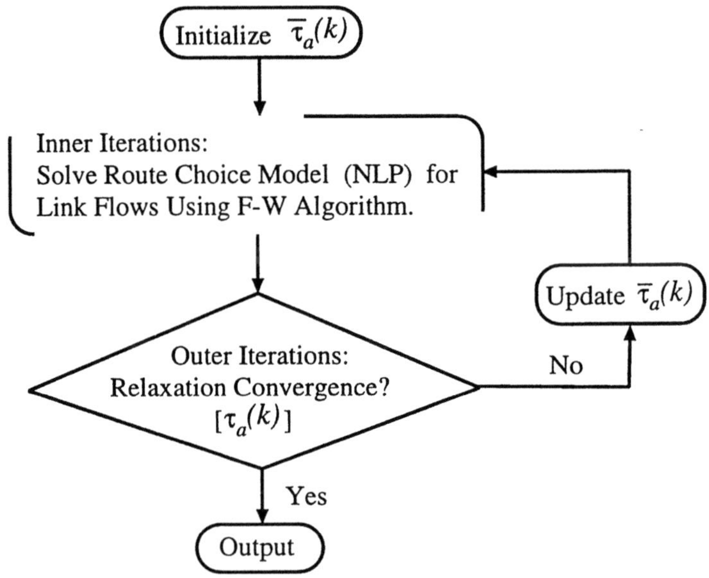

Figure 7.1: Flowchart of the Solution Algorithm

7.4.4 Numerical Example

We illustrate the solution of the link-time-based variational inequality model with the 4-link, 4-node test network shown in Figure 7.2. A symmetrical network is intentionally used to demonstrate that the route travel times in the solution are equal. To convert continuous time VI model (7.45) into an NLP, assignment time period $[0, T]$ is subdivided into $K = 5$ small time intervals, and the OCP is reformulated as a discrete time NLP. We then use an algorithm based on the Frank-Wolfe and relaxation techniques to solve this NLP. This algorithm was coded in FORTRAN and solved on a HP-715 workstation. Some computational results for a small network are given to illustrate the instantaneous DUO traffic flows.

The following link travel time functions were used in the computations:

$$c_a(k) = g_{1a}(k) + g_{2a}(k)$$

$$g_{1a}(k) = \beta_{1a} + \beta_{2a}[u_a(k)]^2 + \beta_{3a}[x_a(k)]^2$$

$$g_{2a}(k) = \beta_{4a} + \beta_{5a}[v_a(k)]^2 + \beta_{6a}[x_a(k)]^2$$

where time interval $k = 1, 2, \cdots, 5$. Parameter values for each link travel time function are given in Table 7.1, and the trip table is given in Table 7.2. The optimal link flows and corresponding optimal link travel times are given in Table 7.3. The optimal route travel times are given in Table 7.4. In this discrete time example, $x_a(k)$ represents vehicles on the link at the beginning of interval k; $u_a(k)$ and $v_a(k)$ represent inflow and exit flow during interval k.

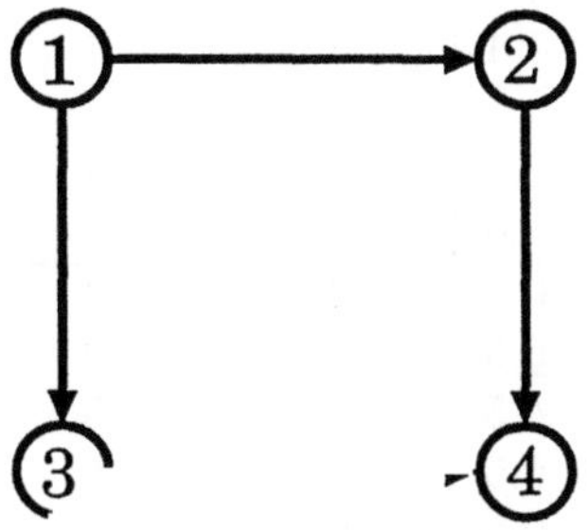

Figure 7.2: Test Network

Table 7.1: Parameters of Link Travel Time Functions

link a	β_{1a}	β_{2a}	β_{3a}	β_{4a}	β_{5a}	β_{6a}
1—2	1.	0.001	0.	0.	0.015	0.003
1—3	1.	0.001	0.	0.	0.015	0.003
2—4	1.	0.001	0.	0.	0.015	0.003
3—4	1.	0.001	0.	0.	0.015	0.003

Table 7.2: Required Flows from Origin 1 to Destination 4

Time Interval k	1	2	3	4	5
Flow/interval	20.	0.	0.	0.	0.

Table 7.3: Optimal Numbers of Vehicles, Inflows, Exit Flows and Travel Times

Interval k	Link a	Vehicles $x_a(k+1)$	Inflow $u_a(k)$	Exit Flow $v_a(k)$	Vehicles $x_a(k)$	Travel Time $c_a(k)$
1	1 — 2	10.0	10.0	0.0	0.0	1.1
2	1 — 2	4.5	0.0	5.5	10.0	1.8
3	1 — 2	0.0	0.0	4.5	4.5	1.4
4	1 — 2	0.0	0.0	0.0	0.0	1.0
5	1 — 2	0.0	0.0	0.0	0.0	1.0
1	1 — 3	10.0	10.0	0.0	0.0	1.1
2	1 — 3	4.4	0.0	5.6	10.0	1.8
3	1 — 3	0.0	0.0	4.4	4.4	1.3
4	1 — 3	0.0	0.0	0.0	0.0	1.0
5	1 — 3	0.0	0.0	0.0	0.0	1.0
1	2 — 4	0.0	0.0	0.0	0.0	1.0
2	2 — 4	5.5	5.5	0.0	0.0	1.0
3	2 — 4	4.5	4.5	5.5	5.5	1.6
4	2 — 4	0.0	0.0	4.5	4.5	1.4
5	2 — 4	0.0	0.0	0.0	0.0	1.0
1	3 — 4	0.0	0.0	0.0	0.0	1.0
2	3 — 4	5.6	5.6	0.0	0.0	1.0
3	3 — 4	4.4	4.4	5.6	5.6	1.6
4	3 — 4	0.0	0.0	4.4	4.4	1.3
5	3 — 4	0.0	0.0	0.0	0.0	1.0

Table 7.4: Dynamic User-Optimal Route Travel Times

Interval	Route Travel Times	
k	Route 1-2-4	Route 1-3-4
1	2.1	2.1
2	2.8	2.8
3	3.0	2.9
4	2.4	2.3
5	2.0	2.0

In this example, note from Table 7.4 that instantaneous travel times on routes 1–2–4 and 1–3–4 are equal during each interval (except for small differences resulting from incomplete convergence). Thus, the results indicate the existence of instantaneous dynamic user optimality in the intuitive sense that the two routes have equal travel times in each interval.

Note from Table 7.3 that the *average* travel time that vehicles incur on link 1–2 from the beginning of interval 2 is 1.8 intervals (some vehicles have smaller travel times and some have larger); 10 vehicles are on the link at the beginning of interval 2. Some vehicles exit during interval 2, which reduces the travel time in subsequent intervals. Thus, 5.5 vehicles exit during interval 2 (decreasing the travel time to 1.4 intervals for interval 3) and 4.5 vehicles exit during interval 3. Note also that 5.5 vehicles are on link 2–4 at the beginning of interval 3, and the average travel time is 1.6 from the beginning of interval 3. Also, 4.5 vehicles enter this link during interval 3, 5.5 vehicles exit during interval 3, and 4.5 vehicles exit during interval 4, reducing the travel time to only 1.0 for interval 5. At the beginning of interval 6 (column $x_a(k+1)$ in rows $k=5$), no vehicle remains on any link.

7.5 Notes

As discussed in Sections 7.1-7.4, route-based and link-based definitions of instantaneous DUO are not necessarily equivalent. Route-based definitions do not provide any routing strategies at intermediate decision nodes and have less applicability compared to the link-based definitions in ATIS systems. Furthermore, link-based DUO definitions imply route-based DUO definitions and are more realistic in terms of users' route choice behavior. Therefore, in this book and in studies of dynamic incident management, we focus on models formulated according to the link-based DUO definitions.

Problems

7.1 Compare the differences of instantaneous link travel time and actual link travel time. Also explain the differences of instantaneous route travel time and actual route travel time.

7.2 Discuss the assumptions of instantaneous DUO route choice. Give three possible scenarios where instantaneous DUO could play an important role.

7.3 For the network shown in Figure 7.3, the travel time functions are

$$c_1(t) = 2u_1(t) + 5, \qquad c_2(t) = 5$$

$$c_3(t) = 7, \qquad c_4(t) = u_4(t) + 7$$

The O-D flow is $f^{14}(t) = 10t$ units. Determine inflows on links 1 and 2 for time period $[0, 10]$, using an instantaneous DUO route choice model. Also determine the optimal route travel times.

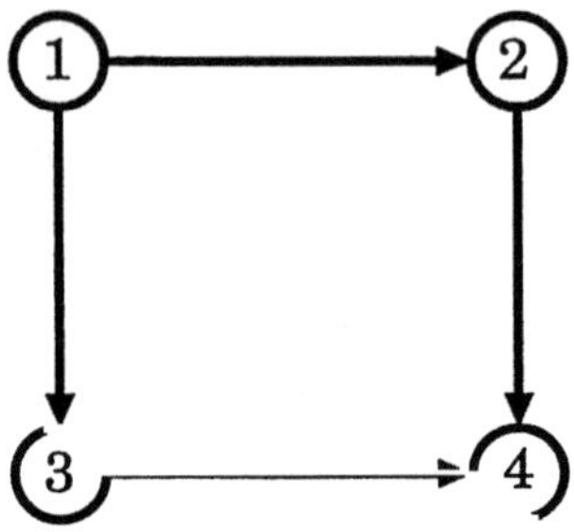

Figure 7.3: Network for Problem 7.3

7.4 For the network shown in Figure 7.3, there are two O-D pairs (1,4) and (2,4). Assume that all links have the same link travel time function.

(a) If O-D flows are $f^{14}(t) = 5t$ units and $f^{24}(t) = 0$ units, are the route departure flows identical for both instantanteous DUO route choice and ideal DUO route choice? Explain why.

(b) If O-D flows are $f^{14}(t) = 5t$ units and $f^{24}(t) = 5$ units, are the route departure flows identical for both instantanteous DUO route choice and ideal DUO route choice? Explain why.

7.5 Discuss when the route-time-based model is more efficient than the link-time-based model. Also discuss when the link-time-based model is more efficient than the route-time-based model.

7.6 Discuss when the link-time-based model and the route-time-based model will produce identical results.

7.7 In the relaxation procedure for solving the VI problem, is it possible to formulate a different optimization subproblem objective function other than equation (7.61)? If it is possible, what variables should be temporarily fixed in the relaxation?

7.8 Provide physical interpretation for the cost terms $t^{r}_{1a}(k)$, $t_{2a}(k)$, $t_{3a}(k)$ in the LP subproblem objective function (7.88).

7.9 Without using the expanded time-space network, redesign the solution algorithm for the instantaneous DUO route choice problem.

Chapter 8

Extensions of Instantaneous Route Choice Models

In this chapter, we first present optimal control formulations for instantaneous dynamic user-optimal route choice problems. In Sections 8.1 and 8.2, we present two instantaneous dynamic user-optimal route choice models. The equivalence of Model 1 with DUO route choice is demonstrated by proving the equivalence of the first order necessary conditions of the model with the instantaneous DUO route choice conditions. The second DUO model employs a different link travel time function assumption; its equivalence with the instantaneous DUO route choice conditions is also demonstrated. In Section 8.3 and 8.4, we generalize the variational inequality model of Chapter 7 to the situation of travel disutility and multi-class travelers instead of travel times for a single class of travelers only. The multi-class route-cost-based variational inequality model is first presented in Section 8.3. Then, the multi-class link-cost-based variational inequality model is presented in Section 8.4.

8.1 Optimal Control Model 1

Given a general transportation network, we define a decision node for each route p of each O-D pair as any node on the route including the origin. The instantaneous route travel time between a decision node and the destination node is calculated using the currently prevailing link travel times. In this section, using optimal control theory, we formulate an alternative model, which is equivalent to the instantaneous DUO route choice conditions. Our objective here is to explore the complexity of the problem. In this section we use more restricted definitions of used links and routes than Chapter 7, as follows. For any link a on any route from origin r to destination s, link a is defined as being used at time t if $u_a^{rs}(t) > 0$ and $v_a^{rs}(t) > 0$. Furthermore, a route p between r and s is defined as being used at time t if $u_{ap}^{rs}(t) > 0$ and $v_{ap}^{rs}(t) > 0$ for all links a on route p from r to s.

We assume that the time-dependent origin-destination trip pattern is known *a priori*. In other words, the departure times of travelers are given. The instantaneous dynamic user-optimal (DUO) route choice problem is to determine vehicle flows at each instant of time on each link resulting from drivers using minimal-time routes under the currently prevailing travel times. We restate the link-based dynamic generalization of the conventional static user-optimal state from Chapter 7:

> **Link-Time-Based Instantaneous DUO State:** *If, for each O-D pair at each decision node at each instant of time, the instantaneous travel times for all routes that are being used equal the minimal instantaneous route travel time, the dynamic traffic flow over the network is in a link-time-based instantaneous dynamic user-optimal state.*

Although the instantaneous user-optimal travel times for all routes being used are equal at each decision node at each instant of time, route flows with the same departure time and the same origin-destination may actually experience somewhat different route travel times. This is because the route time may subsequently change due to changing network traffic conditions, even though at each decision node the flows select the route that is currently the best.

In optimal control theory terminology, the dynamic user-optimal route choice problem is to find the dynamic trajectories of link states and inflow and exit flow control variables, given the time-dependent O-D flow requirements, the network and the link travel time functions. The formulation of the problem in this section is based on the underlying choice criterion that each traveler uses the route that minimizes his/her instantaneous travel time when departing from the origin or any intermediate node to his/her destination. This route choice rule implies that in a travel-time-based instantaneous dynamic user-optimal state, the trajectories of the link flow states and inflows and exit flows are such that the instantaneous travel times at each decision node of all used routes connecting any given O-D pair will be identical and not greater than the instantaneous travel times of routes which are not being used.

8.1.1 Model Formulation

Recall from Chapter 4 that the instantaneous travel time $c_a[x_a(t), u_a(t), v_a(t)]$, or simply $c_a(t)$, over link a is assumed to be dependent on the number of vehicles $x_a(t)$, the inflow $u_a(t)$ and the exit flow $v_a(t)$ on link a at time t. This instantaneous link time is the travel time that would be incurred if traffic conditions on the link remain unchanged while traversing the link. In Model 1, we assume the instantaneous travel time $c_a(t)$ on link a is the sum of two components: 1) an instantaneous flow-dependent running time $g_{1a}[x_a(t), u_a(t)]$ over link a; 2) an instantaneous queuing delay $g_{2a}[x_a(t), v_a(t)]$. It follows that

$$c_a(t) = g_{1a}[x_a(t), u_a(t)] + g_{2a}[x_a(t), v_a(t)]. \tag{8.1}$$

The two components $g_{1a}[x_a(t), u_a(t)]$ and $g_{2a}[x_a(t), v_a(t)]$ of the time-dependent link travel time function $c_a[x_a(t), u_a(t), v_a(t)]$ are assumed to be nonnegative and differentiable with respect to $x_a(t), u_a(t)$ and $x_a(t), v_a(t)$, respectively. The instantaneous travel time function $\psi_p^{rs}(t)$ for each route p between O-D pair (r, s) is

$$\psi_p^{rs}(t) = \sum_{a \in rsp} c_a[x_a(t), u_a(t), v_a(t)] \qquad \forall r, s, p; \tag{8.2}$$

the summation is over all links a in route p from origin r to destination s.

Using optimal control theory, the equivalent optimization model of the instantaneous dynamic user-optimal route choice problem (**Model 1**) is formulated as follows.

$$\min_{u,v,x,e,E} \int_0^T \sum_a \left\{ \int_0^{u_a(t)} g_{1a}[x_a(t), \omega] d\omega + \int_0^{v_a(t)} g_{2a}[x_a(t), \omega] d\omega \right\} dt \tag{8.3}$$

s.t.

Relationship between state and control variables:

$$\frac{dx_{ap}^{rs}(t)}{dt} = u_{ap}^{rs}(t) - v_{ap}^{rs}(t) \qquad \forall a, p, r, s; \tag{8.4}$$

$$\frac{dE_p^{rs}(t)}{dt} = e_p^{rs}(t) \qquad \forall p, r; s \neq r; \tag{8.5}$$

Flow conservation constraints:

$$f^{rs}(t) = \sum_{a \in A(r)} \sum_p u_{ap}^{rs}(t) \qquad \forall r, s; \tag{8.6}$$

$$\sum_{a \in B(j)} v_{ap}^{rs}(t) = \sum_{a \in A(j)} u_{ap}^{rs}(t) \qquad \forall j, p, r, s; j \neq r, s; \tag{8.7}$$

$$\sum_{a \in B(s)} \sum_p v_{ap}^{rs}(t) = e^{rs}(t) \qquad \forall r, s; s \neq r; \tag{8.8}$$

Flow propagation constraints:

$$x_{ap}^{rs}(t) = \sum_{b \in \tilde{p}} \{x_{bp}^{rs}[t + \bar{\tau}_a(t)] - x_{bp}^{rs}(t)\} + \{E_p^{rs}[t + \bar{\tau}_a(t)] - E_p^{rs}(t)\}$$

$$\forall a \in B(j); j \neq r; p, r, s; \tag{8.9}$$

Definitional constraints:

$$\sum_{rsp} u_{ap}^{rs}(t) = u_a(t), \qquad \sum_{rsp} v_{ap}^{rs}(t) = v_a(t), \qquad \forall a; \tag{8.10}$$

$$\sum_{rsp} x_{ap}^{rs}(t) = x_a(t), \qquad \forall a; \qquad (8.11$$

Nonnegativity conditions:

$$x_{ap}^{rs}(t) \geq 0, \quad u_{ap}^{rs}(t) \geq 0, \quad v_{ap}^{rs}(t) \geq 0 \qquad \forall a, p, r, s; \qquad (8.12$$

$$e_p^{rs}(t) \geq 0, \quad E_p^{rs}(t) \geq 0, \qquad \forall p, r, s; \qquad (8.13$$

Boundary conditions:

$$E_p^{rs}(0) = 0, \qquad \forall p, r, s; \qquad (8.14$$

$$x_{ap}^{rs}(0) = 0, \qquad \forall a, p, r, s. \qquad (8.15$$

The objective function terms are similar to the objective function of the wel known static user-optimal (UO) model. The first two constraints (8.4)-(8.5) ar state equations for the flow on each link a and for cumulative arrivals at eac destination. Equations (8.6)-(8.8) are flow conservation constraints at eac node including origins and destinations. The other constraints include flo propagation constraints, definitional constraints, nonnegativity and boundar conditions. In addition, we need several definitional constraints as follows:

$$\sum_{rsp} u_{ap}^{rs}(t) = u_a(t), \qquad \sum_{rsp} v_{ap}^{rs}(t) = v_a(t),$$

$$\sum_{p} x_{ap}^{rs}(t) = x_a^{rs}(t), \qquad \sum_{rs} x_a^{rs}(t) = x_a(t),$$

$$\sum_{p} e_p^{rs}(t) = e^{rs}(t).$$

In summary, the control variables are $u_{ap}^{rs}(t)$, $v_{ap}^{rs}(t)$, and $e_p^{rs}(t)$; the state var ables are $x_{ap}^{rs}(t)$ and $E_p^{rs}(t)$. Model 1 can be solved by discretizing time so th: it becomes an ordinary nonlinear program and by using a fixed estimate of ea(link travel time $\bar{\tau}_a(t)$, which is updated in an iterative relaxation fashion.

8.1.2 Optimality Conditions

The extended Hamiltonian function for the instantaneous DUO route choi program (8.3)-(8.15) is constructed as

$$\begin{aligned}
\mathcal{H} &= \sum_a \left\{ \int_0^{u_a(t)} g_{1a}[x_a(t), \omega] d\omega + \int_0^{v_a(t)} g_{2a}[x_a(t), \omega] d\omega \right\} \\
&+ \sum_{rs} \sum_{ap} \lambda_{ap}^{rs}(t)[u_{ap}^{rs}(t) - v_{ap}^{rs}(t)] + \sum_r \sum_{s \neq r} \sum_p \nu_p^{rs}(t) e_p^{rs}(t) \\
&+ \sum_s \sum_{r \neq s} \sigma_r^{rs}(t)[f^{rs}(t) - \sum_{a \in A(r)} \sum_p u_{ap}^{rs}(t)]
\end{aligned}$$

$$+ \sum_{rs}\sum_{j\neq rs}\sum_{p}\sigma_{jp}^{rs}(t)[\sum_{a\in B(j)} v_{ap}^{rs}(t) - \sum_{a\in A(j)} u_{ap}^{rs}(t)]$$
$$+ \sum_{r}\sum_{s\neq r}\sigma_{s}^{rs}(t)[\sum_{a\in B(s)}\sum_{p} v_{ap}^{rs}(t) - e^{rs}(t)]$$
$$+ \sum_{rsp}\sum_{j\neq r}\sum_{a\in B(j)}\mu_{ap}^{rs}(t)\left\{ x_{ap}^{rs}(t) + \sum_{b\in\bar{p}} x_{bp}^{rs}(t) + E_{p}^{rs}(t)\right.$$
$$\left. - \sum_{b\in\bar{p}} x_{bp}^{rs}[t+\bar{\tau}_a(t)] - E_p^{rs}[t+\bar{\tau}_a(t)]\right\}$$

where $\lambda_{ap}^{rs}(t)$ are Lagrange multipliers associated with the link state equations, $\nu_p^{rs}(t)$ are Lagrange multipliers associated with the destination node state equations, $\sigma_{jp}^{rs}(t)$ are Lagrange multipliers associated with the node flow conservation equations, and $\mu_{ap}^{rs}(t)$ are the Lagrange multipliers associated with the flow propagation equations. For each link a which points from node l to node m, the first order necessary conditions of instantaneous DUO route choice program (8.3)-(8.15) include

$$\frac{\partial\mathcal{H}}{\partial u_{ap}^{rs}(t)} = g_{1a}[x_a(t), u_a(t)] + \lambda_{ap}^{rs}(t) - \sigma_{lp}^{rs}(t) \geq 0, \qquad \forall l; a\in A(l), p, r, s, \quad (8.16)$$

$$\text{and} \qquad u_{ap}^{rs}(t)\frac{\partial\mathcal{H}}{\partial u_{ap}^{rs}(t)} = 0 \qquad \forall a, p, r, s; \quad (8.17)$$

$$\frac{\partial\mathcal{H}}{\partial v_{ap}^{rs}(t)} = g_{2a}[x_a(t), v_a(t)] - \lambda_{ap}^{rs}(t) + \sigma_{mp}^{rs}(t) \geq 0,$$
$$\forall m; a\in B(m), p, r, s, \quad (8.18)$$

$$\text{and} \qquad v_{ap}^{rs}(t)\frac{\partial\mathcal{H}}{\partial v_{ap}^{rs}(t)} = 0 \qquad \forall a, p, r, s; \quad (8.19)$$

$$\frac{\partial\mathcal{H}}{\partial e_p^{rs}(t)} \geq 0, \qquad \forall p, r, s, \quad (8.20)$$

$$\text{and} \qquad e_p^{rs}(t)\frac{\partial\mathcal{H}}{\partial e_p^{rs}(t)} = 0 \qquad \forall p, r, s; \quad (8.21)$$

$$\frac{d\lambda_{ap}^{rs}(t)}{dt} = -\frac{\partial\mathcal{H}}{\partial x_{ap}^{rs}(t)} \qquad \forall a, p, r, s; \quad (8.22)$$

$$\frac{d\nu_p^{rs}(t)}{dt} = -\frac{\partial\mathcal{H}}{\partial E_p^{rs}(t)} \qquad \forall p, r, s; \quad (8.23)$$

$$u_{ap}^{rs}(t)\geq 0, \qquad v_{ap}^{rs}(t)\geq 0, \qquad x_{ap}^{rs}(t)\geq 0, \qquad \forall a, p, r, s; \quad (8.24)$$

$$e_p^{rs}(t)\geq 0, \qquad E_p^{rs}(t)\geq 0, \qquad \forall p, r, s. \quad (8.25)$$

Note that $\sigma_{rp}^{rs}(t) = \sigma_r^{rs}(t)$ when node l equals origin r.

8.1.3 DUO Equivalence Analysis

Since the objective function of optimal control program (8.3)-(8.15) is convex with respect to the control variables, there is a unique optimal solution. We now show that the set of link states, inflows and exit flows that solves this program also satisfies the travel-time-based instantaneous dynamic user-optimal route choice conditions. This equivalence is demonstrated below by proving that the first order necessary conditions for the optimal control program (8.3)-(8.15) are identical to the instantaneous dynamic user-optimal conditions. The equivalence between the instantaneous DUO route choice conditions and the first order necessary conditions of the optimal control program means that the instantaneous DUO route choice conditions are satisfied at the optimal solution of this program.

Combining equations (8.16)-(8.17) with equations (8.18)-(8.19), the following equations can be derived for each link a which points from node l to node m.

$$\frac{\partial \mathcal{H}}{\partial u_{ap}^{rs}(t)} + \frac{\partial \mathcal{H}}{\partial v_{ap}^{rs}(t)} = c_a(t) - \sigma_{lp}^{rs}(t) + \sigma_{mp}^{rs}(t) \geq 0,$$

$$\forall a \in A(l) \cap B(m); p, r, s; \tag{8.26}$$

and

$$u_{ap}^{rs}(t) \frac{\partial \mathcal{H}}{\partial u_{ap}^{rs}(t)} = 0 \qquad \forall a, p, r, s; \tag{8.27}$$

and

$$v_{ap}^{rs}(t) \frac{\partial \mathcal{H}}{\partial v_{ap}^{rs}(t)} = 0 \qquad \forall a, p, r, s. \tag{8.28}$$

For route p between origin node r and destination node s, let i denote node r or any intermediate node on this route. Denote route $\tilde{p}$ as $(i, 1, 2, \cdots, n, s)$. The instantaneous travel time $\psi_{\tilde{p}}^{ri}(t)$ for the remaining route $\tilde{p}$ between i and s is

$$\psi_{\tilde{p}}^{ri}(t) = \sum_{a \in is\tilde{p}} c_a[x_a(t), u_a(t), v_a(t)] \qquad \forall i \in p, r, s. \tag{8.29}$$

Suppose we consider a set of routes p from $r \to i \to s$ and the corresponding set of subroutes $\tilde{p}$. The flow conservation constraint at node i can be revised as

$$\sum_{a \in B(i)} v_{ap}^{rs}(t) = \sum_{a \in A(i)} u_{ap}^{rs}(t) \qquad \forall i, p, r, s; i \neq r, s. \tag{8.30}$$

The fourth term in the Hamiltonian function should be revised as

$$\sum_{rs} \sum_{i \neq rs} \sigma_i^{rs}(t) \sum_p \left[\sum_{a \in B(j)} v_{ap}^{rs}(t) - \sum_{a \in A(j)} u_{ap}^{rs}(t) \right]$$

so that $\sigma_{ip}^{rs}(t) = \sigma_i^{rs}(t)$ for the set of subroutes $\tilde{p}$. Note that all derivations from equation (8.16) to equation (8.28) will follow for this set of subroutes $\tilde{p}$.

Now if route $\tilde{p}$ is being used at time t, $u_a^{rs}(t)$ and $v_a^{rs}(t)$ are both positive by definition. Thus, by (8.26)-(8.28),

$$\begin{aligned}\psi_{\tilde{p}}^{ri}(t) &= [\sigma_i^{rs}(t) - \sigma_1^{rs}(t)] + [\sigma_{1p}^{rs}(t) - \sigma_{2p}^{rs}(t)] + \cdots \\ &+ [\sigma_{n-1,p}^{rs}(t) - \sigma_{np}^{rs}(t)] + [\sigma_{np}^{rs}(t) - \sigma_s^{rs}(t)] \\ &= \sigma_i^{rs}(t) - \sigma_s^{rs}(t)\end{aligned}$$

for every route $\tilde{p}$ being used at time t. Note that $\sigma_{ip}^{rs}(t) = \sigma_i^{rs}(t)$ in the above equation. Thus, routes being used from i to s at time t have travel times equal to $[\sigma_i^{rs}(t) - \sigma_s^{rs}(t)]$. More generally, we obtain the following for each remaining route $\tilde{p}$ between i and s.

$$u_{ap}^{rs}(t)\; v_{ap}^{rs}(t)\; [\psi_{\tilde{p}}^{ri}(t) - \sigma_i^{rs}(t) + \sigma_s^{rs}(t)] = 0 \quad \forall a, i, p, r, s; a \in \tilde{p}, i \in p; \tag{8.31}$$

$$\psi_{\tilde{p}}^{ri}(t) \geq \sigma_i^{rs}(t) - \sigma_s^{rs}(t) \qquad \forall i \in p, r, s; \tag{8.32}$$

$$u_{ap}^{rs}(t) \geq 0, \qquad v_{ap}^{rs}(t) \geq 0, \qquad \forall a, p, r, s; a \in \tilde{p}. \tag{8.33}$$

Conditions (8.31)-(8.33) hold for each remaining route $\tilde{p}$ between i and s, where i is any intermediate node (including the origin) between each O-D pair rs in the network. For route $\tilde{p}$ connecting node i and destination s, if each link is being used at time t, then $u_{ap}^{rs}(t)$ and $v_{ap}^{rs}(t)$ will be positive, so that the quantities in brackets in equation (8.31) will be zero, i.e., equation (8.32) will hold as an equality. Thus, routes which are being used at time t have travel times equal to $[\sigma_i^{rs}(t) - \sigma_s^{rs}(t)]$.

Equation (8.32) states that the difference of Lagrange multipliers $[\sigma_i^{rs}(t) - \sigma_s^{rs}(t)]$ of node-destination pair (i, s) is less than or equal to the instantaneous travel times on all routes connecting this node-destination pair (i, s). Therefore, this difference of Lagrange multipliers $[\sigma_i^{rs}(t) - \sigma_s^{rs}(t)]$ equals the instantaneous minimal route travel time between node i and destination s. For any remaining unused route $\tilde{p}$ between i and s, at least one link a in route $\tilde{p}$ is *not* being used at time t. Thus the inflow $u_{ap}^{rs}(t)$ or the exit flow $v_{ap}^{rs}(t)$ is equal to zero, so that (8.32) may hold as a strict inequality, i.e., the instantaneous travel time $\psi_{\tilde{p}}^{ri}(t)$ on remaining route $\tilde{p}$ will not be less than the instantaneous minimal route travel time $[\sigma_i^{rs}(t) - \sigma_s^{rs}(t)]$.

Furthermore, if we consider any pair of decision nodes i and j between origin r and destination s, the above results also hold for each sub-route between i and j. Thus, at the optimal solution to Model 1, flows always use the instantaneous minimal time subroutes, even if all links on an entire route are not in use at the same time.

Since intermediate node i could be origin node r, the above results also hold for routes from r to s. With the above interpretation, it is now clear that equations (8.31)-(8.33) state the travel-time-based instantaneous dynamic user-optimal conditions. This optimal control program (8.3)-(8.15) can be referred to as an instantaneous DUO route choice program.

8.2 Optimal Control Model 2

8.2.1 Model Formulation

Various DUO route choice models can be formulated with alternative link travel time assumptions. In Model 1, the instantaneous travel time $c_a(t)$ over link a is assumed to depend on the number of vehicles $x_a(t)$, the inflow $u_a(t)$ and the exit flow $v_a(t)$ on link a at time t. In this section, the instantaneous link travel time $c_a(t)$ is assumed to depend only on $x_a(t)$:

$$c_a(t) = c_a[x_a(t)]. \tag{8.34}$$

The time-dependent link travel time function $c_a[x_a(t)]$ is again assumed to be nonnegative and differentiable for all $x_a(t)$. In addition, the actual link travel time $\tau_a(t)$ is introduced, and also assumed to be dependent only on $x_a(t)$, i.e.,

$$\tau_a(t) = \tau_a[x_a(t)]. \tag{8.35}$$

Using optimal control theory, the equivalent optimization program of the instantaneous dynamic user-optimal route choice problem (**Model 2**) is formulated as follows.

$$\min_{u,v,x,e,E} \int_0^T \sum_a \left\{ \int_0^{u_a(t)} c_a[x_a(t)]\, d\omega \right\} dt \tag{8.36}$$

s.t. constraints (8.4)–(8.15).

The objective function is similar to the objective function of the well-known static user-optimal model. The constraints are identical to those of Model 1 except that the flow propagation constraints need to be revised. The flow propagation constraints can be expressed directly based on the actual link travel time function $\tau_a(t)$, instead of its estimate $\bar{\tau}_a(t)$ in a diagonalization fashion. The new flow propagation constraints are

$$x_{ap}^{rs}(t) = \sum_{b \in \bar{p}} \{x_{bp}^{rs}[t + \tau_a(t)] - x_{bp}^{rs}(t)\} + \{E_p^{rs}[t + \tau_a(t)] - E_p^{rs}(t)\}$$

$$\forall a \in B(j); j \neq r; p, r, s. \tag{8.37}$$

The instantaneous link travel time function $c_a(t)$ in Model 2 is a special case of $c_a[x_a(t), u_a(t), v_a(t)]$ in Model 1. We prove in the next section that the unique optimal solution to Model 2 is in a travel-time-based instantaneous DUO state.

8.2.2 Optimality Conditions

We now formulate the extended Hamiltonian function and derive the first-order necessary conditions. The extended Hamiltonian for Model 2 is constructed as

$$\mathcal{H} = \sum_a \int_0^{v_a(t)} c_a[x_a(t), \omega] d\omega$$

$$
\begin{aligned}
&+ \sum_{rs}\sum_{ap} \lambda_{ap}^{rs}(t)[u_{ap}^{rs}(t) - v_{ap}^{rs}(t)] + \sum_{r}\sum_{s\neq r}\sum_{p} \nu_p^{rs}(t) e_p^{rs}(t) \\
&+ \sum_{s}\sum_{r\neq s} \sigma_r^{rs}(t)[f^{rs}(t) - \sum_{a\in A(r)}\sum_{p} u_{ap}^{rs}(t)] \\
&+ \sum_{rs}\sum_{j\neq rs}\sum_{p} \sigma_{jp}^{rs}(t)[\sum_{a\in B(j)} v_{ap}^{rs}(t) - \sum_{a\in A(j)} u_{ap}^{rs}(t)] \\
&+ \sum_{r}\sum_{s\neq r} \sigma_s^{rs}(t)[\sum_{a\in B(s)}\sum_{p} v_{ap}^{rs}(t) - e^{rs}(t)] \\
&+ \sum_{rsp}\sum_{j\neq r}\sum_{a\in B(j)} \mu_{ap}^{rs}(t) \Bigg\{ x_{ap}^{rs}(t) + \sum_{b\in\bar{p}} x_{bp}^{rs}(t) + E_p^{rs}(t) \\
&- \sum_{b\in\bar{p}} x_{bp}^{rs}[t+\tau_a(t)] - E_p^{rs}[t+\tau_a(t)] \Bigg\}
\end{aligned}
$$

where $\lambda_{ap}^{rs}(t)$ are Lagrange multipliers associated with the link state equations, $\nu_p^{rs}(t)$ are Lagrange multipliers associated with the destination node state equations, $\sigma_{jp}^{rs}(t)$ are Lagrange multipliers associated with the node flow conservation equations, and $\mu_{ap}^{rs}(t)$ are Lagrange multipliers associated with the flow propagation equations. For each link a which points from node l to node m, the first order necessary conditions of Model 2 include

$$\frac{\partial \mathcal{H}}{\partial u_{ap}^{rs}(t)} = \lambda_{ap}^{rs}(t) - \sigma_{lp}^{rs}(t) \geq 0, \qquad \forall l; a \in A(l), p, r, s, \tag{8.38}$$

$$\text{and} \qquad u_{ap}^{rs}(t)\frac{\partial \mathcal{H}}{\partial u_{ap}^{rs}(t)} = 0 \qquad \forall a, p, r, s; \tag{8.39}$$

$$\frac{\partial \mathcal{H}}{\partial v_{ap}^{rs}(t)} = c_a[x(t), v(t)] - \lambda_{ap}^{rs}(t) + \sigma_{mp}^{rs}(t) \geq 0, \qquad \forall m; a \in B(m), p, r, s, \tag{8.40}$$

$$\text{and} \qquad v_{ap}^{rs}(t)\frac{\partial \mathcal{H}}{\partial v_{ap}^{rs}(t)} = 0 \qquad \forall a, p, r, s; \tag{8.41}$$

$$\frac{\partial \mathcal{H}}{\partial e_p^{rs}(t)} \geq 0, \qquad \forall p, r, s, \tag{8.42}$$

$$\text{and} \qquad e_p^{rs}(t)\frac{\partial \mathcal{H}}{\partial e_p^{rs}(t)} = 0 \qquad \forall p, r, s; \tag{8.43}$$

$$\frac{d\lambda_{ap}^{rs}(t)}{dt} = -\frac{\partial \mathcal{H}}{\partial x_{ap}^{rs}(t)}, \qquad \forall a, p, r, s; \tag{8.44}$$

$$\frac{d\nu_p^{rs}(t)}{dt} = -\frac{\partial \mathcal{H}}{\partial E_p^{rs}(t)} \qquad \forall p, r, s; \tag{8.45}$$

$$u_{ap}^{rs}(t) \geq 0, \qquad v_{ap}^{rs}(t) \geq 0, \qquad x_{ap}^{rs}(t) \geq 0, \qquad \forall a, p, r, s; \tag{8.46}$$

$$e_p^{rs}(t) \geq 0, \qquad E_p^{rs}(t) \geq 0, \qquad \forall p, r, s. \tag{8.47}$$

Since the actual link travel time $\tau_a(t)$ is also assumed to be dependent only on $x_a(t)$, the terms in flow propagation constraints (8.37) have no impact on the partial derivative of the Hamiltonian function with respect to control variables $u_{ap}^{rs}(t)$ and $v_{ap}^{rs}(t)$.

The equivalence of the instantaneous dynamic user-optimal conditions and Model 2 is demonstrated by showing that the unique trajectories of link states, inflows and exit flows that solve Model 2 also satisfy the instantaneous dynamic user-optimal conditions. As with Model 1, this equivalence is demonstrated by proving that the first order necessary conditions for Model 2 are identical to the instantaneous dynamic user-optimal conditions.

Combining equations (8.38)-(8.39) with equations (8.40)-(8.41), the same equations as equations (8.31)-(8.33) for Model 1 can be derived for each link a which points from node l to node m. Thus, we follow the derivation of optimality conditions for Model 1 and obtain the same equations (8.26)-(8.28) for each remaining route $\tilde{p}$ between intermediate node i and destination s. Similarly, we can obtain the same interpretation of the optimality conditions for Model 2 as those for Model 1.

It is now clear that the optimality conditions of Model 2 state the instantaneous dynamic user-optimal route choice conditions. Therefore, Model 2 is an equivalent instantaneous DUO route choice program. Since the flow propagation constraints are expressed directly using the actual link travel time function $\tau_a(t)$, instead of its estimate $\bar{\tau}_a(t)$ in a diagonalization fashion, Model 2 is a complete optimal control model. It can be shown that it is equivalent to a variational inequality.

8.3 A Multi-Class Route-Cost-Based Model

In this section, we define an instantaneous dynamic user-optimal (DUO) model based on travel costs or disutilities instead of travel times. To be consistent with Chapter 5, we stratify travelers into M classes for each O-D pair according to the socio-economic characteristics of each traveler. When $M = 1$, the following definition and VI model reduces to a single class model, but one which is different from the above VI model based on instantaneous travel times.

8.3.1 Multi-Class Route-Cost-Based Conditions

The instantaneous route disutility function is dependent on the instantaneous route travel time, fuel consumption enroute, automobile operating cost, etc. Denote $\tilde{c}_{ma}(t)$ as the instantaneous disutility function for travelers of class m entering link a at time t. It follows that

$$\tilde{c}_{ma}(t) = \alpha_{ma} + \beta_{ma}\, c_a(t) \qquad \forall m, a \tag{8.48}$$

where α_{ma} is a fixed instantaneous disutility parameter for class m travelers on link a and β_{ma} is a parameter to transform instantaneous link travel time $c_a(t)$ into the disutility of class m travelers. Denote $\tilde{\psi}_{mp}^{rs}(t)$ as the instantaneous route travel disutility for class m travelers from origin r to destination s at time t, and $\tilde{\sigma}_m^{rs}(t)$ as the minimal instantaneous route travel disutility for class m travelers from origin r to destination s at time t. The instantaneous route travel disutility for all allowable routes is computed as follows:

$$\tilde{\psi}_{mp}^{rs}(t) = \sum_{a \in rsp} \tilde{c}_{ma}(t) \qquad \forall m, p, r, s \tag{8.49}$$

where the summation is over all links a on route p. The minimal instantaneous route travel disutility for each O-D pair rs is

$$\tilde{\sigma}_m^{rs}(t) = \min_p \tilde{\psi}_{mp}^{rs}(t) \qquad \forall m, r, s. \tag{8.50}$$

We then present the definition of the multi-class route-cost-based instantaneous DUO state as follows.

Multi-Class Route-Cost-Based Instantaneous DUO State: *If, for each class m and each O-D pair at each instant of time, the instantaneous travel* disutilities *for all routes that are being used equal the minimal instantaneous route travel* disutility, *the dynamic traffic flow over the network is in a multi-class route-cost-based instantaneous dynamic user-optimal state.*

Denote $f_{mp}^{rs}(t)$ as the route inflow of class m from origin r to destination s over route p at time t. The equivalent route-based multi-class instantaneous DUO route choice conditions can be summarized as follows:

$$\tilde{\psi}_{mp}^{rs^*}(t) - \tilde{\sigma}_m^{rs^*}(t) \geq 0 \qquad \forall m, p, r, s; \tag{8.51}$$

$$f_{mp}^{rs^*}(t) \left[\tilde{\psi}_{mp}^{rs^*}(t) - \tilde{\sigma}_m^{rs^*}(t)\right] = 0 \qquad \forall m, p, r, s; \tag{8.52}$$

$$f_{mp}^{rs}(t) \geq 0 \qquad \forall m, p, r, s. \tag{8.53}$$

The above definition of the multi-class instantaneous DUO state and the corresponding route-based route choice conditions are defined for each origin-destination pair. They are not defined for each decision node-destination pair as in Section 8.1, which is similar to the link-based definitions in the next section. Thus, this definition does not provide any rerouting strategies for travelers at any intermediate intersection.

8.3.2 Dynamic Network Constraints

The dynamic network constraints are written for each class of travelers, designated by index m. The constraint set for this problem is summarized as follows.

Relationship between state and control variables:

$$\frac{dx_{map}^{rs}}{dt} = u_{map}^{rs}(t) - v_{map}^{rs}(t) \qquad \forall m, a, p, r, s; \tag{8.54}$$

$$\frac{dE_{mp}^{rs}(t)}{dt} = e_{mp}^{rs}(t) \qquad \forall m, p, r; s \neq r; \tag{8.55}$$

Flow conservation constraints:

$$f_m^{rs}(t) = \sum_{a \in A(r)} \sum_p u_{map}^{rs}(t) \qquad \forall m, r, s; \tag{8.56}$$

$$\sum_{a \in B(j)} v_{map}^{rs}(t) = \sum_{a \in A(j)} u_{map}^{rs}(t) \qquad \forall j, m, p, r, s; j \neq r, s; \tag{8.57}$$

$$\sum_{a \in B(s)} \sum_p v_{map}^{rs}(t) = e_m^{rs}(t) \qquad \forall m, r, s; s \neq r; \tag{8.58}$$

Flow propagation constraints:

$$x_{ap}^{rs}(t) = \sum_{b \in \tilde{p}} \{x_{bp}^{rs}[t + \tau_a(t)] - x_{bp}^{rs}(t)\} + \{E_p^{rs}[t + \tau_a(t)] - E_p^{rs}(t)\}$$

$$\forall a \in B(j); j \neq r; p, r, s; \tag{8.59}$$

Definitional constraints:

$$\sum_{mrsp} u_{map}^{rs}(t) = u_a(t), \qquad \sum_{mrsp} v_{map}^{rs}(t) = v_a(t), \qquad \forall a; \tag{8.60}$$

$$\sum_{mrsp} x_{map}^{rs}(t) = x_a(t), \qquad \forall a; \tag{8.61}$$

Nonnegativity conditions:

$$x_{map}^{rs}(t) \geq 0, \quad u_{map}^{rs}(t) \geq 0, \quad v_{map}^{rs}(t) \geq 0 \quad \forall m, a, p, r, s; \tag{8.62}$$

$$e_{mp}^{rs}(t) \geq 0, \qquad E_{mp}^{rs}(t) \geq 0, \qquad \forall m, p, r, s; \tag{8.63}$$

Boundary conditions:

$$E_{mp}^{rs}(0) = 0, \qquad \forall m, p, r, s; \tag{8.64}$$

$$x_{map}^{rs}(0) = 0, \qquad \forall m, a, p, r, s. \tag{8.65}$$

8.3.3 The Variational Inequality Problem

Then, the equivalent variational inequality formulation of multi-class route-cost-based instantaneous DUO route choice conditions (8.51)-(8.53) may be stated as follows.

> **Theorem 8.1.** *The dynamic traffic flow pattern satisfying network constraint set (8.54)-(8.65) is in a multi-class route-cost-based instantaneous DUO route choice state if and only if it satisfies the variational inequality problem:*

$$\int_0^T \sum_{rs} \sum_{mp} \tilde{\psi}_{mp}^{rs*}(t) \left[f_{mp}^{rs}(t) - f_{mp}^{rs*}(t) \right] dt \geq 0 \tag{8.66}$$

The proofs of necessity and sufficiency for variational inequality (8.66) follow in the same way as in Section 7.2.3 for the single class route-time-based case.

8.4 A Multi-Class Link-Cost-Based Model

In this section, we consider a DUO multi-class link-based VI model for the instantaneous dynamic user-optimal problem. The dynamic network constraints have to be written for each class of travelers designated by index m. The constraint set for this problem is identical to the constraint set (8.54)-(8.65) for the multi-class route-based VI model in Section 8.3. As before, the basic difference between the two models is that the variational inequality of the multi-class link-cost-based model is formulated using link-based flow variables instead of route-based variables in the multi-class route-cost-based VI model.

8.4.1 Multi-Class Link-Cost-Based Conditions

The instantaneous route disutility function is dependent on the instantaneous route travel time, fuel consumption enroute and automobile operating cost, etc. Denote $\tilde{c}_{ma}(t)$ as the instantaneous disutility function for class m travelers entering link a at time t. It follows that

$$\tilde{c}_{ma}(t) = \alpha_{ma} + \beta_{ma}\, c_a(t) \qquad \forall m, a \tag{8.67}$$

where α_{ma} is a fixed instantaneous disutility parameter for class m travelers on link a and β_{ma} is a parameter to transform instantaneous link travel time $c_a(t)$ into the disutility of class m travelers. Denote $\tilde{\psi}_{mp}^{rs}(t)$ as the instantaneous route travel disutility for class m travelers from origin r to destination s at time t. The instantaneous route travel disutility for all allowable routes p between origin r and node i is computed as follows:

$$\tilde{\psi}_{mp}^{ri}(t) = \sum_{a \in isp} \tilde{c}_{ma}(t) \qquad \forall m, p, r, i \tag{8.68}$$

where the summation is over all links a on route p. We then define a multi-class link-cost-based instantaneous DUO state as follows.

Multi-Class Link-Cost-Based Instantaneous DUO State: *If, for any departure flow of class m from each decision node to each destination node at each instant of time, the instantaneous travel* disutilities *equal the minimal instantaneous route travel* disutility, *the dynamic traffic flow over the network is in a multi-class link-cost-based instantaneous dynamic user-optimal state.*

Define $\tilde{\sigma}_m^{ri^*}(t)$ as the minimal instantaneous route travel disutility for class m travelers from origin r to node i at time t. The asterisk denotes that the travel disutility is computed using multi-class link-cost-based instantaneous DUO traffic flows. For class m travelers on link $a = (i, j)$, the minimal instantaneous travel disutility $\tilde{\sigma}_m^{rj^*}(t)$ from origin r to node j should be equal to or less than the minimal instantaneous travel disutility $\tilde{\sigma}_m^{ri^*}(t)$ from origin r to node i plus the instantaneous link travel disutility $\tilde{c}_{ma}^*(t)$ at time instant t. It follows that

$$\tilde{\sigma}_m^{ri^*}(t) + \tilde{c}_{ma}^*(t) \geq \tilde{\sigma}_m^{rj^*}(t) \qquad \forall m, a = (i, j), r. \tag{8.69}$$

If any departure flow of class m from origin r to node i enters link a at time t, or $u_{ma}^r(t) > 0$, then the multi-class link-cost-based instantaneous DUO route choice conditions require that link a is on the minimal instantaneous travel disutility route. In other words, the instantaneous minimal travel disutility $\tilde{\sigma}_m^{rj^*}(t)$ for vehicles departing origin r toward node j at time t should equal the minimal instantaneous travel disutility $\tilde{\sigma}_m^{ri^*}(t)$ for vehicles departing origin r to node i plus the instantaneous link travel disutility $\tilde{c}_{ma}^*(t)$ at time instant t. It follows that

$$\tilde{\sigma}_m^{ri^*}(t) + \tilde{c}_{ma}^*(t) = \tilde{\sigma}_m^{rj^*}(t), \qquad \text{if } u_{ma}^{r^*}(t) > 0 \quad \forall m, a = (i, j), r. \tag{8.70}$$

The above equations are also equivalent to the following:

$$\left[\tilde{\sigma}_m^{ri^*}(t) + \tilde{c}_{ma}^*(t) - \tilde{\sigma}_m^{rj^*}(t)\right] u_{ma}^{r^*}(t) = 0 \qquad \forall m, a = (i, j), r. \tag{8.71}$$

Denote $\tilde{\Theta}_{ma}^{r^*}(t)$ as the difference between the minimal instantaneous travel disutility from origin r to node j and the instantaneous travel disutility from origin r to node i plus the instantaneous travel disutility on link a for class m at time t. It follows that

$$\tilde{\Theta}_{ma}^{r^*}(t) = \tilde{\sigma}_m^{ri^*}(t) + \tilde{c}_{ma}^*(t) - \tilde{\sigma}_m^{rj^*}(t) \qquad \forall m, a, r; a = (i, j). \tag{8.72}$$

Thus, the multi-class link-cost-based instantaneous DUO route choice conditions can be summarized as follows:

$$\tilde{\Theta}_{ma}^{r^*}(t) \geq 0 \qquad \forall m, a = (i, j), r; \tag{8.73}$$

$$u_{ma}^{r^*}(t)\,\tilde{\Theta}_{ma}^{r^*}(t) = 0 \qquad \forall m, a = (i,j), r; \tag{8.74}$$

$$u_{ma}^{r}(t) \geq 0 \qquad \forall m, a = (i,j), r. \tag{8.75}$$

Similar to Lemma 7.1, we can prove that multi-class link-cost-based instantaneous DUO route choice conditions (8.73)-(8.75) imply route-based multi-class instantaneous DUO route choice conditions (8.51)-(8.53).

8.4.2 The Variational Inequality Problem

The equivalent variational inequality formulation of multi-class link-cost-based instantaneous DUO route choice conditions (8.73)-(8.75) may then be stated as follows.

> **Theorem 8.2.** *The dynamic traffic flow pattern satisfying network constraint set (8.54)-(8.65) is in a multi-class link-cost-based instantaneous DUO route choice state if and only if it satisfies the variational inequality problem:*

$$\int_0^T \sum_r \sum_{ma} \tilde{\Theta}_{ma}^{r^*}(t)\; \left[u_{ma}^{r}(t) - u_{ma}^{r^*}(t)\right]\; dt \;\geq\; 0 \tag{8.76}$$

The proofs of necessity and sufficiency follow in the same way as for the single class case.

8.5 Notes

There are many possible dynamic generalizations of the static user-optimal route choice model. Some of them may be suitable for providing instantaneous information and advising vehicles of their best routes in a dynamic route guidance system. Since our instantaneous DUO route choice models have a direct correspondence to the static UO assignment model, other static UO formulations can be expected to have their DUO counterparts. Various optimal control formulations should be investigated and their corresponding solution characteristics, such as the uniqueness of the optimal control strategies, should be studied.

The definition of instantaneous DUO is similar to that given in Friesz et al (1989). However, in their model, the route travel costs are equal at every instant of time; in contrast, in our models, instantaneous route travel times equal the minimum instantaneous travel time only at each decision node that has flows to the destination. We show that our alternative definition allows the resulting models to equilibrate flows using route travel times based only on link driving times. This is a fundamental difference in the definition of instantaneous route travel time between their model and our models. We discuss this point further in the following.

Moreover, we discuss some fundamental differences among existing dynamic route choice models. These models diverge at two points: 1) system-optimal vs. user-optimal; and 2) the interpretation of route travel time. Following Merchant and Nemhauser (1978a), Carey (1987) presented improved dynamic system-optimal models which considered the minimization of cumulative instantaneous route travel times. However, the dynamic *user-optimal* state was not considered in his models. Friesz et al (1989) presented an instantaneous dynamic user-optimal route choice model which did consider the equilibration of instantaneous unit route travel cost. In their model, however, the instantaneous unit route travel cost is defined as

$$\psi_p^{rs}(t) = \sum_{a \in rsp} \left[c_a[x_a(t)] + \frac{\dot{\lambda}_a(t)}{g'_a[x_a(t)]} \right] \qquad \forall r, s, p, \tag{8.77}$$

where the numerator of the second term is the time variation (rate of change) of the Lagrange multiplier $\lambda_a(t)$ and the denominator is the derivative of exit flow function $g_a[x_a(t)]$ with respect to link state variable $x_a(t)$. Since both of these could change from problem to problem, their presence may present a difficulty for the physical interpretation of the model. In contrast, our models only use the summation of instantaneous link travel times as the instantaneous route travel time:

$$\psi_p^{rs}(t) = \sum_{a \in rsp} c_a[x_a(t), u_a(t), v_a(t)] \qquad \forall r, s, p. \tag{8.78}$$

This explicit definition of instantaneous route travel time reflects the dynamic route choice behavior and decision criterion of travelers. It is also consistent with the definition of instantaneous route travel time in the DSO route choice models of Merchant-Nemhauser (1978a) and Carey (1987). We note that it is not known how to formulate an optimization program which is consistent with our definition of instantaneous route travel time if exit flow functions instead of exit flow variables are used.

Problems

8.1 Assume that link travel time function has the following simplified form

$$c_a(t) = c_a[x_a(t), v_a(t)]$$

Design an optimal control problem for instantaneous DUO route choice. Demonstrate the equivalence of the first-order necessary conditions with the link-time-based instantaneous DUO route choice conditions.

8.2 Design an optimal control problem for route-time-based instantaneous DUO route choice. Demonstrate the equivalence of the first-order necessary conditions

conditions with the route-time-based instantaneous DUO route choice conditions.

8.3 In implementing the output from the optimal control formulation (8.3)-(8.15), do we need to control both inflow and exit flow for a highway link?

8.4 Discuss the advantages and disadvantages of the optimal control formulations over the variational inequality formulation for instantaneous DUO route choice problems.

8.5 A two-link network is shown in Figure 8.1. Assume the instantaneous link travel time functions are

$$c_1(t) = 4 + u_1(t), \qquad c_2(t) = 3 + 2u_2(t)$$

The instantaneous disutility functions $\tilde{c}_a(t)$ are

$$\tilde{c}_1(t) = 3 + c_1(t), \qquad \tilde{c}_2(t) = 1 + 1.5c_2(t)$$

The O-D flow is 5 units for time period $[0, 5]$. Determine route departure flows and route travel times for time period $[0, 5]$ based on the link-time-based and the link-cost-based instantaneous DUO route choice criteria, respectively. Compare the differences of the two cases.

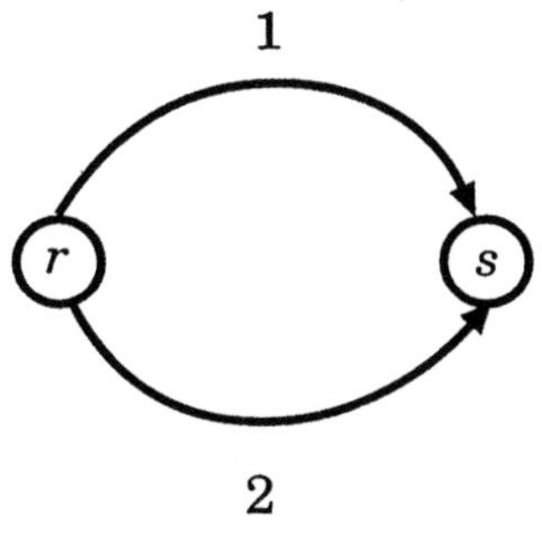

Figure 8.1: Network of Problems 8.5 and 8.6

8.6 For the two-link network shown in Figure 8.1. There are two classes of travelers. The O-D departure flows for class 1 and 2 are 5 units and 10 units for time period $[0, 5]$, respectively. For both classes ($m = 1, 2$), the instantaneous disutility functions $\tilde{c}_{ma}(t)$ are

$$\tilde{c}_{11}(t) = 3 + u_1(t), \qquad \tilde{c}_{21}(t) = 5 + u_1(t)$$

$$\tilde{c}_{12}(t) = 2 + 2u_2(t), \qquad \tilde{c}_{22}(t) = 3 + 2u_2(t)$$

Determine route departure flows and route travel times for time period $[0, 5]$ based on the multi-class link-cost-based instantaneous DUO route choice criterion. [Hint: the solution may not be unique.]

Chapter 9

Ideal Stochastic Dynamic Route Choice Models

Dynamic route guidance systems are being developed to inform and guide drivers regarding their best departure times and routes so as to avoid congestion delay. However, some drivers may not rely on the information provided by the route guidance system to adjust their departure times and routes. Furthermore, drivers without navigation systems do not have perfect information on the network traffic and must use their own experience and perception of current traffic conditions to make travel decisions. Thus, there is a need to develop dynamic route choice models under imperfect as well as perfect information.

To date, several dynamic route choice models have been proposed for both route and departure time choices; see Chapter 1 for an overview. However, few existing models can be applied to the situation with imperfect traffic information or indeterministic traffic conditions. Moreover, deterministic dynamic user-optimal (DUO) route choice models assign flows to minimal travel time routes only. Those deterministic models can not represent realistically the dispersion of traffic flow across different travel time routes. In response to this gap, this chapter presents an ideal stochastic dynamic user-optimal (SDUO) route choice model which is a stochastic generalization of our ideal deterministic DUO route choice model presented in Chapter 5. This model represents an alternative approach to stratifying vehicles into different classes and using class-specific travel disutilities to determine dynamic user-optimal routes. That approach was discussed in Chapters 5 and 8.

Perceptions of travel times have been explored extensively under the assumption of static traffic conditions (Daganzo and Sheffi, 1977; Fisk, 1980; Sheffi and Powell, 1982, etc.). Mirchandani and Soroush (1988) generalized these models to include stochastic link travel times as well as random perception errors. Recently, Cassetta (1991) studied the variation of dynamic route choice from day to day. Vythoulkas (1990) also extended stochastic static route choice models into the dynamic route choice framework.

The SDUO route choice problem is to find dynamic trajectories of link states and inflow and exit flow control variables, given time-dependent O-D departure rates, the network, link travel time functions and some assumptions about the randomness of traffic information. Basically, SDUO problems can be classified according to the following:

1. random components in traveler's perceptions of travel times;
2. randomness of origin-destination flows;
3. randomness of the link traffic flow states and link travel times.

The travelers' perceptions of the travel time of a route may vary from traveler to traveler due to the perception error of each traveler. In other words, each traveler perceives his/her own travel time. This perceived travel time is a random variable distributed across the population of travelers. Due to the variation in their perceptions of travel times, travelers may choose different routes (not necessarily the shortest, nor do the routes necessarily have the same actual travel time) between a given O-D pair in selecting their perceived shortest routes. Thus, there is a need to develop route choice models under imperfect information as well as perfect information.

Stochastic variation of O-D demand is another contributing factor to stochastic network conditions. It arises in several ways. First, the estimates or forecasts of O-D demand may include large errors. These errors can be considered as stochastic input to the route choice models. Second, the day-to-day variation of O-D demands may be another stochastic factor. Current route choice models do not consider day to day variations.

Third, the stochastic events may occur on the traffic network. In reality, many stochastic factors affect link traffic flow. For example, double parking or a vehicle breakdown can reduce the capacity of a link. Random variations in link times due to signals are another important factor. Accidents can reduce the link capacity to a larger extent. Route choice models should be able to handle these stochastic variations of the traffic state. However, these last two types of stochastic dynamic route choice model are not considered further here.

In this chapter, we include only the random component resulting from travelers' perceptions of travel times. In this case, route choice is a process of selecting among alternative routes, for which the perceived time-dependent travel times include a random error. A general probit-type variational inequality model is first formulated by assuming a general distribution of route choice dispersion. The stochastic route choice assumption is introduced into dynamic route choice problems so that our SDUO models are dynamic generalizations of conventional stochastic static user-optimal (SUO) models under the assumption of dispersed travel choice.

The proposed model is a better approximation of real world conditions because O-D departure flows are dispersed across different travel time routes, whereas deterministic DUO route choice models assign O-D departure flows to minimal travel time routes only. Furthermore, the SDUO model provides a better representation of travelers' route choice behavior than their static counterparts because time-dependent traffic flows and travel times are explicitly taken into account.

We first present definitions of travel times in stochastic dynamic problems in Section 9.1. In Section 9.2, we present an ideal SDUO route choice model based on stochastic flow loading with a general distribution function of *mean* actual travel times experienced by drivers over alternative routes for each O-D pair. This model is formulated as a variational inequality. Our ideal deterministic DUO route choice model in Chapter 5 is a special case of the ideal SDUO route choice model when the variance of the perceived actual route travel time is zero.

To solve the ideal SDUO model for large networks, we need to develop solution algorithms avoiding route enumeration. To this end, the continuous variational inequality is converted into a discrete variational inequality. The relaxation method is used to solve this discrete VI. A nonlinear programming problem (NLP) is formulated under relaxation so that the Method of Successive Averages can be used to solve this NLP. For a general multinomial probit, Monte Carlo sampling is proposed to estimate the dynamic route choice probabilities. On the other hand, for a multinomial logit, a multiple-route dynamic route choice algorithm (DYNASTOCH1) similar to Dial's efficient-route algorithm (STOCH) is suggested to solve the LP subproblem, which is a *flow-independent* ideal SDUO route choice problem. In the DYNASTOCH1 algorithm, explicit route enumeration can be avoided and dynamic route choice probabilities can be computed accurately. A numerical example is presented in Section 9.4 to illustrate the solution of the proposed algorithm.

9.1 Redefinition of Dynamic Travel Times

Define $\tau_a(t)$ as the mean actual travel time over link a for vehicles entering link a at time t. In this model, we assume the mean actual travel time $\tau_a(t)$ on link a is the sum of two components: 1) a flow-dependent cruise time $g_{1a}[x_a(t), u_a(t)]$ over link a; and 2) a queuing delay $g_{2a}[x_a(t), v_a(t)]$. It follows that

$$\tau_a(t) = g_{a1}[x_a(t), u_a(t)] + g_{2a}[x_a(t), v_a(t)]. \tag{9.1}$$

The two components $g_{1a}[x_a(t), u_a(t)]$ and $g_{2a}[x_a(t), v_a(t)]$ are assumed to be nonnegative and differentiable with respect to $x_a(t), u_a(t)$ and $x_a(t), v_a(t)$, respectively.

Denote the minimal free flow travel time from any node i to destination node s as σ_o^{is}. An *efficient route* from origin r to destination s is redefined to include only links $a = (i, j)$ such that node j is closer to destination s than node i, i.e., $\sigma_o^{is} > \sigma_o^{js}$. In the following, efficient routes are used in the logit-based stochastic dynamic network loading. These routes need not be explicitly enumerated to solve our models; in Section 9.3, we show that these routes can be implicitly generated.

Let $T_a(t)$ denote the perceived actual travel time for flows entering link a at time t. $T_a(t)$ is a random variable with mean $\tau_a(t)$. Similarly, define $\eta_p^{rs}(t)$ as the mean actual travel time experienced over route p by vehicles departing origin r toward destination s at time t. Once the mean actual link travel time $\tau_a(t)$ is determined, the mean actual route travel time $\eta_p^{rs}(t)$ can be computed using the following recursive formula:

$$\eta_p^{rj}(t) = \eta_p^{ri}(t) + \tau_a[t + \eta_p^{ri}(t)] \qquad \forall r, s, p; i, j \in p \tag{9.2}$$

where link $a = (i, j)$ is on route p.

9.2 Formulation of the Model

The stochastic dynamic route choice model attempts to account for dispersion of travelers onto multiple routes between an O-D pair by assuming errors in travelers' perception of network travel times. The dynamic user-optimal route choice model yields some dispersion by considering congestion effects, but assumes that travelers are completely knowledgeable about network travel conditions and therefore encounter no perception errors. It is reasonable to assume that travelers do perceive travel conditions differently from what would actually be measured and that a more behaviorally realistic model would consider perceived travel time, different from the measured time by the amount of a random error term, and congestion effects. A model which accounts for both of these route dispersion factors is the Stochastic Dynamic User-Optimal (SDUO) route choice model. Similar to the DUO route choice model, the SDUO model is derived by stating a VI problem whose solution characterizes a set of pre-specified SDUO route choice conditions.

9.2.1 Network Constraints

In this section, the constraint set for our stochastic dynamic route choice problem is first summarized as follows:

Relationship between state and control variables:

$$\frac{dx_{ap}^{rs}}{dt} = u_{ap}^{rs}(t) - v_{ap}^{rs}(t) \qquad \forall a, p, r, s; \tag{9.3}$$

$$\frac{dE_p^{rs}(t)}{dt} = e_p^{rs}(t) \qquad \forall p, r; s \neq r; \tag{9.4}$$

Flow conservation constraints:

$$f^{rs}(t) = \sum_{a \in A(r)} \sum_p u_{ap}^{rs}(t) \qquad \forall r, s; \tag{9.5}$$

$$\sum_{a \in B(j)} v_{ap}^{rs}(t) = \sum_{a \in A(j)} u_{ap}^{rs}(t) \qquad \forall j, p, r, s; j \neq r, s; \tag{9.6}$$

$$\sum_{a \in B(s)} \sum_{p} v_{ap}^{rs}(t) = e^{rs}(t) \qquad \forall r, s; s \neq r; \tag{9.7}$$

Flow propagation constraints:

$$x_{ap}^{rs}(t) = \sum_{b \in \tilde{p}} \{x_{bp}^{rs}[t + \tau_a(t)] - x_{bp}^{rs}(t)\} + \{E_p^{rs}[t + \tau_a(t)] - E_p^{rs}(t)\}$$

$$\forall a \in B(j); j \neq r; p, r, s; \tag{9.8}$$

Definitional constraints:

$$\sum_{rsp} u_{ap}^{rs}(t) = u_a(t), \qquad \sum_{rsp} v_{ap}^{rs}(t) = v_a(t), \qquad \forall a; \tag{9.9}$$

$$\sum_{rsp} x_{ap}^{rs}(t) = x_a(t), \qquad \forall a; \tag{9.10}$$

Nonnegativity conditions:

$$x_{ap}^{rs}(t) \geq 0, \quad u_{ap}^{rs}(t) \geq 0, \quad v_{ap}^{rs}(t) \geq 0 \quad \forall a, p, r, s; \tag{9.11}$$

$$e_p^{rs}(t) \geq 0, \qquad E_p^{rs}(t) \geq 0, \qquad \forall p, r, s; \tag{9.12}$$

Boundary conditions:

$$E_p^{rs}(0) = 0, \qquad \forall p, r, s; \tag{9.13}$$

$$x_{ap}^{rs}(0) = 0, \qquad \forall a, p, r, s. \tag{9.14}$$

In summary, in this problem the control variables are $u_{ap}^{rs}(t)$, $v_{ap}^{rs}(t)$ and $e_p^{rs}(t)$; the state variables are $x_{ap}^{rs}(t)$ and $E_p^{rs}(t)$; and the functionals are $\tau_a(t)$. The stochastic dynamic route choice model has similar constraints as the deterministic ideal DUO route choice model in Chapter 5. Before considering the VI formulation, some background information regarding discrete choice theory in route choice modeling and the definition of ideal SDUO is presented.

9.2.2 Stochastic Route Choice and the Ideal SDUO State

The fundamental assumption in discrete choice theory is that individual's preferences for alternatives in a choice set can be represented with a utility measure. Given the set of possible utility measures, and assuming that travelers are preference maximizers, the probability that some alternative is the most preferred can be calculated. This probability, along with the number of individual decision makers, determines the number of individuals preferring each alternative.

For each origin-destination pair, there are many alternative efficient routes p, each with some mean actual travel time for flows departing from each origin to each destination at each time t. Due to variations in perception and exogenous factors, actual route travel times are perceived differently by each traveler. We assume the perceived actual route travel time to be a random variable for each instant of time t. Given his or her perception of actual route travel time at each time instant t, each traveler is assumed to choose the route with minimal perceived actual travel time. Let $\Omega_p^{rs}(t)$ denote the perceived actual route travel time for flows departing origin r toward destination s on route p at each time t. Then, we assume that

$$\Omega_p^{rs}(t) = \eta_p^{rs}(t) + \xi_p^{rs}(t) \qquad \forall r, s, p, \tag{9.15}$$

where $\eta_p^{rs}(t)$ is an observable or deterministic measure, while $\xi_p^{rs}(t)$ is a random error term associated with the route p under consideration. The attributes, both observable and unobservable, which influence the decision process, may be alternative specific, as well as model specific. In other words, the random term may account for errors due to lack of information about the alternative, differences in individual tastes, or model specification errors associated with the assumptions required to make the model solvable.

Given the distribution of the error terms $\xi_p^{rs}(t)$, the distribution of the perceived route travel times may be calculated. In general, we assume that $E[\xi_p^{rs}(t)] = 0$, i.e.,

$$E[\Omega_p^{rs}(t)] = \eta_p^{rs}(t) \qquad \forall r, s, p, \tag{9.16}$$

In other words, the average perceived actual route travel time is the mean actual route travel time.

Given this assumption, the costs, $\Omega_p^{rs}(t)$ and $\eta_p^{rs}(t)$ are often referred to as perceived and measured costs, respectively. Based on a probability density function for the actual route travel time at each time t, the stochastic dynamic route choice problem is to determine how many travelers use each route at each origin node at each instant of time. If travelers are cost minimizers, then the probability that alternative p is chosen by an individual, or the route choice function, can be stated as:

$$P_p^{rs}(t) = \text{Prob}[\Omega_p^{rs}(t) \leq \Omega_q^{rs}(t),\ \forall \text{ routes } q \text{ between } r \text{ and } s] \quad \forall r, s, p. \tag{9.17}$$

where Prob is the choice function representing the proportion of individuals who choose alternative p. Therefore, the probability that a given efficient route is chosen at time t is the probability that its actual route travel time is perceived to be the lowest of all the alternative routes. Once the distribution of the perceived route travel times is specified, the probability of selecting each alternative route can be calculated and the time-dependent route flow can be assigned as follows:

$$f_p^{rs}(t) = f^{rs}(t) P_p^{rs}(t) \qquad \forall r, s, p. \tag{9.18}$$

As shown in Sheffi (1985), a satisfaction function can be defined as follows:

$$S^{rs}(t) = E[\min_p \Omega_p^{rs}(t)] \qquad \forall r, s, p. \tag{9.19}$$

This satisfaction function has the following property:

$$\frac{\partial S^{rs}(t)}{\partial \eta_q^{rs}(t)} = P_p^{rs}(t) \qquad \forall r, s, p. \tag{9.20}$$

This property will be used in the following to derive the solution algorithm and the objective function of a reduced nonlinear programming problem.

We next define the ideal SDUO state, which is the stochastic generalization of the deterministic *ideal* DUO state in Chapter 5.

> **Ideal SDUO State:** *If, for each O-D pair at each instant of time, the actual travel times perceived by travelers departing at the same time over used routes are equal and minimal, the dynamic traffic flow over the network is in an ideal stochastic dynamic user-optimal state.*

This definition represents another kind of perception of travel times by travelers. It can also be called a predictive (or anticipatory) SDUO since the mean actual route travel time is a predicted mean route travel time. Sometimes, it is termed a stochastic dynamic user-equilibrium, which is the dynamic generalization of stochastic user-equilibrium (Daganzo and Sheffi, 1977).

9.2.3 Two Popular Route Choice Functions

In static network models, the stochastic route choice literature is dominated by two route choice functions, the multinomial logit and multinomial probit. In general, these two discrete choice models are distinguished by the random error component in the disutility (cost or impedance) functions, which is determined by the modeler's specification of the random error term's distribution. In practice, these two choice functions are particularly suited to route choice applications because of the difficulty of explicitly identifying the large number of route choice alternatives in a large-scale transportation network. Similar arguments apply to stochastic dynamic route choice. In the following, the logit and probit route choice functions are discussed.

The logit route choice function is investigated first. We assume that

$$\Omega_p^{rs}(t) = \eta_p^{rs}(t) - \frac{1}{\theta}\epsilon_p^{rs}(t) \qquad \forall r, s, p, \tag{9.21}$$

where θ is a nonnegative parameter that scales the perceived actual route travel time, and $\epsilon_p^{rs}(t)$ is a random error term associated with the route p under consideration. We assume that $\epsilon_p^{rs}(t)$ are identically and independently distributed (IID) Gumbel variates for each time instant with location parameter equal to zero and scale parameter θ (see Ben-Akiva and Lerman, 1985). The multinomial logit choice function is generally defined as follows for a given set of alternatives :

$$P_p^{rs}(t) = \frac{\exp[-\theta\eta_p^{rs}(t)]}{\sum_q \exp[-\theta\eta_q^{rs}(t)]} \qquad \forall r, s, p. \tag{9.22}$$

Thus, for each route p being used at time t, the logit route flow is

$$f_p^{rs}(t) = f^{rs}(t)\frac{\exp[-\theta\eta_p^{rs}(t)]}{\sum_q \exp[-\theta\eta_q^{rs}(t)]} \qquad \forall r,s,p. \tag{9.23}$$

When $\theta \rightarrow \infty$, $Var[\Omega_p^{rs}(t)] \rightarrow 0$, and the perceived actual travel time between O-D equals the mean actual travel time. In this case, travelers choose the minimal actual travel time route at each time instant, as in the ideal deterministic DUO route choice model. The assumption of independently and identically distributed Gumbel variates restricts the random error components to have equal scale parameters θ. The fundamental properties related to the maximum and to the difference of two Gumbel variates with equal scale parameters result in the simple expression for the logit model.

In practice, the logit model must be applied with full consideration of its simplifying assumptions. The independence assumption for random error terms requires that two alternatives be considered independently by an individual regardless of how correlated the decision attributes are. The other simplifying assumption that the random error components of the cost function are identically distributed results in equal variance among the alternatives, regardless of the relative differences in the deterministic portion of the cost function. The limitations of the logit model for computing choice proportions among alternatives with highly correlated random cost components, such as found in route choice problems, are well known. The logit model, however, has been used in practice because its simplicity and because an efficient solution method was found by Dial (1971).

Although the logit model is computationally attractive, the limitations for route choice problems due to the IID assumptions may cause it to produce questionable results. A model which accounts for the correlation of attributes among alternatives is the multinomial probit model. While more theoretically sound, it is also more difficult to solve than the logit model.

The multinomial probit model can be derived from discrete choice theory and follows from equations (9.15) through (9.17) where the $\xi_p^{rs}(t)$ in (9.15) are assumed to be multivariate normal distributed with mean given by a vector of zeros and a covariance matrix for routes p at each time t. Unlike the logit model, equation (9.17) cannot be expressed in closed form because the minimum of the set of random variables describing the alternative's travel time involves the calculation of a multiple integral of a cumulative normal density function which cannot be evaluated in closed form. Also unlike the logit model, there are no restrictions of independence or identical distributions for the random error components.

The multinomial probit route choice function is characterized by a covariance matrix with positive non-diagonal elements when two or more route choices share one or more network links. The route choice attributes of these routes are dependent, unlike the multinomial logit model, where the covariance matrix is diagonal. The properties of the normal distribution simplify the derivation of the multinomial probit model just as the properties of the

Gumbel distribution did for the multinomial logit model. The probit choice function is represented by equation (9.17) where the $\Omega_p^{rs}(t)$ are multivariate normal distributed at each time t and represent the travel time of alternative route p.

To apply the probit choice model to a route choice problem, it is necessary to specify the parameters of the perceived route travel time distributions. As noted previously, the mean of the random error components in the perceived route travel time functions are assumed to be zero, which yields:

$$E[\Omega_p^{rs}(t)] = \eta_p^{rs}(t) \qquad \forall r, s, p. \tag{9.24}$$

The variance of these perceived travel times represents the error associated with perception, individual tastes, functional specification, and nonquantifiable attributes. In the dynamic route choice problem, the mean actual route travel time can be computed by using the mean actual link travel time via a recursive formula as given in Chapter 4. Assume route p consists of nodes $(r, 1, 2, \cdots, i, \cdots, s)$. Denote $\eta_p^{ri}(t)$ as the travel time *actually* experienced over route p from origin r to node i by vehicles departing origin r at time t. Then, a recursive formula for mean actual route travel time $\eta_p^{rs}(t)$ is:

$$\eta_p^{ri}(t) = \eta_p^{r(i-1)}(t) + \tau_a[t + \eta_p^{r(i-1)}(t)] \qquad \forall p, r, i; i = 1, 2, \cdots, s;$$

where link $a = (i-1, i)$ and $\tau_a(t)$ is the mean actual link travel time. However, the variances for the perceived actual route travel times are difficult to compute by using a similar recursive formula. This point is demonstrated by the following example.

Suppose that there are only two links 1 and 2 on one route p linking O-D pair rs. The nodes are $r, 1, s$, respectively. The perceived route travel times can be computed iteratively as follows:

$$\Omega_p^{r1}(t) = T_1(t) = \tau_1(t) + \xi_1(t) \tag{9.25}$$

$$\begin{aligned} \Omega_p^{rs}(t) &= \Omega_p^{r1}(t) + T_2[t + \Omega_p^{r1}(t)] \\ &= T_1(t) + T_2[t + \Omega_p^{r1}(t)] \\ &= \tau_1(t) + \xi_1(t) + \tau_2[t + \Omega_p^{r1}(t)] + \xi_2[t + \Omega_p^{r1}(t)] \end{aligned} \tag{9.26}$$

If some distribution is assumed for the perceived link travel time, a closed-form distribution function for the perceived route travel time is not easy to find. In the above example, if a normal distribution is assumed for the perceived link travel time, i.e. $T_1(t)$ and $T_2[t + \Omega_p^{r1}(t)]$ have normal distributions, their summation (the route travel time) has a normal distribution. However, its distribution parameters, mean and variance, do not have a closed form, because $\Omega_p^{r1}(t)$ in $T_2[t + \Omega_p^{r1}(t)]$ is a random variable. Subsequently, for any two routes, the covariance has no closed form as well. To overcome this problem, a probit distribution is assumed for the perceived actual route travel time directly. For example, we assume a normal distribution for $\Omega_p^{rs}(t)$, i.e.

$$\Omega_p^{rs}(t) \sim N(\mu_p^{rs}(t), [\sigma_p^{rs}(t)]^2) \qquad \forall\, r, s, p \tag{9.27}$$

where the mean $\mu_p^{rs}(t) = \eta_p^{rs}(t)$. In the above two-link example, the perceived route travel time can be decomposed into the summation of perceived link travel times. However, a closed form distribution for perceived link travel times is not available, because $\Omega_p^{r1}(t)$ in $T_2[t + \Omega_p^{r1}(t)]$ is a random variable. Subsequently, for any two routes p and q with shared links, the covariance $Cov[\Omega_p^{rs}(t), \Omega_q^{rs}(t)]$ has no closed form as well. For these reasons, the calculation of the probit choice function for dynamic route choice applications can be performed by using a Monte Carlo sampling procedure, similar to the static route choice application by Sheffi and Powell (1981).

9.2.4 Ideal Route Choice Conditions and VI Problem

The general route choice probability of selecting each alternative route can be calculated and the time-dependent route flow can be determined as follows:

$$f_p^{rs}(t) = f^{rs}(t) P_p^{rs}(t) \qquad \forall r, s, p. \tag{9.28}$$

The above equality states the SDUO route choice conditions. In general, the mean actual route travel time $\eta_p^{rs}(t)$ is increasing with route departure flow $f_p^{rs}(t)$, i.e.,

$$\frac{\partial \eta_p^{rs}(t)}{\partial f_p^{rs}(t)} > 0 \qquad \forall r, s, p. \tag{9.29}$$

For each route p and each O-D pair rs, define a cost term as follows:

$$G_p^{rs}(t) = [f_p^{rs}(t) - f^{rs}(t) P_p^{rs}(t)] \frac{\partial \eta_p^{rs}(t)}{\partial f_p^{rs}(t)} = 0 \qquad \forall r, s, p. \tag{9.30}$$

Clearly, the above equality contains the SDUO route choice conditions (9.28), since $\partial \eta_p^{rs}(t) / \partial f_p^{rs}(t) > 0$.

The ideal stochastic dynamic user-optimal (SDUO) route choice problem is to determine vehicle flows at each instant of time on each link resulting from drivers using minimal perceived actual travel time routes. Now, we present a VI formulation for this ideal SDUO route choice model.

Theorem 9.1. *The dynamic traffic flow pattern satisfying network constraint set (9.3)-(9.14) is in an ideal SDUO route choice state if and only if it satisfies the variational inequality problem:*

$$G_p^{rs}(t) \left\{ f_p^{rs}(t) - f_p^{rs^*}(t) \right\} \geq 0 \tag{9.31}$$

for each time instant $t \in [0, +\infty)$.

Proof.
It is necessary and sufficient to prove that equation (9.30) is equivalent to inequality (9.31). As shown in Nagurney (1993), the system of equations (9.30) is equivalent to variational inequality (9.31).

9.2.5 Analysis of Dispersed Route Choice

Next, we use the logit-based ideal SDUO route choice model to illustrate the relationship of the ideal SDUO route choice model and the ideal DUO route choice model. The logit-based stochastic route choice function is

$$f_p^{rs}(t) = f^{rs}(t) \frac{\exp[-\theta \eta_p^{rs}(t)]}{\sum_q \exp[-\theta \eta_q^{rs}(t)]} \qquad \forall r, s, p. \tag{9.32}$$

We assume θ to be nonnegative. In the following we show that as $\theta \to \infty$, the solution approaches that of deterministic dynamic user-optimal route choice model based on actual route travel times in Chapter 5. Rewriting the logit-based route flow constraints (9.32), we have

$$\frac{f_p^{rs}(t)}{\exp[-\theta \eta_p^{rs}(t)]} = \frac{f^{rs}(t)}{\sum_m \exp[-\theta \eta_m^{rs}(t)]} \qquad \forall r, s, p. \tag{9.33}$$

Now assume that there is another route q from r to s which has flow $f_q^{rs}(t)$ at each instant of time. Thus, it follows that

$$\frac{f_q^{rs}(t)}{\exp[-\theta\ \eta_q^{rs}(t)]} = \frac{f^{rs}(t)}{\sum_m \exp[-\theta\ \eta_m^{rs}(t)]} \qquad \forall r, s, q. \tag{9.34}$$

Comparing the above two equations, we have

$$\frac{f_p^{rs}(t)}{\exp[-\theta\ \eta_p^{rs}(t)]} = \frac{f_q^{rs}(t)}{\exp[-\theta\ \eta_q^{rs}(t)]} \qquad \forall p, q, r, s. \tag{9.35}$$

Assume that flows $f_p^{rs}(t)$ and $f_q^{rs}(t)$ are positive. Taking the logarithms of the above equation, it follows that

$$\ln f_p^{rs}(t) + \theta\ \eta_p^{rs}(t) = \ln f_q^{rs}(t) + \theta\ \eta_q^{rs}(t) \qquad \forall p, q, r, s. \tag{9.36}$$

Dividing the above equation by θ, we have

$$\frac{1}{\theta} \ln f_p^{rs}(t) + \eta_p^{rs}(t) = \frac{1}{\theta} \ln f_q^{rs}(t) + \eta_q^{rs}(t) \qquad \forall p, q, r, s. \tag{9.37}$$

As $\theta \to \infty$, the first terms of both the left-hand-side and the right-hand-side will approach zero. Thus, we have

$$\eta_p^{rs}(t) = \eta_q^{rs}(t) \qquad \forall p, q, r, s. \tag{9.38}$$

The above equation demonstrates that for any O-D pair rs, any route carrying positive O-D departure flow has equal mean actual travel time.

9.3 Solution Algorithm

9.3.1 The Discrete Variational Inequality Problem

To convert our continuous time VI problem (9.31) into a discrete time VI problem, the time period $[0, T]$ is subdivided into K small time intervals. Each time interval is regarded to be one unit of time. Then, $u_a(k)$ represents the inflow into link a during interval k, and $v_a(k)$ represents the exit flow from link a during interval k. To simplify the formulation, we round the estimated actual travel time on each link in the following way so that each estimated travel time is equal to a multiple of the time interval:

$$\bar{\tau}_a(k) = i \quad \text{if} \quad i - 0.5 \leq \bar{\tau}_a(k) < i + 0.5,$$

where i is an integer and $0 \leq i \leq K$. We note that this round-off method is used only in the flow propagation constraints. More accurate flow propagation constraints can be obtained by making the time intervals smaller. The evaluation of the link/route travel time functions and the subsequent objective function does not have this round-off error, so these round-off errors do not accumulate in the subsequent minimal-cost route search.

By discretizing time, the continuous VI problem (9.31) is converted into the following VI problem:

$$G_p^{rs}(k) \left\{ f_p^{rs}(k) - f_p^{rs^*}(k) \right\} \geq 0 \tag{9.39}$$

To solve the above VI problem, it is necessary to solve an inequality of its summation over all time intervals k, all O-D pairs rs, and all routes p. It follows that

$$\sum_{k=1}^{K} \sum_{rs} \sum_{p} G_p^{rs}(k) \left\{ f_p^{rs}(k) - f_p^{rs^*}(k) \right\} \geq 0 \tag{9.40}$$

The above discrete VI can be solved by using a combination of the relaxation method and the Method of Successive Averages (MSA).

9.3.2 The Relaxation Method

For each relaxation iteration, we temporarily fix:

1. the actual travel time $\tau_a(k)$ in link flow propagation constraints as $\bar{\tau}_a(k)$;

2. the route flow variables in the route travel time function so that route travel time $\eta_p^{rs^*}(k)$ is fixed as $\bar{\eta}_p^{rs}[f_p^{rs}(k)]$ for each route;

3. the route choice probability $P_p^{rs}(k)$ as $\bar{P}_p^{rs}$, which depends on $\bar{\eta}_p^{rs}[f_p^{rs}(k)]$.

Thus, the discrete VI (9.40) becomes

$$\sum_{k=1}^{K}\sum_{rs}\sum_{p}\bar{G}_p^{rs}(k)\ \left\{f_p^{rs}(k)-f_p^{rs^*}(k)\right\}\ \geq\ 0 \tag{9.41}$$

where

$$\bar{G}_p^{rs}(t)=[f_p^{rs}(t)-f^{rs}(t)\bar{P}_p^{rs}(t)]\frac{\partial\bar{\eta}_p^{rs}[f_p^{rs}(k)]}{\partial f_p^{rs}(t)}\qquad\forall r,s,p. \tag{9.42}$$

Next, an objective function of a nonlinear programming program or discrete-time optimal control problem can be constructed as follows:

$$\begin{aligned}\min\quad Z(u,v,x,E,f)\quad &=\quad \sum_{k=1}^{K}\sum_{rs}\left\{-f^{rs}(k)\ S^{rs}(k)+\sum_{p}f_p^{rs}(k)\ \bar{\eta}_p^{rs}[f_p^{rs}(k)]\right.\\ &-\quad \left.\sum_{p}\int_0^{f_p^{rs}(k)}\bar{\eta}_p^{rs}[\omega]d\omega\right\}\end{aligned} \tag{9.43}$$

where the satisfaction function $S^{rs}(k)$ is defined as follows:

$$S^{rs}(k)=E[\min_p\ \Omega_p^{rs}(k)]\qquad\forall r,s,p,k. \tag{9.44}$$

This satisfaction function has the following property:

$$\frac{\partial S^{rs}(k)}{\partial\eta_q^{rs}(k)}=P_p^{rs}(k)\qquad\forall r,s,p,k. \tag{9.45}$$

Note that this nonlinear programming program has the same set of constraints as the discrete VI. Now, we need to verify that the above nonlinear programming program is equivalent to the VI under the relaxation. Based on Theorem 3.3.3, we need to demonstrate that the gradient of the objective function is equivalent to cost function term of the VI, i.e.,

$$\nabla_f Z(u,v,x,E,f)=\bar{G}_p^{rs}(k) \tag{9.46}$$

In the following, we demonstrate that the above equation holds. Taking the partial derivative of Z with respect to route departure flow $f_l^{mn}(k)$, it follows that

$$\begin{aligned}\frac{\partial Z}{f_l^{mn}(k)}\quad &=\quad -f^{mn}(k)\ \frac{\partial S^{mn}(k)}{\bar{\eta}_l^{mn}(k)}\frac{\partial\bar{\eta}_p^{mn}[f_l^{mn}(k)]}{f_l^{mn}(k)}\\ &+\quad \bar{\eta}_l^{mn}[f_l^{mn}(k)]+f_l^{mn}(k)\ \frac{\partial\bar{\eta}_p^{mn}[f_l^{mn}(k)]}{f_l^{mn}(k)}\\ &-\quad \bar{\eta}_l^{mn}[f_l^{mn}(k)]\\ &=\quad \left\{f_l^{mn}(k)-f^{mn}(k)\ \bar{P}_l^{mn}(k)\right\}\ \frac{\partial\bar{\eta}_p^{mn}[f_l^{mn}(k)]}{f_l^{mn}(k)}\\ &=\quad \bar{G}_l^{mn}(k)\end{aligned} \tag{9.47}$$

Because the cost term $\bar{G}_p^{rs}(k)$ of the variational inequality (9.41) has a symmetric and semidefinite Jacobian matrix, an optimization program (NLP) can be formulated, which is equivalent to the above discrete VI (9.41). The resulting NLP is stated as follows:

$$\min \ Z(f) = \sum_{k=1}^{K}\sum_{rs}\left\{-f^{rs}(k)\, S^{rs}(k) + \sum_{p} f_p^{rs}(k)\, \bar{\eta}_p^{rs}[f_p^{rs}(k)] - \sum_{p}\int_0^{f_p^{rs}(k)} \bar{\eta}_p^{rs}[\omega]d\omega\right\} \tag{9.48}$$

s.t.

Relationship between state and control variables:

$$x_{ap}^{rs}(k+1) = x_{ap}^{rs}(k) + u_{ap}^{rs}(k) - v_{ap}^{rs}(k) \qquad \forall a,p,r,s; k=1,\cdots,K; \tag{9.49}$$

$$F_p^{rs}(k+1) = F_p^{rs}(k) + f_p^{rs}(k) \qquad \forall r,s,p; k=1,\cdots,K; \tag{9.50}$$

$$E_p^{rs}(k+1) = E_p^{rs}(k) + e_p^{rs}(k) \qquad \forall r,s,p; k=1,\cdots,K; \tag{9.51}$$

Flow conservation constraints:

$$f_p^{rs}(k) = \sum_{a\in A(r)} \delta_{ap}^{rs}\, u_{ap}^{rs}(k) \qquad \forall p,r,s; k=1,\cdots,K; \tag{9.52}$$

$$\sum_{a\in B(j)} v_{ap}^{rs}(k) - \sum_{a\in A(j)} u_{ap}^{rs}(k) = 0 \qquad \forall j,p,r,s; j\neq r,s; k=1,\cdots,K; \tag{9.53}$$

$$e_p^{rs}(k) = \sum_{a\in B(s)} \delta_{ap}^{rs}\, v_{ap}^{rs}(k) \qquad \forall p,r,s\neq r; k=1,\cdots,K; \tag{9.54}$$

Flow propagation constraints:

$$x_{ap}^{rs}(k) = \sum_{b\in\tilde{p}}\{x_{bp}^{rs}[k+\bar{\tau}_a(k)] - x_{bp}^{rs}(k)\} + \{E_p^{rs}[k+\bar{\tau}_a(k)] - E_p^{rs}(k)\} \tag{9.55}$$

$$\forall a\in B(j); j\neq r; p,r,s; k=1,\cdots,K+1;$$

Definitional constraints:

$$\sum_p u_{ap}^{rs}(k) = u_a^{rs}(k), \qquad \sum_{rs} u_a^{rs}(k) = u_a(k), \qquad \forall a; k=1,\cdots,K; \tag{9.56}$$

$$\sum_p v_{ap}^{rs}(k) = v_a^{rs}(k), \qquad \sum_{rs} v_a^{rs}(k) = v_a(k), \qquad \forall a; k=1,\cdots,K; \tag{9.57}$$

$$\sum_p x_{ap}^{rs}(k) = x_a^{rs}(k), \qquad \sum_{rs} x_a^{rs}(k) = x_a(k), \qquad \forall a; k=1,\cdots,K+1; \tag{9.58}$$

$$\sum_p e_p^{rs}(k) = e^{rs}(k), \quad \sum_p E_p^{rs}(k) = E^{rs}(k), \quad \forall r, s; k = 1, \cdots, K+1; \tag{9.59}$$

$$\sum_p f_p^{rs}(k) = f^{rs}(k), \quad \sum_p F_p^{rs}(k) = F^{rs}(k), \quad \forall r, s; k = 1, \cdots, K+1; \tag{9.60}$$

Nonnegativity conditions:

$$x_{ap}^{rs}(k+1) \geq 0, \quad u_{ap}^{rs}(k) \geq 0, \quad v_{ap}^{rs}(k) \geq 0, \quad \forall a, p, r, s; k = 1, \cdots, K; \tag{9.61}$$

$$e_p^{rs}(k) \geq 0, \quad f_p^{rs}(k) \geq 0, \quad E_p^{rs}(k+1) \geq 0, \quad F_p^{rs}(k+1) \geq 0, \quad \forall p, r, s; k = 1, \cdots, K; \tag{9.62}$$

Boundary conditions:

$$E_p^{rs}(1) = 0, \qquad F_p^{rs}(1) = 0, \qquad \forall p, r, s; \qquad x_{ap}^{rs}(1) = 0, \qquad \forall a, p, r, s. \tag{9.63}$$

9.3.3 Method of Successive Averages

The above NLP problem can be solved by using a solution procedure termed the Method of Successive Averages (MSA), as proposed by Sheffi and Powell (1982). The algorithm and convergence measures used in this section to determine the NLP solution follow directly from their work. Like the F-W method, the MSA is a descent method, determining a search direction and applying a prespecified step size during each iteration to find a new solution. In the F-W method, the solution to an LP determines the search direction, and the step size is the one which minimizes the objective function along the search direction. The MSA similarly determines a search direction by solving a subproblem, but the step size is predetermined.

For the stochastic dynamic user-optimal objective function (9.48), determining an optimal step size is difficult, if not impossible, due to the computational difficulty of evaluating the objective function or the lack of an analytic expression for the objective function. If the route choice probability distribution is multinomial logit, an analytic expression for this choice function may exist, and the objective function for the equivalent optimization problem and associated partial derivatives can be expressed in closed form. To evaluate this objective function and its partial derivatives, however, requires explicit enumeration of routes which is intractable for real transportation networks.

If the route choice probability distribution is multinomial probit, the choice function involves the Normal density function and does not have a closed form expression. The stochastic dynamic user-optimal objective function (9.48) in this case can only be evaluated approximately by means of Monte Carlo sampling. The predetermined step sizes are thus used in the absence of a closed form objective function.

Define subproblem variables as $p, q, y, \bar{E}, \bar{f}$ in the MSA subproblem, corresponding to main problem variables u, v, x, E, f. Note that route departure flow is the only set of independent variables. Once the route departure flow is determined, all link flow variables can be calculated by using the flow conservation equations and flow propagation equations. Applying a decent algorithm to the minimization of each discretized SDUO nonlinear program requires, at each iteration, the algorithmic step can be written as follows:

$$\mathbf{f}^{n+1}(k) = \mathbf{f}^{n}(k) + \alpha_n \mathbf{d}^{n}(k) \qquad \forall p; k = 1, \cdots, K; \tag{9.64}$$

where $\mathbf{f}^{n+1}(k)$ denotes the value of the decision variable vector $\{f_p^{rs}(k)\}$ at the nth inner iteration, α_n is a scalar representing the move size, and $\mathbf{d}^n(k)$ is a decent direction vector computed at $\mathbf{f}^n(k)$. The gradient vector of the objective function is given by equation (9.46). For each O-D pair rs, the departure flow in the MSA subproblem can be calculated as follows:

$$\bar{f}_{pn}^{rs}(k) \;=\; f^{rs}(k)\; P_{pn}^{rs}(k) \qquad \forall r, s.p; k = 1, \cdots, K. \tag{9.65}$$

The components of the decent direction can be written as

$$d_{pn}^{rs}(k) \;=\; \bar{f}_{pn}^{rs}(k) - f_{pn}^{rs}(k) \tag{9.66}$$

$$\;=\; f^{rs}(k)\; P_{pn}^{rs}(k) - f_{pn}^{rs}(k) \qquad \forall p; k = 1, \cdots, K. \tag{9.67}$$

where $P_{pn}^{rs}(k)$ is the probability of choosing route p between zones r and s given the current route travel times $\eta^{rs}(k)$ at iteration n. The MSA subproblem link flows can be determined by using flow conservation and propagation constraints once route departure flow $f_{pn}^{rs}(k)$ is determined.

The subproblem link flows can be determined in the case where $P_{pn}^{rs}(k)$ is a multinomial logit function by using a dynamic generalization of Dial's algorithm and implicitly performing the assignment described by equation (9.65). If the route choice function is multinomial probit, an unbiased estimate for the subproblem link flows can be determined by a Monte Carlo sampling for the $P_{pn}^{rs}(k)$. Given the stochastic dynamic network loading subproblem solution, the unbiased estimate of the random search direction vector, $(\bar{\mathbf{f}}^n - \mathbf{f}^n)$, may be combined with the predetermined step size to yield a new MSA main problem solution:

$$f_{p(n+1)}^{rs}(k) = f_{pn}^{rs}(k) + \frac{1}{n}[\bar{f}_{pn}^{rs}(k) - f_{pn}^{rs}(k)] \qquad \forall r, s, p; k = 1, \cdots, K. \tag{9.68}$$

The resulting MSA main problem link flow solutions are

$$u_a^{n+1}(k) = u_a^n(k) + \frac{1}{n}[p_p^n(k) - u_a^n(k)] \qquad \forall a; k = 1, \cdots, K; \tag{9.69}$$

$$v_a^{n+1}(k) = v_a^n(k) + \frac{1}{n}[q_p^n(k) - u_a^n(k)] \qquad \forall a; k = 1, \cdots, K; \tag{9.70}$$

$$x_a^{n+1}(k) = x_a^n(k) + \frac{1}{n}[y_p^n(k) - x_a^n(k)] \qquad \forall a; k = 2, \cdots, K+1; \tag{9.71}$$

The proof that this algorithm converges to the optimal link flow vector as $n \to \infty$ is based Blum (1954) and is examined in the context of stochastic route choice by Powell and Sheffi (1982). The step size in equations (9.68)-(9.71), is one possible step size determined from a more general expression:

$$\alpha^n = \frac{k_1}{k_2 + n} \tag{9.72}$$

where k_1 is any positive constant and k_2 is any nonnegative constant. Sheffi and Powell (1982) discuss the derivation of this expression. It is possible to use other step sizes by using other values of k_1 and k_2, thus changing the rate of convergence of the MSA algorithm. To choose the values which yield faster convergence, one must experiment with the performance of the algorithm with various values, a process known as tuning.

For a probit based stochastic dynamic user-optimal problem, the MSA involves an iterative procedure of performing a stochastic dynamic network loading by using a Monte Carlo sampling followed by a combination of the previous solution with the subproblem solution. The Monte Carlo procedure involves sampling actual route traversal times from (9.27), where the parameters may be given by specific assumptions. Given the actual route travel times, a dynamic shortest-route problem can be solved to find the minimal travel time route between each O-D pair during time interval k. Then, another sampling and dynamic shortest-route sequence is performed, noting which route was selected at each sequence. If L is the number of sampling and route selection iterations performed and L_p is the number of times route p was chosen as the minimal travel time route for flows departing during time interval k, then the probit choice probability of alternative p, $P_p(t)$, during time interval k is given by:

$$P_p(k) \approx \frac{L_p}{L}, \qquad \forall\, p, k \tag{9.73}$$

Equation (9.73) becomes an equality as $L \to \infty$. Note that in the above analysis, the same numbers of sampling and route selection iterations are assumed for each time interval k.

For the dynamic route choice problem, rather than storing the routes for each iteration, the dynamic shortest-route solution can be followed by an all-or-nothing loading of all O-D flows to the new dynamic minimal travel time route. The O-D flow is assigned to links on the dynamic minimal time route and saved in a set of temporary flow vectors, $\bar{y}_a(k)$, $\bar{p}_a(k)$, $\bar{q}_a(k)$, for each time interval k. This set of temporary flow vectors is then averaged with the current flow vector, $y_a^l(k)$, $p_a^l(k)$, $q_a^l(k)$, from the previous iteration:

$$y_a^l(k) = \frac{(l-1)y_a^{l-1}(k) + \bar{y}_a^l(k)}{n}, \qquad \forall\, p, k \tag{9.74}$$

$$p_a^l(k) = \frac{(l-1)p_a^{l-1}(k) + \bar{p}_a^l(k)}{n}, \qquad \forall\, p, k \tag{9.75}$$

$$q_a^l(k) = \frac{(l-1)q_a^{l-1}(k) + \bar{q}_a^l(k)}{n}, \qquad \forall\, p, k \tag{9.76}$$

As iteration counter $l \rightarrow \infty$, $y_a^l(k)$, $p_a^l(k)$, $q_a^l(k)$ converge to the link flow variables determined by the stochastic dynamic network loading of the O-D flows according to the specified multinomial probit route choice density function.

The route departure flow at iteration l is given by:

$$f_p^l(k) = f(k)P_p^l(k) \qquad \forall p, k \tag{9.77}$$

where $P_p^l(k)$ is the approximate probit choice probability function at iteration l and as $L \rightarrow \infty$,

$$f_p(k) = f(k)P_p(k) \qquad \forall p, k \tag{9.78}$$

where $P_p(k)$ is the probit route choice probability function given the parameters determined by the specification of the route travel time distributions. Note that in equations (9.77)-(9.78), O-D pair rs is suppressed for clarity of showing iteration superscript l.

Each MSA iteration involves Monte Carlo sampling to determine the link flow variables resulting from apportioning the O-D flows to the network based on the route choice probability distribution. In the case of a multinomial probit route choice model, link flow variables are determined by sampling link travel times from the Normal distribution in equation (9.27). Dynamic minimum travel time routes are determined followed by an all-or-nothing loading to these routes. This procedure is repeated some predetermined number of times, and the link flow vectors averaged, resulting in the multinomial probit stochastic assignment described by equations (9.74)- (9.76).

A detailed analysis of the effect of the number of Monte Carlo iterations on the convergence of the algorithm is necessary; however, some experience drawn from solving static SUO problems could be useful. For example, one to ten Monte Carlo sampling iterations could be enough to estimate the multinomial probit stochastic network loading. Furthermore, the effect of route choice probability distribution specification and implementation on the MSA algorithm and on the relaxation algorithm require much more extensive analysis.

While variation as a convergence indicator does not provide information on how close the solution is to the optimum, it does provide a measure of stability in link flow variables. In order to perform a comparative analysis for several values of model parameters, it is necessary to solve the models to consistently accurate solutions.

Convergence measures and stopping conditions for the Method of Successive Averages present a problem. Lower bound measures cannot be calculated in the manner of F-W lower bounds since typically neither objective function nor gradient values can be computed. For this reason, Sheffi and Powell (1982) devised a convergence indicator based on the change in link flow vector components over the three to five most recent iterations. The idea is that the variation in link flows in recent iterations should dissipate as the link flow vectors converge to the optimal solution.

Their measure is referred to here as the *variation*. A similar measure is used for the SDUO problem. This measure is computed as the ratio of the sum of the standard deviations of individual link flow vector components over the last M iterations to the sum of the means of the individual link flow vector components over the last M iterations. The variation decreases as the link flow vector component values show decreasing change over successive iterations. We use the number of vehicles $x_a^i(k)$ as the major indicator for this convergence check, where the index i ranges from 1 to I, the number of recent iterations to be considered. At iteration n of the MSA algorithm, the variation V^n is:

$$V^n = \frac{\sum_a \sum_k \left\{ \frac{1}{I} \left[\sum_{i=n-I+1}^{n} [x_a^i(k)]^2 - I[\bar{x}_a^n(k)]^2 \right] \right\}^{\frac{1}{2}}}{\sum_a \sum_k \bar{x}_a^n(k)} \qquad n \geq I \qquad (9.79)$$

where

$$\bar{x}_a^n(k) = \frac{1}{I} \sum_{i=n-I+1}^{n} x_a^i(k) \qquad \forall a, k \qquad (9.80)$$

9.3.4 Summary of the Solution Algorithm

In this combined algorithm, we define the travel time approximation procedure (relaxation) as the outer iteration and the MSA procedure as the inner iteration. The flowchart of our solution algorithm for solving our ideal SDUO route choice model is shown in Figure 9.1 and the algorithm is summarized as follows.

Step 0: Initialization.
Find an initial feasible solution $\{x_a^1(k)\}$, $\{u_a^1(k)\}$, $\{v_a^1(k)\}$, $\{E^1(k)\}$. Set the outer iteration counter m = 1.

Step 1: Relaxation. (Outer Iteration)
Find a new estimate of actual link travel time $\tau_a^m(k)$ for the flow propagation constraints and solve the ideal SDUO route choice program (NLP) by *MSA method* as follows. Set the inner iteration counter n = 1.

[Step 1.1]: *Update.* Calculate the link travel time functions using the temporarily fixed link flow variables.

[Step 1.2]: *Direction Finding.* Implement a stochastic dynamic network loading procedure based on the temporarily fixed actual link travel time and find an auxiliary set of link flow variables.

[Step 1.3]: *Move.* Find a new solution by combining current solution $\{u_a(k)\}$, $\{v_a(k)\}$, $\{x_a(k)\}$, $\{E^{rs}(k)\}$ and previous solution $\{p_a(k)\}$, $\{q_a(k)\}$, $\{y_a(k)\}$, $\{\bar{E}^{rs}(k)\}$ as equations (9.69)-(9.71).

[Step 1.4]: *Convergence Test for Inner Iterations (MSA).* If n equals a prespecified number or equation (9.79) is satisfied, go to step 2; otherwise, set $n = n + 1$ and go to step 1.1.

Step 2: Convergence Test for Outer Iterations.
If $\tau_a^m(k) \simeq \tau_a^{m+1}(k)$, stop. The current solution, $\{u_a(k)\}$, $\{v_a(k)\}$, $\{x_a(k)\}$, $\{E^{rs}(k)\}$, is in a near ideal SDUO state; otherwise, set $m = m + 1$ and go to step 1.

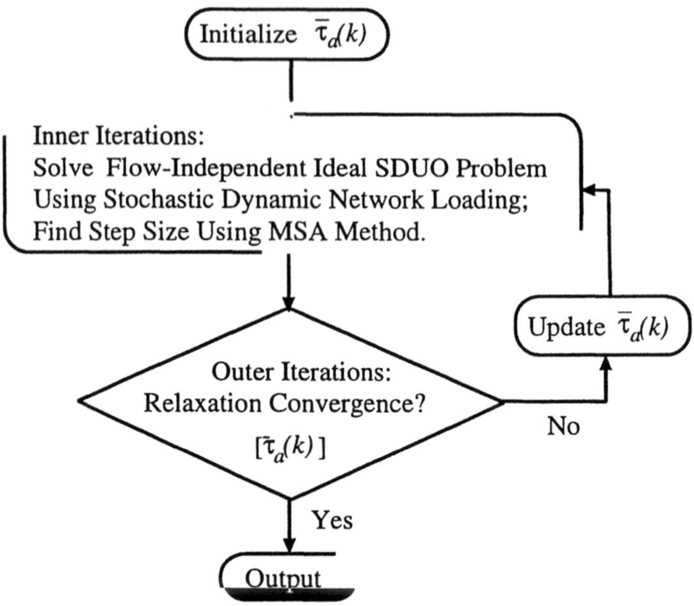

Figure 9.1: Flowchart of the Ideal Solution Algorithm

The number of inner iterations n and the number of outer iterations m are co-related. If we set m large, then n should be set small and vice versa. Often, only a few inner iterations are needed for convergence.

9.3.5 A Logit-Based Ideal Stochastic Loading

In the above algorithm, a stochastic dynamic network loading procedure is needed to find the MSA subproblem solution. For a multinomial probit, a Monte Carlo sampling is needed to estimate the route choice probability. For a multinominal logit, a dynamic generalization of Dial's algorithm can be designed to find the accurate route choice probability by implicitly enumerate routes. This problem is in fact a *flow-independent* logit-based ideal SDUO route choice problem. The flow-independent ideal SDUO route choice problem is to assign vehicle flows between each O-D pair based on a logit route choice probability calculated using the given actual route travel times for each time interval.

The solution algorithm given here is similar to Dial's algorithm for static flow-independent stochastic network assignment. This algorithm effectively implements a logit-type ideal dynamic route choice model at the network level. As in the instantaneous model, it assigns probabilities and flows to efficient routes connecting each O-D pair for each time interval. This algorithm also includes a preliminary phase which identifies the set of efficient routes connecting each O-

D pair. The O-D departure flows are then assigned only to these routes during each interval k, using the logit formula based on actual route travel times. In this algorithm, all probabilities $\text{Prob}_{(i,j)|i}(t)$ for assigning inflows to link (i, j) are evaluated during time interval t instead of time interval k. Here, interval t is the arrival time interval at node I.

Statement of the Algorithm

The steps of this algorithm for one O-D pair rs are outlined below, and are repeated for each O-D pair in the network. The following procedure is called the DYNASTOCH1 algorithm to represent its dynamic version for the flow-independent ideal SDUO route choice problem.

Step 0: Initialization.

Compute the minimal actual travel time $\pi^{js}(t)$ from node j to destination s for vehicles departing node j during interval t. Calculate the likelihood for each link (i, j) during each interval t:

$$L_{(i,j)}(t) = \begin{cases} \exp\{\theta\,[\pi^{is}(t) - \pi^{js}[t + \tau_a(t)] - \tau_a(t)]\}, & \text{if } \sigma_o^{is} > \sigma_o^{js} \\ 0, & \text{otherwise} \end{cases}$$

In this expression, $\tau_a(t)$ is the mean actual travel time on link $a = (i, j)$ during time interval t.

Step 1: Backward Pass.

By examining all nodes j in ascending sequence with respect to $\pi^{js}(t)$, calculate the weight for each link $a = (i, j)$ during each time interval t:

$$W_{(i,j)}(t) = \begin{cases} L_{(i,j)}(t), & \text{if } j = s \text{ (destination)} \\ L_{(i,j)}(t) \displaystyle\sum_{(j,m)\in A(j)} W_{(j,m)}[t + \tau_a(t)], & \text{otherwise} \end{cases}$$

When the origin r is reached, this step is completed.

Step 2: Forward Pass.

Consider all nodes i in descending sequence with respect to $\pi^{is}(t)$, starting with origin r. When each node i is considered during each time interval t, compute the inflow of link (i, j) during each interval t using the following formula:

$$u_{(i,j)}(t) =$$

$$\begin{cases} f^{rs}(t) \quad \dfrac{W_{(i,j)}(t)}{\sum_{(i,l)\in A(i)} W_{(i,l)}(t)}, & \text{if } i = r \text{ (origin)} \\ \left\{\displaystyle\sum_{(n,i)\in B(i)} v_{(n,i)}(t)\right\} \quad \dfrac{W_{(i,j)}(t)}{\sum_{(i,l)\in A(i)} W_{(i,l)}(t)}, & \text{otherwise} \end{cases}$$

This step is implemented iteratively until destination s is reached. Note that the sum in the denominator includes all links emanating from the upstream node of the link under consideration. The sum of the exit flow variables is taken over all links arriving at the upstream node of the link under consideration.

The flow generated by this algorithm is equivalent to a logit-based flow-independent route assignment between each O-D pair, given that only reasonable routes are considered. The proof for this algorithm is similar to that for the instantaneous SDUO route choice model.

9.3.6 Proof of the Algorithm

We now prove that the algorithm does generate logit-based flow-independent ideal SDUO route choices between each O-D pair. We note that each link likelihood $L_{(i,j)}(t)$, is proportional to the logit probability that link $a = (i, j)$ is used during time interval t by a traveler chosen at random from among the population of trip-makers between r and s, given that the traveler is at node i during interval t. The probability that a given route will be used is proportional to the product of all the likelihoods of the links comprising this route. Suppose route l consists of nodes $(r, 1, 2, \cdots, i-1, i, \cdots, n, s)$ and links $(1, 2, \cdots, h)$. The probability of using route l between r and s, $P_l^{rs}(k)$, for vehicles departing r during interval k and arriving at node i during interval t, is then

$$P_l^{rs}(k) = G(k) \prod_{a \in l} \{L_{(i,j)}(t)\}^{\delta_{al}^{rs}} \tag{9.81}$$

where $G(k)$ is a proportionality constant for each time interval k and the product is taken over all links in the network. Here, $t = k + \eta_l^{ri}(k)$. The incidence variable δ_{al}^{rs} ensures that $P_l^{rs}(k)$ includes only those links in route l between r and s. Substituting the expression for the likelihood $L_{(i,j)}(t)$ in the above equation, the choice probability of choosing a particular efficient route becomes

$$\begin{aligned} P_l^{rs}(k) &= G(k) \prod_{a \in l} \exp\left\{\theta \left\{\pi^{is}(t) - \pi^{js}[t + \tau_a(t)] - \tau_a(t)\right\} \delta_{al}^{rs}\right\} \\ &= G(k) \exp\left\{\theta \sum_{a \in l} \{\pi^{is}(t) - \pi^{js}[t + \tau_a(t)] - \tau_a(t)\} \delta_{al}^{rs}\right\} \\ &= G(k) \exp\left\{\theta \left[\pi^{rs}(k) - \eta_l^{rs}(k)\right]\right\} \end{aligned} \tag{9.82}$$

The last equality results from the following summations:

$$\begin{aligned} \sum_{a \in l} \{\pi^{is}(t) - \pi^{js}[t + \tau_a(t)]\} &= \pi^{rs}(k) - \pi^{1s}[k + \eta_l^{r1}(k)] \\ &+ \pi^{1s}[k + \eta_l^{r1}(k)] - \pi^{2s}[k + \eta_l^{r2}(k)] \\ &+ \vdots \end{aligned}$$

$$\begin{aligned} &+ \quad \pi^{ns}[k+\eta_l^{rn}(k)] - \pi^{ss}[k+\eta_l^{rs}(k)] \\ &= \quad \pi^{rs}(k) - \pi^{ss}[k+\eta_l^{rs}(k)] \\ &= \quad \pi^{rs}(k) \end{aligned} \tag{9.83}$$

and

$$\sum_{a \in l} \tau_a(t) = \tau_1(k) + \tau_2[k+\eta_l^{r1}(k)] + \cdots + \tau_h[k+\eta_l^{rn}(k)] = \eta_l^{rs}(k) \tag{9.84}$$

Since

$$\sum_q P_q^{rs}(k) = 1.0, \tag{9.85}$$

the proportionality constant must equal

$$G(k) = \frac{1}{\sum_q \exp\left\{\theta\left[\pi^{rs}(k) - \eta_q^{rs}(k)\right]\right\}} \tag{9.86}$$

Thus,

$$\begin{aligned} P_l^{rs}(k) &= \frac{\exp\left\{\theta\left[\pi^{rs}(k) - \eta_l^{rs}(k)\right]\right\}}{\sum_q \exp\left\{\theta\left[\pi^{rs}(k) - \eta_q^{rs}(k)\right]\right\}} \\ &= \frac{\exp\left\{-\theta\ \eta_l^{rs}(k)\right]\}}{\sum_q \exp\left\{-\theta\ \eta_q^{rs}(k)\right\}} \end{aligned} \tag{9.87}$$

Equation (9.87) depicts a logit model of route choice during each interval k among the efficient routes connecting O-D pair rs. This algorithm does generate a route choice probability using actual route travel times.

The DYNASTOCH1 algorithm does not require route enumeration. It does require the calculation of time-dependent minimal actual travel time routes for every O-D pair in the network during each time interval k. In the following, we show that the forward pass of the algorithm generates a time-dependent logit-based route flow assignment. To do so, we only need to show that the calculated link inflows are obtained in a manner consistent with the expression in the forward pass. This is done by proving the algorithm diverts trips from each node i during each time interval t (arrival time interval at node I) according to appropriate conditional link probabilities. A conditional link probability during interval t is the probability that a trip between r and s uses a particular link $a = (i,j)$ during interval t, given it goes through the link's tail node i during interval t. This probability is stated as

$$\text{Prob}_{(i,j)|i}(t) = \frac{\text{Prob}_{(i,j),i}(t)}{\text{Prob}_i(t)} = \frac{\text{Prob}_{(i,j)}(t)}{\text{Prob}_i(t)} = \frac{\text{Prob}_{(i,j)}(t)}{\sum_j \text{Prob}_{(i,j)}(t)} \tag{9.88}$$

The probability of using link $a = (i, j)$ during interval t is the summation of probabilities of departure trips using route P at origin r during interval k, where link $a = (i, j)$ is on route p, given that the traveler reaches node i during interval $t = k + \eta_p^{ri}(k)$. It follows that

$$\text{Prob}_{(i,j)}(t) = \sum_{p:(i,j)} \text{Prob}_p(t) \tag{9.89}$$

where $p : (i, j)$ means those routes p which include link (i, j). It is useful to write (9.89) in a more elaborate form, so as to facilitate cancellation of common factors in the numerator and denominator of equation (9.88). To do so, an efficient route through link (i,j) can be partitioned into three sets of links:

1. $P_i = \{\text{all links topologically preceding link } (i, j)\}$;
2. link $(i, j) = \{(i, j)\}$;
3. $P_j = \{\text{all links topologically following link } (i, j)\}$.

Denote $\mathcal{P}_i$ as the set of P_i and $\mathcal{P}_j$ as the set of P_j. Then

$$\sum_{p:(i,j)} \text{Prob}_p(t) = h(t)\, L_{(i,j)}(t) \{\sum_{p\in\mathcal{P}_i} \prod_{(m,n)} L_{(m,n)}(t')\}\{\sum_{p\in\mathcal{P}_j} \prod_{(m,n)} L_{(m,n)}(t'')\} \tag{9.90}$$

Note that $t' + \eta_P^{mi}(t') = t$ where $t' < t$, and $t + \eta_P^{im}(t) = t''$ where $t'' > t$. Equation (9.90) follows from the fact that all efficient routes can be constructed by independently choosing a member from each of $\mathcal{P}_i$ and $\mathcal{P}_j$ and putting link (i, j) in between, given that flows arrive at node i during interval t. Such combinations constitute efficient routes. Substituting (9.89) and (9.90) into (9.88), it follows that

$$\text{Prob}_{(i,j)|i}(t)$$

$$= \frac{h(t)\, L_{(i,j)}(t) \left\{\sum_{p\in\mathcal{P}_i} \prod_{(m,n)} L_{(m,n)}(t')\right\} \left\{\sum_{p\in\mathcal{P}_j} \prod_{(m,n)} L_{(m,n)}(t'')\right\}}{\sum_l h(t)\, L_{(i,l)}(t) \left\{\sum_{p\in\mathcal{P}_i} \prod_{(m,n)} L_{(m,n)}(t')\right\} \left\{\sum_{p\in\mathcal{P}_j} \prod_{(m,n)} L_{(m,n)}(t'')\right\}}$$

$$= \frac{h(t)\, L_{(i,j)}(t) \left\{\sum_{p\in\mathcal{P}_i} \prod_{(m,n)} L_{(m,n)}(t')\right\} \left\{\sum_{p\in\mathcal{P}_j} \prod_{(m,n)} L_{(m,n)}(t'')\right\}}{h(t) \left\{\sum_{p\in\mathcal{P}_i} \prod_{(m,n)} L_{(m,n)}(t')\right\} \sum_l \left\{L_{(i,l)}(t) \left\{\sum_{p\in\mathcal{P}_j} \prod_{(m,n)} L_{(m,n)}(t'')\right\}\right\}}$$

$$= \frac{L_{(i,j)}(t)\left\{\sum\limits_{p\in\mathcal{P}_j}\prod\limits_{(m,n)} L_{(m,n)}(t'')\right\}}{\sum\limits_{l}\left\{L_{(i,l)}(t)\left\{\sum\limits_{p\in\mathcal{P}_j}\prod\limits_{(m,n)} L_{(m,n)}(t'')\right\}\right\}} = \frac{W_{(i,j)}(t)}{\sum\limits_{(i,l)} W_{(i,l)}(t)} \tag{9.91}$$

where the numerator of (9.91) is the definition of $W_{(i,j)}(t)$. By the link weights calculated in the Backward Pass (Step 1), the quotient in Forward Pass (Step 2) is equal to the right-hand-side of equation (9.91). Thus, we complete the proof that the diversion inflows are indeed those implied by the probabilities defined in equation (9.81). We also note the flow over each link should satisfy the flow propagation constraints.

9.4 Numerical Example

We illustrate the solutions of ideal SDUO route choice model with the 4-link, 4-node test network shown in Figure 9.2. Time period $[0, T]$ is subdivided into $K = 8$ small time intervals. The algorithms were coded in FORTRAN and solved on a HP workstation.

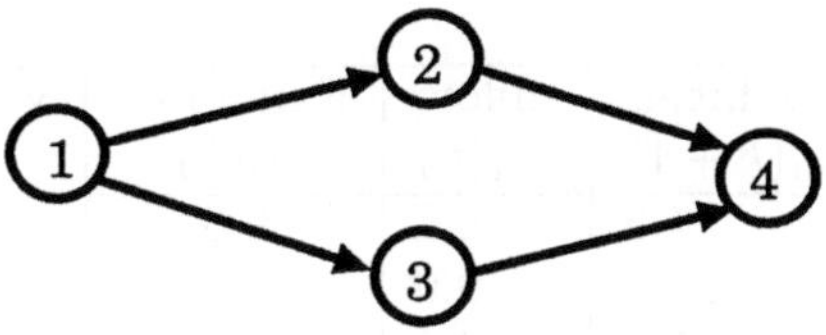

Figure 9.2: Test Network

The following link travel time functions were used in the computations:

$$\tau_a(k) = g_{1a}(k) + g_{2a}(k)$$

$$g_{1a}(k) = \beta_{1a} + \beta_{2a}[u_a(k)]^2 + \beta_{3a}[x_a(k)]^2$$
$$g_{2a}(k) = \beta_{4a} + \beta_{5a}[v_a(k)]^2 + \beta_{6a}[x_a(k)]^2$$

where time interval $k = 1, 2, \cdots, 8$. The trip table is given in Table 9.1. The parameters for each link travel time function are given in 9.2. The route choice dispersion parameter θ is given as 1. The optimal link flows and the corresponding optimal link travel times are given in 9.3. The optimal route travel times are given in 9.4. In this discrete time example, $x_a(k)$ represents vehicles on the link at the beginning of interval k; $u_a(k)$ and $v_a(k)$ represent inflow and exit flow during interval k.

In this example, note from Table 9.4 that the mean actual travel times on routes 1–2–4 and 1–3–4 are not equal during each interval. These travel times are a property of the logit-type route flows in each interval.

Table 9.1: Required Flows from Origin 1 to Destination 4

Time Interval k	1	2	3	4	5	6	7	8
Flow/interval	40.	0.	0.	0.	0.	0.	0.	0.

Table 9.2: Parameters of Link Travel Time Functions for Ideal Model

link a	β_{1a}	β_{2a}	β_{3a}	β_{4a}	β_{5a}	β_{6a}
1—2	1.	0.001	0.	0.	0.015	0.002
1—3	2.	0.001	0.	0.	0.015	0.002
2—4	1.	0.001	0.	0.	0.015	0.002
3—4	1.	0.001	0.	0.	0.015	0.002

Table 9.3: Optimal Link Flows and Travel Times for Ideal Model

Interval k	Link a	Vehicles $x_a(k+1)$	Inflow $u_a(k)$	Exit Flow $v_a(k)$	Vehicles $x_a(k)$	Travel Time $c_a(k)\&\tau_a(k)$
1	1 — 2	24.5	24.5	0.0	0.0	1.60
2	1 — 2	21.5	0.0	3.0	24.5	2.33
3	1 — 2	5.5	0.0	16.0	21.5	5.77
4	1 — 2	0.0	0.0	5.5	5.5	1.52
1	1 — 3	15.5	15.5	0.0	0.0	2.24
2	1 — 3	15.5	0.0	0.0	15.5	2.48
3	1 — 3	0.0	0.0	15.5	15.5	6.08
4	1 — 3	0.0	0.0	0.0	0.0	2.00
1	2 — 4	0.0	0.0	0.0	0.0	1.00
2	2 — 4	3.0	3.0	0.0	0.0	1.01
3	2 — 4	16.0	16.0	3.0	3.0	1.41
4	2 — 4	18.6	5.5	2.9	16.0	1.67
5	2 — 4	5.5	0.0	13.1	18.6	4.26
6	2 — 4	5.5	0.0	0.0	5.5	1.06
7	2 — 4	0.0	0.0	5.5	5.5	1.52
1	3 — 4	0.0	0.0	0.0	0.0	1.00
2	3 — 4	0.0	0.0	0.0	0.0	1.00
3	3 — 4	15.5	15.5	0.0	0.0	1.24
4	3 — 4	12.6	0.0	2.9	15.5	1.60
5	3 — 4	0.0	0.0	12.6	12.6	3.70

Table 9.4: Ideal Route Travel Times

Interval	Mean Actual Route Travel Times	
k	Route 1-2-4	Route 1-3-4
1	3.01	3.48
2	6.59	4.08

9.5 Notes

Stochastic route choice models and solution algorithms have been studied extensively under the assumption of static traffic conditions. Dial (1971) presented the STOCH method to perform a logit-based, flow-independent stochastic traffic assignment. Daganzo and Sheffi (1977) presented a probit-based stochastic user-optimal route choice model. Subsequently, a Monte Carlo sampling approach to solving this problem were presented by Sheffi and Powell (1982). Fisk (1980) proposed a stochastic user-optimal (SUO) route choice model based on logit-type route flow method. The Method of Successive Averages (MSA) was suggested to solve this model (Sheffi, 1985). Recently, two improved algorithms were proposed by Chen and Alfa (1991) for solving the logit-type SUO route choice model. Their algorithms use information in the objective function to speed up the convergence of the one-dimensional search.

In this chapter, we have formulated an ideal stochastic dynamic user-optimal route choice problem as a variational inequality. The solution of the VI model results in dispersed dynamic route choice governed by a probit distribution incorporating mean actual route travel times. In stochastic dynamic route choice model, the O-D departure flows are dispersed across different travel time routes so that stochastic dynamic models represent more realistic travel choice behavior than deterministic models. This is one of the major merits of stochastic dynamic route choice models over deterministic dynamic route choice models.

Other SDUO route choice models should be investigated based on different types of distributions for travel time perception errors. However, more realistic distributions of travel time perception errors than the normal distribution should be considered, such as the Gamma distribution. This issue provides a challenge to both mathematical modeling and estimation of parameters of the distribution of perception errors from data. Nevertheless, realistic distributions of travel time perception errors should depend on real data collected from ITS operational tests. Our models only provide a theoretical approach toward a better understanding of stochastic dynamic travel choices and travel time predictions. Our future research also includes calibration of the route choice dispersion parameters and development of more general models incorporating stochastic mode and departure time choice. Using realistic link travel time functions in Chapter 14, we plan to implement our models on larger networks.

Problems

9.1 Discuss the major stochastic factors for dynamic route choice. Provide a rating for these factors based on their importance and modeling feasibility.

9.2 *Flow dispersion* is usually referred to the phenomenon that O-D flows are distributed among different travel time routes. Explain the similarities and differences of *flow dispersion* and *stochastic route choice.*

9.3 In reality, some travelers do not switch routes regardless of whether real-time traffic information is provided or not. If a stochastic dynamic route choice model is used to depict their route choice, what kind of route choice probability distribution functions should be used?

9.4 Discuss the relationship of ideal SDUO route choice conditions and ideal DUO route choice conditions. Explain how ideal SDUO route choice conditions reduce to ideal DUO route choice conditions under certain conditions.

9.5 The time-dependent route departure flow can be assigned by using the route choice probability as follows:

$$f_p^{rs}(t) - f^{rs}(t)P_p^{rs}(t) = 0 \qquad \forall r, s, p.$$

In addition to logit and probit distributions, what could be a possible distribution function for the route choice probability.

9.6 A two-link network is shown in Figure 9.3. Assume the perceived route travel time functions are

$$\Omega_1(t) = 5 + \xi_1(t)$$
$$\Omega_2(t) = 10 + \xi_2(t)$$

where $\xi_p^{rs}(t)$ is a random error term associated with the route p under consideration. Both error terms follow uniform distributions between [-2,2], [-5,5], respectively. Determine the route choice probability $P_1(t)$.

9.7 For the network shown in Figure 9.2, the mean link travel time functions are

$$\tau_1(t) = 2u_1(t) + 5, \qquad \tau_2(t) = 10$$
$$\tau_3(t) = 15, \qquad \tau_4(t) = u_4(t) + 10.$$

Assume that the following logit-based stochastic route choice function is used to assign traffic from node 1 to node 4:

$$P_1(t) = \frac{\exp[-\eta_1^{14}(t)]}{\exp[-\eta_1^{14}(t)] + \exp[-\eta_2^{14}(t)]}$$

The O-D flow is $f^{14}(t) = 10t$ units. Determine mean inflows on links 1 and 2 for time period $[0, 10]$ and find the optimal mean route travel times. [Hint: assume no flow dispersion in the link flow propagation.]

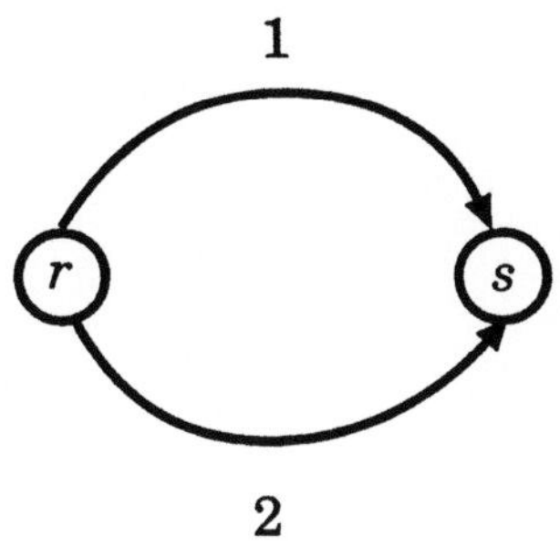

Figure 9.3: Network of Problem 9.6

Chapter 10

Instantaneous Stochastic Dynamic Route Choice Models

This chapter presents an instantaneous stochastic dynamic user-optimal route choice model which is a stochastic generalization of our deterministic DUO route choice model presented in Chapter 8. In this chapter, we include only a random component in travelers' perception of travel times. Then route choice is a process of selection among alternative routes, for which the perceived time-dependent travel times include a random error. The general probit-type variational inequality model is first formulated by assuming a general distribution of route choice dispersion. The stochastic route choice assumption is introduced into dynamic route choice problems so that our SDUO models are dynamic generalizations of conventional stochastic static user-optimal (SUO) models under the assumption of dispersed route choice.

In Section 10.1, the instantaneous SDUO model is presented. The formulation of the instantaneous SDUO route choice problem is based on the underlying choice criterion that each traveler uses the route that minimizes his/her perceived instantaneous travel time when departing from the origin or any intermediate node to his/her destination. The solution of this instantaneous SDUO route choice model results in instantaneous stochastic network flows at each decision node based on a general distribution function of mean instantaneous travel times of alternative routes. It is also shown that our instantaneous DUO route choice model in Chapter 8 is a particular case of the instantaneous SDUO route choice model when the variance of the instantaneous route travel time perception errors is zero.

To solve the instantaneous SDUO model for large networks, we need solution algorithms that avoid route enumeration, as proposed in Section 10.2. The continuous variational inequality is converted into a discrete variational inequality. The relaxation method is used to solve this discrete VI. A nonlinear programming problem (NLP) or a discrete optimal control program is formulated under relaxation so that the Method of Successive Averages can be used to solve this NLP. For a general multinominal probit, Monte Carlo sampling is proposed to compute the dynamic route choice probabilities. For the multinomial logit case, a multiple-route dynamic route choice algorithm (DYNASTOCH2) similar to Dial's efficient-route algorithm (STOCH) is suggested in Section 10.3 to solve the LP subproblem, which is a *flow-independent* instantaneous SDUO route choice problem. In the DYNASTOCH2 algorithm, explicit route enumeration can be avoided and the dynamic route choice probabilities can be computed.

In Section 10.4, an optimal control formulation is presented for a logit-based instantaneous SDUO route choice problem. The optimality conditions are used to demonstrate that the optimal control model is equivalent to the instantaneous SDUO route choice conditions. Finally, a numerical example is presented in Section 10.5 to illustrate the solution of the proposed models with the solution algorithm.

10.1 Formulation of the Model

The *instantaneous* travel time at time t is defined as the travel time that is experienced by vehicles traversing link a when prevailing traffic conditions remain unchanged. Let $C_a(t)$ denote the instantaneous travel time on link a at time t as perceived by a traveler randomly chosen from the population of travelers. $C_a(t)$ is a random variable that is assumed to have mean $c_a(t)$. The *mean* or *measured* (as opposed to *perceived*) *instantaneous* travel time $c_a(t)$ over link a at time t is assumed to be dependent on the number of vehicles $x_a(t)$, the inflow $u_a(t)$ and the exit flow $v_a(t)$ on link a at time t. As described in Chapter 13, an actual link travel time function has a similar form to an instantaneous link travel time function. Refer to Section 13.1 for a detailed discussion of the difference between instantaneous and actual link travel times.

Consider the flow which originates at node r at time t and is destined for node s. There is a set of routes $\{p\}$ between O-D pair rs. Denote the minimal free flow travel time from node i to destination node s as σ_o^{is}. In this chapter, an *efficient route* between rs is redefined to include only links $a = (i, j)$ such that node j is closer to destination s than node i, i.e., $\sigma_o^{is} > \sigma_o^{js}$ (Dial, 1971). In the following, all route related constraints are defined using efficient routes. By using efficient routes, the cyclic flow problem in some dynamic assignment models can be prevented. The mean or measured instantaneous travel time $\psi_p^{rs}(t)$ for each route p between rs is defined as

$$\psi_p^{rs}(t) = \sum_{a \in rsp} c_a[x_a(t), u_a(t), v_a(t)] \qquad \forall r, s, p. \tag{10.1}$$

The summation is over all links a in route p from origin r to destination s. Thus, the mean instantaneous route travel time is that experienced by vehicles if prevailing traffic conditions do not vary until vehicles reach their destination.

This instantaneous route travel time provides a first approximation to the time-dependent vehicle travel time. Finally, the minimal mean instantaneous route travel time $\sigma^{rs}(k) = \min \{\psi_p^{rs}(k)|p\}$.

10.1.1 Network Constraints

In this section, the constraint set for our stochastic dynamic route choice model is summarized as follows:

Relationship between state and control variables:

$$\frac{dx_{ap}^{rs}}{dt} = u_{ap}^{rs}(t) - v_{ap}^{rs}(t) \qquad \forall a, p, r, s; \tag{10.2}$$

$$\frac{dE_p^{rs}(t)}{dt} = e_p^{rs}(t) \qquad \forall p, r; s \neq r; \tag{10.3}$$

Flow conservation constraints:

$$f^{rs}(t) = \sum_{a \in A(r)} \sum_p u_{ap}^{rs}(t) \qquad \forall r, s; \tag{10.4}$$

$$\sum_{a \in B(j)} v_{ap}^{rs}(t) = \sum_{a \in A(j)} u_{ap}^{rs}(t) \qquad \forall j, p, r, s; j \neq r, s; \tag{10.5}$$

$$\sum_{a \in B(s)} \sum_p v_{ap}^{rs}(t) = e^{rs}(t) \qquad \forall r, s; s \neq r; \tag{10.6}$$

Flow propagation constraints:

$$x_{ap}^{rs}(t) = \sum_{b \in \tilde{p}} \{x_{bp}^{rs}[t + \tau_a(t)] - x_{bp}^{rs}(t)\} + \{E_p^{rs}[t + \tau_a(t)] - E_p^{rs}(t)\}$$

$$\forall a \in B(j); j \neq r; p, r, s; \tag{10.7}$$

Definitional constraints:

$$\sum_{rsp} u_{ap}^{rs}(t) = u_a(t), \qquad \sum_{rsp} v_{ap}^{rs}(t) = v_a(t), \qquad \forall a; \tag{10.8}$$

$$\sum_{rsp} x_{ap}^{rs}(t) = x_a(t), \qquad \forall a; \tag{10.9}$$

Nonnegativity conditions:

$$x_{ap}^{rs}(t) \geq 0, \quad u_{ap}^{rs}(t) \geq 0, \quad v_{ap}^{rs}(t) \geq 0 \quad \forall a, p, r, s; \tag{10.10}$$

$$e_p^{rs}(t) \geq 0, \quad E_p^{rs}(t) \geq 0, \quad \forall p, r, s; \tag{10.11}$$

Boundary conditions:

$$E_p^{rs}(0) = 0, \qquad \forall p, r, s; \tag{10.12}$$

$$x_{ap}^{rs}(0) = 0, \qquad \forall a, p, r, s. \tag{10.13}$$

In this problem the control variables are $u_{ap}^{rs}(t)$, $v_{ap}^{rs}(t)$ and $e_p^{rs}(t)$; the state variables are $x_{ap}^{rs}(t)$ and $E_p^{rs}(t)$; and the functionals are $\tau_a(t)$. The stochastic dynamic route choice model has similar constraints as the deterministic instantaneous DUO route choice model in Chapter 8. Before considering the VI formulation, the definition of instantaneous SDUO is presented.

10.1.2 Definition of an Instantaneous SDUO State

The formulation of the instantaneous SDUO route choice problem is based on the underlying choice criterion that each traveler uses the route that minimizes his/her perceived instantaneous travel time when departing from the origin or any intermediate node to his/her destination. In the following, we extend a similar analysis for stochastic static route choice given by Sheffi (1985) to the case of instantaneous stochastic dynamic route choice.

For each origin-destination pair, there are many alternative routes p, each with some instantaneous travel time for each time instant. Due to variations in perceptions and exogenous factors, these instantaneous route travel times are perceived differently by each traveler. We represent the perceived instantaneous route travel time as a random variable for each time instant t. Given his or her perception of instantaneous route travel time at each time instant t, each traveler is assumed to choose the route with minimal perceived instantaneous travel time. Given a probability density function for the instantaneous route travel time at each time t, the instantaneous stochastic dynamic user-optimal route choice problem is to determine how many travelers will use each route at each decision node at each instant of time.

Let $\Psi_p^{rs}(t)$ denote the perceived instantaneous route travel time on route p between origin r and destination s at each time t. Then, we assume that

$$\Psi_p^{rs}(t) = \psi_p^{rs}(t) + \xi_p^{rs}(t) \qquad \forall r, s, p, \tag{10.14}$$

where $\xi_p^{rs}(t)$ is a random error term associated with the route p under consideration. Furthermore, assume that the mean of the error $E[\xi_p^{rs}(t)] = 0$, or $E[\Psi_p^{rs}(t)] = \psi_p^{rs}(t)$. In other words, the mean perceived instantaneous route travel time is the mean instantaneous route travel time.

The share of travelers choosing route p at each time t, $P_p^{rs}(t)$, is

$$P_p^{rs}(t) = Pr[\Psi_p^{rs}(t) \leq \Psi_q^{rs}(t), \ \forall \text{ routes } q \text{ between } r \text{ and } s] \quad \forall r, s, p. \tag{10.15}$$

In other words, the probability that a given route for each O-D pair is chosen at time t is the probability that its instantaneous route travel time is perceived to be the lowest of all the alternative routes. Once the distribution of perceived instantaneous route travel times is specified, the probability of selecting each alternative route can be calculated and the route flow can be assigned as follows:

$$f_p^{rs}(t) = f^{rs}(t) P_p^{rs}(t) \qquad \forall r, s, p \tag{10.16}$$

where $f_p^{rs}(t)$ is a control variable specifying the departure rate on route p from r to s at time t.

Analogous to the deterministic instantaneous DUO route choice model, we define the used links and routes. For any link a and any O-D pair rs, link a is defined as being used at time t if $u_a^{rs}(t) > 0$. Furthermore, a route p between r and s is defined as being used at time t if route departure flow $f_p^{rs}(t) > 0$. This definition is used in the general variational inequality model for the instantaneous SDUO route choice problem.

The instantaneous stochastic dynamic user-optimal (SDUO) route choice problem is to determine vehicle flows at each instant of time on each link resulting from drivers using perceived minimal-time routes under the currently prevailing travel times. The instantaneous stochastic dynamic user-optimal state is the following dynamic generalization of the conventional stochastic static user-optimal (SUO) state.

> **Instantaneous SDUO State:** *If, for each O-D pair at each decision node at each instant of time, the perceived instantaneous travel times for all routes that are being used equal the minimal perceived instantaneous route travel time, the dynamic traffic flow over the network is in an instantaneous stochastic dynamic user-optimal state.*

This definition is the stochastic generalization of the one given in Chapter 8. It assumes that all drivers make their route choice decisions using their perceptions of the current prevailing O-D travel times so that traffic flows are more dispersed over alternative routes than the deterministic dynamic route choice models.

Similar to Chapter 10, the two most often used probability distribution function, logit and probit, are discussed in the following. First, the logit model is discussed. In this chapter, we assume that

$$\Psi_p^{rs}(t) = \psi_p^{rs}(t) - \frac{1}{\theta}\epsilon_p^{rs}(t) \qquad \forall r, s, p, \tag{10.17}$$

where θ is a nonnegative parameter that scales the perceived instantaneous route travel time, and $\epsilon_p^{rs}(t)$ is a random error term associated with the route p under consideration (Sheffi, 1985). We assume that the errors $\epsilon_p^{rs}(t)$ are identically and independently distributed (IID) Gumbel variates for each time instant t. Other distributions of the perceived instantaneous route travel times are discussed in general probit models. Based on random utility theory, the logit route choice probability can be expressed as

$$P_p^{rs}(t) = \frac{\exp[-\theta\psi_p^{rs}(t)]}{\sum_l \exp[-\theta\psi_l^{rs}(t)]} \qquad \forall r, s, p. \tag{10.18}$$

As $\theta \rightarrow \infty$, $var[\Psi_p^{rs}(t)] \rightarrow 0$ and the perceived instantaneous travel time between the O-D equals the mean instantaneous travel time. In this case, travelers will choose the minimal-time route at each time instant, as in the deterministic instantaneous DUO route choice model in Chapter 8. We note that the above analysis for each O-D pair also applies for each decision node-destination pair.

The multinomial probit model can be derived from discrete choice theory and follows from equations (10.14) through (10.15) where the $\xi_p^{rs}(t)$ in (10.14) are assumed to be multivariate normal distributed with mean given by a vector of zeros and a covariance matrix at each time t. Unlike the logit model, equation (10.15) cannot be expressed in closed form because the minimum of the set of random variables describing the alternative's travel time involves the calculation of a multiple integral of a cumulative normal density function which cannot be evaluated in closed form. Also unlike the logit model, there are no restrictions of independence or identical distributions for the random error components.

The multinomial probit route choice function is characterized by a covariance matrix with positive non-diagonal elements when two or more route choices share one or more network links. The route choice attributes of these routes are dependent, unlike the multinomial logit model, where the covariance matrix is diagonal due to the IID assumptions. The properties of the normal distribution simplify the derivation of the multinomial probit model, just as the properties of the Gumbel distribution did for the multinomial logit model. The probit choice function is represented by equation (10.15) where the $\Psi_p^{rs}(t)$ are multivariate normal distributed at each time t and represent the travel time of alternative route p.

10.1.3 Instantaneous Route Choice Conditions and VI Problem

The general route choice probability of selecting each alternative route can be calculated and the time-dependent route flow can be assigned as follows:

$$f_p^{rs}(t) - f^{rs}(t)P_p^{rs}(t) = 0 \qquad \forall r,s,p. \tag{10.19}$$

The above equality states the instantaneous SDUO route choice conditions. In general, the mean instantaneous route travel time $\psi_p^{rs}(t)$ is increasing with route departure flow $f_p^{rs}(t)$, i.e.,

$$\frac{\partial \psi_p^{rs}(t)}{\partial f_p^{rs}(t)} > 0 \qquad \forall r,s,p. \tag{10.20}$$

For each route p and each O-D pair rs, define a cost term as follows:

$$G_p^{rs}(t) = [f_p^{rs}(t) - f^{rs}(t)P_p^{rs}(t)]\frac{\partial \psi_p^{rs}(t)}{\partial f_p^{rs}(t)} = 0 \qquad \forall r,s,p. \tag{10.21}$$

Clearly, the above equality states the instantaneous SDUO route choice conditions (10.19), since $\partial\psi_p^{rs}(t)/\partial f_p^{rs}(t) > 0$.

The instantaneous stochastic dynamic user-optimal (SDUO) route choice problem is to determine vehicle flows at each instant of time on each link resulting from drivers using minimal perceived instantaneous travel time routes. In the following, we present the VI formulation for the instantaneous SDUO route choice model.

Theorem 10.1. *The dynamic traffic flow pattern satisfying network constraint set (10.2)-(10.13) is in an instantaneous SDUO route choice state if and only if it satisfies the variational inequality problem:*

$$G_p^{rs}(t) \left\{ f_p^{rs}(t) - f_p^{rs^*}(t) \right\} \geq 0 \tag{10.22}$$

for each time instant $t \in [0, +\infty)$.

Proof.
It is necessary and sufficient to prove that equation (10.21) is equivalent to the inequality (10.22). As shown in Nagurney (1993), the system of equations (10.21) is equivalent to a variational inequality (10.22).

10.2 Solution Algorithm

10.2.1 The Discrete Variational Inequality Problem

To convert our continuous time VI problem (10.22) into a discrete time VI problem, the time period $[0, T]$ is subdivided into K small time intervals. Each time interval is regarded to be one unit of time. Then, $u_a(k)$ represents the inflow into link a during interval k, and $v_a(k)$ represents the exit flow from link a during interval k. To simplify the formulation, we round the estimated actual travel time on each link in the following way so that each estimated travel time is equal to a multiple of the time interval:

$$\bar{\tau}_a(k) = i \quad \text{if} \quad i - 0.5 \leq \bar{\tau}_a(k) < i + 0.5,$$

where i is an integer and $0 \leq i \leq K$. We note that this round-off method is used only in the flow propagation constraints. More accurate flow propagation constraints can be obtained by making the time intervals smaller. The evaluation of the link/route travel time functions and the subsequent objective function does not have this round-off error, so these round-off errors do not accumulate in the subsequent minimal-cost route search.

By discretizing time, the continuous VI problem (10.22) is converted into the following VI problem:

$$G_p^{rs}(k) \left\{ f_p^{rs}(k) - f_p^{rs^*}(k) \right\} \geq 0 \tag{10.23}$$

To solve the above VI problem, it is necessary to solve an inequality of its summation over all time intervals k, all O-D pairs rs, and all routes p. It follows that

$$\sum_{k=1}^{K}\sum_{rs}\sum_{p} G_p^{rs}(k)\ \left\{f_p^{rs}(k) - f_p^{rs^*}(k)\right\}\ \geq 0 \tag{10.24}$$

The above discrete VI can be solved by using a combination of the relaxation method and the Method of Successive Averages (MSA).

10.2.2 The Relaxation Method

For each relaxation (or relaxation) iteration, we temporarily fix:

1. the actual travel time $\tau_a(k)$ in link flow propagation constraints as $\bar{\tau}_a(k)$;
2. the route flow variables in route travel time function so that route travel time $\psi_p^{rs^*}(k)$ is fixed as $\bar{\psi}_p^{rs}[f_p^{rs}(k)]$ for each route;
3. the route choice probability $P_p^{rs}(k)$ as $\bar{P}_p^{rs}$, which depends on $\bar{\psi}_p^{rs}[f_p^{rs}(k)]$.

Thus, the discrete VI (10.24) becomes

$$\sum_{k=1}^{K}\sum_{rs}\sum_{p} \bar{G}_p^{rs}(k)\ \left\{f_p^{rs}(k) - f_p^{rs^*}(k)\right\}\ \geq 0 \tag{10.25}$$

where

$$\bar{G}_p^{rs}(t) = [f_p^{rs}(t) - f^{rs}(t)\bar{P}_p^{rs}(t)]\frac{\partial\bar{\psi}_p^{rs}[f_p^{rs}(k)]}{\partial f_p^{rs}(t)} \qquad \forall r,s,p. \tag{10.26}$$

Next, an objective function of a nonlinear programming program or discrete-time optimal control problem can be constructed as follows:

$$\begin{aligned}\min\quad Z(u,v,x,E,f) \quad &= \quad \sum_{k=1}^{K}\sum_{rs}\Bigg\{-f^{rs}(k)\ S^{rs}(k) + \sum_{p} f_p^{rs}(k)\ \bar{\psi}_p^{rs}[f_p^{rs}(k)] \\ &\quad - \sum_{p}\int_0^{f_p^{rs}(k)} \bar{\psi}_p^{rs}[\omega]d\omega\Bigg\}\end{aligned} \tag{10.27}$$

where the satisfaction function $S^{rs}(k)$ is defined as follows:

$$S^{rs}(k) = E[\min_{p}\ \Omega_p^{rs}(k)] \qquad \forall r,s,p,k. \tag{10.28}$$

This satisfaction function has the following property:

$$\frac{\partial S^{rs}(k)}{\partial\psi_q^{rs}(k)} = P_p^{rs}(k) \qquad \forall r,s,p,k. \tag{10.29}$$

Note that this nonlinear programming program has the same set of constraints as the discrete VI. Now, we need to verify that the above nonlinear programming program is equivalent to the VI under the relaxation. Based on Theorem 3.3.3, we need to demonstrate that the gradient of the objective function is equivalent to cost function term of the VI, i.e.,

$$\nabla_f Z(u, v, x, E, f) = \bar{G}_p^{rs}(k) \tag{10.30}$$

In the following, we demonstrate that the above equation holds. Taking the partial derivative of Z with respect to route departure flow $f_l^{mn}(k)$, it follows that

$$\begin{aligned}
\frac{\partial Z}{f_l^{mn}(k)} &= -f^{mn}(k)\, \frac{\partial S^{mn}(k)}{\bar{\psi}_l^{mn}(k)} \frac{\partial \bar{\psi}_p^{mn}[f_l^{mn}(k)]}{f_l^{mn}(k)} \\
&+ \bar{\psi}_l^{mn}[f_l^{mn}(k)] + f_l^{mn}(k)\, \frac{\partial \bar{\psi}_p^{mn}[f_l^{mn}(k)]}{f_l^{mn}(k)} \\
&- \bar{\psi}_l^{mn}[f_l^{mn}(k)] \\
&= \left\{f_l^{mn}(k) - f^{mn}(k)\, \bar{P}_l^{mn}(k)\right\} \frac{\partial \bar{\psi}_p^{mn}[f_l^{mn}(k)]}{f_l^{mn}(k)} \\
&= \bar{G}_l^{mn}(k)
\end{aligned} \tag{10.31}$$

Because the cost term $\bar{G}_p^{rs}(k)$ of the variational inequality (10.25) has a symmetric and semidefinite Jacobian matrix, an optimization program (NLP) can be formulated, which is equivalent to the above discrete VI (10.25). The resulting NLP is stated as follows:

$$\begin{aligned}
\min_f \quad Z(f) &= \sum_{k=1}^{K} \sum_{rs} \Bigg\{ -f^{rs}(k)\, S^{rs}(k) + \sum_p f_p^{rs}(k)\, \bar{\psi}_p^{rs}[f_p^{rs}(k)] \\
&- \sum_p \int_0^{f_p^{rs}(k)} \bar{\psi}_p^{rs}[\omega] d\omega \Bigg\}
\end{aligned} \tag{10.32}$$

s.t.

Relationship between state and control variables:

$$x_{ap}^{rs}(k+1) = x_{ap}^{rs}(k) + u_{ap}^{rs}(k) - v_{ap}^{rs}(k) \qquad \forall a, p, r, s; k = 1, \cdots, K; \tag{10.33}$$

$$F_p^{rs}(k+1) = F_p^{rs}(k) + f_p^{rs}(k) \qquad \forall r, s, p; k = 1, \cdots, K; \tag{10.34}$$

$$E_p^{rs}(k+1) = E_p^{rs}(k) + e_p^{rs}(k) \qquad \forall r, s, p; k = 1, \cdots, K; \tag{10.35}$$

Flow conservation constraints:

$$f_p^{rs}(k) = \sum_{a \in A(r)} \delta_{ap}^{rs}\, u_{ap}^{rs}(k) \qquad \forall p, r, s; k = 1, \cdots, K; \tag{10.36}$$

$$\sum_{a\in B(j)} v_{ap}^{rs}(k) - \sum_{a\in A(j)} u_{ap}^{rs}(k) = 0 \qquad \forall j,p,r,s; j\neq r,s; k=1,\cdots,K; \quad (10.37)$$

$$e_p^{rs}(k) = \sum_{a\in B(s)} \delta_{ap}^{rs}\, v_{ap}^{rs}(k) \qquad \forall p,r,s\neq r; k=1,\cdots,K; \quad (10.38)$$

Flow propagation constraints:

$$x_{ap}^{rs}(k) = \sum_{b\in\tilde{p}} \{x_{bp}^{rs}[k+\bar{\tau}_a(k)] - x_{bp}^{rs}(k)\} + \{E_p^{rs}[k+\bar{\tau}_a(k)] - E_p^{rs}(k)\} \quad (10.39)$$

$$\forall a\in B(j); j\neq r; p,r,s; k=1,\cdots,K+1;$$

Definitional constraints:

$$\sum_p u_{ap}^{rs}(k) = u_a^{rs}(k), \qquad \sum_{rs} u_a^{rs}(k) = u_a(k), \qquad \forall a; k=1,\cdots,K; \quad (10.40)$$

$$\sum_p v_{ap}^{rs}(k) = v_a^{rs}(k), \qquad \sum_{rs} v_a^{rs}(k) = v_a(k), \qquad \forall a; k=1,\cdots,K; \quad (10.41)$$

$$\sum_p x_{ap}^{rs}(k) = x_a^{rs}(k), \qquad \sum_{rs} x_a^{rs}(k) = x_a(k), \qquad \forall a; k=1,\cdots,K+1; \quad (10.42)$$

$$\sum_p e_p^{rs}(k) = e^{rs}(k), \quad \sum_p E_p^{rs}(k) = E^{rs}(k), \quad \forall r,s; k=1,\cdots,K+1; \quad (10.43)$$

$$\sum_p f_p^{rs}(k) = f^{rs}(k), \quad \sum_p F_p^{rs}(k) = F^{rs}(k), \quad \forall r,s; k=1,\cdots,K+1; \quad (10.44)$$

Nonnegativity conditions:

$$x_{ap}^{rs}(k+1)\geq 0, \quad u_{ap}^{rs}(k)\geq 0, \quad v_{ap}^{rs}(k)\geq 0, \qquad \forall a,p,r,s; k=1,\cdots,K; \quad (10.45)$$

$$e_p^{rs}(k)\geq 0, \quad f_p^{rs}(k)\geq 0, \quad E_p^{rs}(k+1)\geq 0, \quad F_p^{rs}(k+1)\geq 0,$$
$$\forall p,r,s; k=1,\cdots,K; \quad (10.46)$$

Boundary conditions:

$$E_p^{rs}(1)=0, \qquad F_p^{rs}(1)=0, \qquad \forall p,r,s; \qquad x_{ap}^{rs}(1)=0, \qquad \forall a,p,r,s. \quad (10.47)$$

10.2.3 Method of Successive Averages

The above NLP problem can be solved by using the Method of Successive Averages (MSA), as proposed by Sheffi and Powell (1982). The algorithm and convergence measures used in this section to determine the NLP solution follow directly from the discussion in Chapter 9. Define subproblem variables as $p, q, y, \bar{E}, \bar{f}$ in the MSA subproblem, corresponding to main problem variables u, v, x, E, f. The resulting MSA main problem link flow solutions are

$$u_a^{n+1}(k) = u_a^n(k) + \frac{1}{n}[p_p^n(k) - u_a^n(k)] \qquad \forall a; k=1,\cdots,K; \quad (10.48)$$

$$v_a^{n+1}(k) = v_a^n(k) + \frac{1}{n}[q_p^n(k) - u_a^n(k)] \qquad \forall a; k = 1, \cdots, K; \tag{10.49}$$

$$x_a^{n+1}(k) = x_a^n(k) + \frac{1}{n}[y_p^n(k) - x_a^n(k)] \qquad \forall a; k = 2, \cdots, K+1; \tag{10.50}$$

For a probit-based stochastic dynamic user-optimal problem, the MSA involves an iterative procedure of performing a stochastic dynamic network loading by using a Monte Carlo sampling followed by a combination of the previous solution with the subproblem solution. The Monte Carlo sampling is performed by sampling random network travel times, computing minimum perceived actual travel time routes from the perceived instantaneous travel times, and assigning all trips to these routes during each Monte Carlo iteration. Let the temporary link flow vector resulting from the all-or-nothing loading be called $\bar{y}_a(k)$, $\bar{p}_a(k)$, $\bar{q}_a(k)$, for each time interval k at Monte Carlo iteration r. Note that the current link flow vectors are $y_a^l(k)$, $p_a^l(k)$, $q_a^l(k)$. Then the stochastic dynamic network loading is solved by the recursive equation:

$$y_a^l(k) = \frac{(l-1)y_a^{l-1}(k) + \bar{y}_a^l(k)}{n}, \qquad \forall\, p, k \tag{10.51}$$

$$p_a^l(k) = \frac{(l-1)p_a^{l-1}(k) + \bar{p}_a^l(k)}{n}, \qquad \forall\, p, k \tag{10.52}$$

$$q_a^l(k) = \frac{(l-1)q_a^{l-1}(k) + \bar{q}_a^l(k)}{n}, \qquad \forall\, p, k \tag{10.53}$$

As iteration counter $l \to \infty$, $y_a^l(k)$, $p_a^l(k)$, $q_a^l(k)$ converge to the link flow variables determined by the stochastic dynamic network loading of the O-D flows according to the specified multinomial probit route choice density function.

Each MSA iteration involves Monte Carlo sampling, which determines the link flow variables resulting from an apportionment of O-D flows to the network based on the route choice probability distribution. A convergence criterion similar to Chapter 9 can be established to check the convergence of the MSA iterations.

10.2.4 Summary of the Solution Algorithm

In this combined algorithm, we define the travel time approximation procedure (relaxation) as the outer iteration and the MSA procedure as the inner iteration. The flowchart of our solution algorithm for solving our instantaneous SDUO route choice model is shown in Figure 10.1 and the algorithm is summarized as follows.

Step 0: Initialization.

Find an initial feasible solution $\{x_a^1(k)\}$, $\{u_a^1(k)\}$, $\{v_a^1(k)\}$, $\{E^1(k)\}$. Set the outer iteration counter m = 1.

Step 1: Relaxation. (Outer Iteration)
Find a new estimate of actual link travel time $\tau_a^m(k)$ for the flow propagation constraints and solve the instantaneous SDUO route choice program (NLP) by *MSA method* as follows. Set the inner iteration counter n = 1.

[Step 1.1]: *Update.* Calculate the link travel time functions using the temporarily fixed link flow variables.

[Step 1.2]: *Direction Finding.* Implement a stochastic dynamic network loading procedure based on the temporarily fixed actual link travel time and find an auxiliary set of link flow variables.

[Step 1.3]: *Move.* Find a new solution by combining current solution $\{u_a(k)\}$, $\{v_a(k)\}$, $\{x_a(k)\}$, $\{E^{rs}(k)\}$ and previous solution $\{p_a(k)\}$, $\{q_a(k)\}$, $\{y_a(k)\}$, $\{\bar{E}^{rs}(k)\}$ as equations (10.48)-(10.50).

[Step 1.4]: *Convergence Test for Inner Iterations (MSA).* If n equals a prespecified number is satisfied, go to step 2; otherwise, set $n = n + 1$ and go to step 1.1.

Step 2: Convergence Test for Outer Iterations.
If $\tau_a^m(k) \simeq \tau_a^{m+1}(k)$, stop. The current solution, $\{u_a(k)\}$, $\{v_a(k)\}$, $\{x_a(k)\}$ $\{E^{rs}(k)\}$, is in a near ideal SDUO state; otherwise, set $m = m + 1$ and go to step 1.

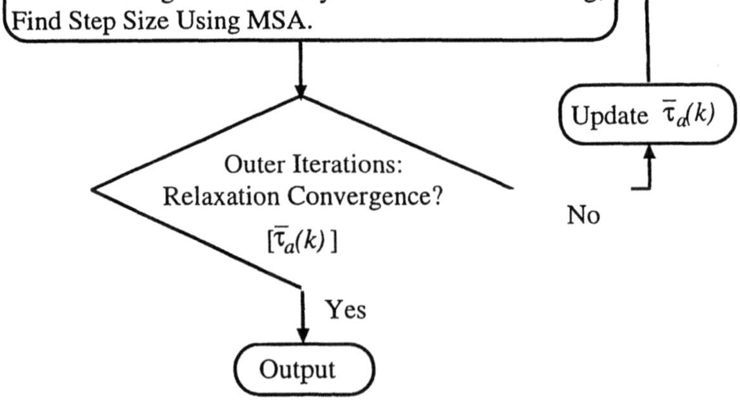

Figure 10.1: Flowchart of the Instantaneous Solution Algorithm

The number of inner iterations n and the number of outer iterations m are co-related. If we set m large, then n should be set small and vice versa. Often, only a few inner iterations are needed for convergence.

According to Powell and Sheffi (1982), the inner iteration procedure will converge. Since the convergence of outer iteration (relaxation) is also robust (Florian and Spiess, 1982), we expect that our algorithm will converge to our desired instantaneous SDUO route choice solutions. In order to speed up convergence, an incremental assignment technique is suggested for finding a good starting solution before the relaxation procedure. Since the subproblem can be decomposed by each artificial origin-destination pair, this problem is also a good candidate for solution with parallel computing techniques.

10.2.5 A Logit-Based Instantaneous Stochastic Loading

In the above solution algorithm, an instantaneous stochastic dynamic network loading is needed to perform the Monte Carlo sampling. In this section, we will show that an accurate stochastic dynamic network loading can be implemented for a logit-based problem. In other words, we discuss a *flow-independent* instantaneous SDUO route choice model and its solution algorithm. The flow-independent instantaneous SDUO route choice problem is to assign time-dependent vehicle flows between each O-D pair based on a route choice probability calculated using the fixed instantaneous route travel times for each time interval. We note that in this flow-independent problem, the instantaneous link travel times $c_a(k)$ do not depend on the link flow variables $u_a(k)$, $v_a(k)$ and $x_a(k)$.

The solution algorithm given here is similar to the algorithm proposed by Dial (1971) for static flow-independent stochastic network assignment. Our algorithm effectively implements a logit-type instantaneous route choice model at the network level. It does not assign probabilities and flows to all routes connecting each O-D pair. Instead, it is assumed that many of these routes constitute unreasonable travel choices that would not be actually considered. Consequently, our algorithm includes a preliminary phase which identifies the set of efficient routes connecting each O-D pair. The O-D departure flows are then assigned only to these routes during each time interval k, using the logit formula based on fixed instantaneous route travel times.

The steps of this algorithm for one O-D pair rs are outlined below, and are repeated for each O-D pair in the network. In view of the correspondence to Dial's STOCH algorithm (1971), the following is called the DYNASTOCH2 algorithm to represent a dynamic version of STOCH algorithm for the flow-independent instantaneous SDUO route choice problem.

Step 0: Initialization.

Compute the minimal instantaneous travel time $\sigma^{ri}(k)$ from origin r to all other nodes for vehicles departing origin r during time interval k.

Calculate the likelihood for each link $a = (i,j)$ during time interval k:

$$L_{(i,j)}(k) = \begin{cases} \exp\{\theta\,[\sigma^{rj}(k) - \sigma^{ri}(k) - c_a(k)]\}, & \text{if } \sigma_o^{is} > \sigma_o^{js} \\ 0, & \text{otherwise} \end{cases}$$

In this expression, $c_a(k)$ is the mean instantaneous travel time on link (i,j) during time interval k.

Step 1: Backward Pass.
By examining all nodes j in ascending sequence with respect to $\sigma^{js}(k)$, calculate the weight for each link $a = (i,j)$ during each time interval k:

$$W_{(i,j)}(k) = \begin{cases} L_{(i,j)}(k), & \text{if } j = s \text{ (destination)} \\ L_{(i,j)}(k) \sum_{(j,m)\in A(j)} W_{(j,m)}(k), & \text{otherwise} \end{cases}$$

When the origin r is reached, this step is completed.

Step 2: Forward Pass.
Consider all nodes i in descending sequence with respect to $\sigma^{is}(k)$, starting with origin r. When each node i is considered during each time interval k, compute the inflow of link (i,j) during each time interval k using the following formula:

$$u_{(i,j)}(k) = \begin{cases} f^{rs}(k)\dfrac{W_{(i,j)}(k)}{\sum_{(i,m)\in A(i)} W_{(i,m)}(k)}, & \text{if } i = r \text{ (origin)} \\ \left\{\sum_{(n,i)\in B(i)} v_{(n,i)}(k)\right\}\dfrac{W_{(i,j)}(k)}{\sum_{(i,m)\in A(i)} W_{(i,m)}(k)}, & \text{otherwise} \end{cases}$$

This step is implemented iteratively until destination s is reached. Note that the sum in each denominator includes all links emanating from the upstream node of the link under consideration. The sum of the exit flow variables is taken over all links arriving at the upstream node of the link under consideration.

The flow generated by this algorithm is equivalent to a logit-based flow-independent route assignment between each O-D pair, given that only efficient routes are considered. We next present a proof for this algorithm.

10.2.6 Proof of the Algorithm

We now prove that the above algorithm does generate logit-based flow-independent instantaneous SDUO route choices between each O-D pair. We note that each link likelihood $L_{(i,j)}(k)$, is proportional to the logit probability that link $a = (i,j)$ is used during interval k by a traveler chosen at random from among the population of trip-makers between r and s, given that the traveler is departing from origin r during interval k. The probability that a given route will be used is proportional to the product of all the likelihoods of the links comprising this route. The probability of using route l between r and s, $P_l^{rs}(k)$, for vehicles departing r during interval k, is then

$$P_l^{rs}(k) = G(k) \prod_{a \in l} \left\{ L_{(i,j)}(k) \right\}^{\delta_{al}^{rs}} \tag{10.54}$$

where $G(k)$ is a proportionality constant for each time interval k and the product is taken over all links in the network. The incidence variable δ_{al}^{rs} ensures that $P_l^{rs}(k)$ will include only those links in the lth route between r and s. Substituting the expression for the likelihood $L_{(i,j)}(k)$ in the above equation, the choice probability of choosing a particular efficient route becomes

$$P_l^{rs}(k) = G(k) \prod_{a \in l} \exp \left\{ \theta \left\{ \sigma^{rj}(k) - \sigma^{ri}(k) - c_a(k) \right\} \delta_{al}^{rs} \right\} \tag{10.55}$$

$$= G(k) \exp \left\{ \theta \sum_{a \in l} \left\{ \sigma^{rj}(k) - \sigma^{ri}(k) - c_a(k) \right\} \delta_{al}^{rs} \right\} \tag{10.56}$$

$$= G(k) \exp \left\{ \theta \left[\sigma^{rs}(k) - \psi_l^{rs}(k) \right] \right\} \tag{10.57}$$

The last equality results from the following summations:

$$\sum_{a \in l} \left\{ \sigma^{rj}(k) - \sigma^{ri}(k) \right\} = \sigma^{rs}(k) - \sigma^{rr}(k) = \sigma^{rs}(k) \tag{10.58}$$

and

$$\sum_{a \in l} c_a(k) = \psi_l^{rs}(k) \tag{10.59}$$

Since

$$\sum_q P_q^{rs}(k) = 1.0, \tag{10.60}$$

the proportionality constant must equal

$$G(k) = \frac{1}{\sum_q \exp \left\{ \theta \left[\sigma^{rs}(k) - \psi_q^{rs}(k) \right] \right\}} \tag{10.61}$$

Thus,

$$P_l^{rs}(k) = \frac{\exp \left\{ \theta \left[\sigma^{rs}(k) - \psi_l^{rs}(k) \right] \right\}}{\sum_q \exp \left\{ \theta \left[\sigma^{rs}(k) - \psi_q^{rs}(k) \right] \right\}} \tag{10.62}$$

$$= \frac{\exp \left\{ -\theta \, \psi_l^{rs}(k) \right] \right\}}{\sum_q \exp \left\{ -\theta \, \psi_q^{rs}(k) \right\}} \tag{10.63}$$

The above equation (10.63) depicts a logit model of route choice during each time interval k among the efficient routes connecting O-D pair rs. Thus, the algorithm does generate a logit-type route choice probability formula using instantaneous route travel times.

The DYNASTOCH2 algorithm does not require explicit route enumeration. It does require the calculation of time-dependent minimal instantaneous travel time routes for every O-D pair in the network during each time interval k. In the following, we show that the forward pass of the algorithm will generate a time-dependent logit-based assignment. To do so, we only need to show that the calculated link inflows are obtained in a manner consistent with the expression in the forward pass. This is done by proving the algorithm diverts trips from each node i during each interval k according to appropriate conditional link probabilities. A conditional link probability during interval k is the probability that a trip between r and s will use a particular link $a = (i,j)$ during interval k, given that it goes through the link's tail node i. This probability is stated as

$$\text{Prob}_{(i,j)|i}(k) = \frac{\text{Prob}_{(i,j),i}(k)}{\text{Prob}_i(k)} = \frac{\text{Prob}_{(i,j)}(k)}{\text{Prob}_i(k)} = \frac{\text{Prob}_{(i,j)}(k)}{\sum_l \text{Prob}_{(i,l)}(k)} \tag{10.64}$$

Denote p as an efficient route from r to s. The probability of using link $a = (i,j)$ during interval k is the summation of probabilities of using route P during interval k, where link $a = (i,j)$ is on route p. It follows that

$$\text{Prob}_{(i,j)}(k) \quad = \quad \sum_{p:(i,j)} \text{Prob}_p(k) \tag{10.65}$$

where $p{:}(i,j)$ means those routes p which include link (i,j). It is useful to write (10.65) in a more elaborate form, so as to facilitate cancellation of common factors in the numerator and denominator of equation (10.64). To do so, an efficient route through link (i,j) can be partitioned into three sets of links:

1. $P_i = \{\text{all links topologically preceding link } (i,j)\}$;
2. link $(i,j) = \{(i,j)\}$;
3. $P_j = \{\text{all links topologically following link } (i,j)\}$.

Denote $\mathcal{P}_i$ as the set of P_i and $\mathcal{P}_j$ as the set of P_j. Then

$$\sum_{p:(i,j)} \text{Prob}_p(k) =$$
$$h(k)\; L_{(i,j)}(k)\; \{\sum_{p\in\mathcal{P}_i} \prod_{(m,n)} L_{(m,n)}(k)\}\{\sum_{p\in\mathcal{P}_j} \prod_{(m,n)} L_{(m,n)}(k)\} \tag{10.66}$$

Equation (10.66) follows from the fact that all the efficient routes can be constructed by independently choosing a member from each of $\mathcal{P}_i$ and $\mathcal{P}_j$ and putting link (i,j) in between. Such combinations constitute efficient routes. Substituting (10.65) and (10.66) into (10.64), it follows that

$$
\begin{aligned}
&\mathrm{Prob}_{(i,j)|i}(k) \\
&= \frac{h(k)\, L_{(i,j)}(k) \left\{ \sum\limits_{p\in\mathcal{P}_i} \prod\limits_{(m,n)} L_{(m,n)}(k) \right\} \left\{ \sum\limits_{p\in\mathcal{P}_j} \prod\limits_{(m,n)} L_{(m,n)}(k) \right\}}{\sum\limits_{l} h(k)\, L_{(i,l)}(k) \left\{ \sum\limits_{p\in\mathcal{P}_i} \prod\limits_{(m,n)} L_{(m,n)}(k) \right\} \left\{ \sum\limits_{p\in\mathcal{P}_j} \prod\limits_{(m,n)} L_{(m,n)}(k) \right\}} \\
&= \frac{h(k)\, L_{(i,j)}(k) \left\{ \sum\limits_{p\in\mathcal{P}_i} \prod\limits_{(m,n)} L_{(m,n)}(k) \right\} \left\{ \sum\limits_{p\in\mathcal{P}_j} \prod\limits_{(m,n)} L_{(m,n)}(k) \right\}}{h(k) \left\{ \sum\limits_{p\in\mathcal{P}_i} \prod\limits_{(m,n)} L_{(m,n)}(k) \right\} \sum\limits_{l} \left\{ L_{(i,l)}(k) \left\{ \sum\limits_{p\in\mathcal{P}_j} \prod\limits_{(m,n)} L_{(m,n)}(k) \right\} \right\}} \\
&= \frac{L_{(i,j)}(k) \left\{ \sum\limits_{p\in\mathcal{P}_j} \prod\limits_{(m,n)} L_{(m,n)}(k) \right\}}{\sum\limits_{l} \left\{ L_{(i,l)}(k) \left\{ \sum\limits_{p\in\mathcal{P}_j} \prod\limits_{(m,n)} L_{(m,n)}(k) \right\} \right\}} = \frac{W_{(i,j)}(k)}{\sum\limits_{l} W_{(i,l)}(k)}
\end{aligned}
\tag{10.67}
$$

By the link weights calculated in the Backward Pass (Step 1), the quotient in Forward Pass (Step 2) is equal to the right-hand-side of equation (10.67). Thus, we complete the proof that the diversion inflows are indeed those implied by the probability defined in equation (10.54). Therefore, our algorithm does generate a logit-type route flow assignment using instantaneous route travel times. The proof is complete. We also note that the flow over each link should satisfy the flow propagation constraints. For a detailed discussion of these constraints, refer to Chapter 4.

10.3 An Instantaneous Optimal Control Model

In this section, we formulate an optimal control model which is equivalent to the instantaneous SDUO route choice conditions. This optimal control model is an alternative formulation to the general variational inequality model under the assumption of logit route choice probability. In the optimal control model, we use the following relaxed definition of used links and routes. For a link a on any route from origin r to destination s, link a is being used at time t if $u_a^{rs}(t) > 0$ and $v_a^{rs}(t) > 0$. Furthermore, a route p between r and s is being used at time t if $u_a^{rs}(t) > 0$ and $v_a^{rs}(t) > 0$ for all links $a \in rsp$. We note that this definition only applies to this instantaneous SDUO route choice model.

Similar to deterministic instantaneous DUO route choice models in Chapter 9, we write the flow propagation constraints using *estimates* of the mean actual link travel times. These link time estimates must be updated in an iterative relaxation solution procedure. Denote $\bar{\tau}_a(t)$ as the estimated mean actual travel time over link a for flows entering link a at time t. Then, the flow propagation constraints are as follows:

$$x_{ap}^{rs}(t) = \sum_{b \in \tilde{p}} \{x_{bp}^{rs}[t + \bar{\tau}_a(t)] - x_{bp}^{rs}(t)\} + \{E_p^{rs}[t + \bar{\tau}_a(t)] - E_p^{rs}(t)\}$$

$$\forall a \in p; p, r, s. \tag{10.68}$$

Using optimal control theory, an equivalent optimization model of the instantaneous stochastic dynamic user-optimal route choice problem is formulated as follows.

$$\min_{u,v,x,e,E} \int_0^T \Bigg\{ \sum_a \left[\int_0^{u_a(t)} g_{1a}[x_a(t),\omega] d\omega + \int_0^{v_a(t)} g_{2a}[x_a(t),\omega] d\omega \right]$$

$$+ \frac{1}{\theta} \Bigg[\sum_{rs} \sum_p \sum_a u_{ap}^{rs}(t) \ln u_{ap}^{rs}(t)$$

$$- \sum_{rs} \sum_p \sum_{a \notin B(s)} v_{ap}^{rs}(t) \ln v_{ap}^{rs}(t) \Bigg] \Bigg\} dt \tag{10.69}$$

s.t.

Relationship between state and control variables:

$$\frac{dx_{ap}^{rs}}{dt} = u_{ap}^{rs}(t) - v_{ap}^{rs}(t) \qquad \forall a, p, r, s; \tag{10.70}$$

$$\frac{dE_p^{rs}(t)}{dt} = e_p^{rs}(t) \qquad \forall p, r; s \neq r; \tag{10.71}$$

Flow conservation constraints:

$$f^{rs}(t) = \sum_{a \in A(r)} \sum_p u_{ap}^{rs}(t) \qquad \forall r \neq s; s; \tag{10.72}$$

$$\sum_{a \in B(j)} v_{ap}^{rs}(t) = \sum_{a \in A(j)} u_{ap}^{rs}(t) \qquad \forall j, p, r, s; j \neq r, s; \tag{10.73}$$

$$e^{rs}(t) = \sum_{a \in B(s)} \sum_p v_{ap}^{rs}(t) \qquad \forall r, s \neq r; \tag{10.74}$$

Flow propagation constraints:

$$x_{ap}^{rs}(t) = \sum_{b \in \tilde{p}} \{x_{bp}^{rs}[t + \bar{\tau}_a(t)] - x_{bp}^{rs}(t)\} + \{E_p^{rs}[t + \bar{\tau}_a(t)] - E_p^{rs}(t)\}$$

$$\forall a \in p; p, r, s; \tag{10.75}$$

Definitional constraints:

$$\sum_p u_{ap}^{rs}(t) = u_a^{rs}(t), \qquad \sum_{rs} u_a^{rs}(t) = u_a(t), \qquad \forall a; \tag{10.76}$$

$$\sum_p v_{ap}^{rs}(t) = v_a^{rs}(t), \qquad \sum_{rs} v_a^{rs}(t) = v_a(t), \qquad \forall a; \tag{10.77}$$

$$\sum_p x_{ap}^{rs}(t) = x_a^{rs}(t), \qquad \sum_{rs} x_a^{rs}(t) = x_a(t), \qquad \forall a; \tag{10.78}$$

$$\sum_p e_p^{rs}(t) = e^{rs}(t), \qquad \sum_p E_p^{rs}(t) = E^{rs}(t), \qquad \forall r, s; \tag{10.79}$$

Nonnegativity conditions:

$$x_{ap}^{rs}(t) \geq 0, \quad u_{ap}^{rs}(t) \geq 0, \quad v_{ap}^{rs}(t) \geq 0 \quad \forall a, p, r, s; \tag{10.80}$$

$$e_p^{rs}(t) \geq 0, \quad E_p^{rs}(t) \geq 0, \qquad \forall p, r, s; \tag{10.81}$$

Boundary conditions:

$$E_p^{rs}(0) = 0, \quad \forall p, r, s; \qquad x_{ap}^{rs}(0) = 0, \quad \forall a, p, r, s; \tag{10.82}$$

where the expressions $u(t) \ln u(t)$ and $v(t) \ln v(t)$ are assigned the value zero at $u(t) = 0$ and $v(t) = 0$, respectively. The first two terms of the objective function are similar to the objective function of the well-known stochastic static UO model. The third and fourth terms of the objective function are also similar to the route flow entropy terms in the objective function of the stochastic UO model. In the last term of the objective function, the entropy function applies to all except the last link on every route between every O-D pair.

The first two constraints (10.70)-(10.71) are state equations for flows on each link a and for the arrivals at destinations. Equations (10.72)-(10.74) are flow conservation constraints at each node including origins and destinations. The other constraints include flow propagation constraints, definitional constraints, nonnegativity, and boundary conditions. In summary, the control variables are $u_{ap}^{rs}(t)$, $v_{ap}^{rs}(t)$, $e_p^{rs}(t)$; the state variables are $x_{ap}^{rs}(t)$, $E_p^{rs}(t)$. We prove below that the optimal solution to the model is in an instantaneous SDUO state.

Generally we assume the route choice dispersion parameter θ to be nonnegative. As $\theta \to \infty$, the third and fourth terms in the objective function will vanish and the solution will approach the deterministic instantaneous user-optimal solution.

To solve the above optimal control model, the same algorithm as in Section 10.3 can be used. As discussed in the following, the first-order necessary conditions of the continuous time model fit a logit-type instantaneous route choice model. This flow-independent subproblem can be efficiently solved using the DYNASTOCH2 algorithm discussed in Section 10.3.3 without route enumeration. In other words, the Monte Carlo sampling should be replaced by the logit-based stochastic dynamic network loading method (DYNASTOCH2). We note that the DYNASTOCH2 algorithm is much different from Dial's STOCH algorithm in that DYNASTOCH2 has to be implemented on an expanded time-space network presented in Chapter 6 and the flow propagation constraints ensure the flow progression over links.

Optimality Conditions

We construct the extended Hamiltonian as follows.

$$\begin{aligned}
\mathcal{H} &= \sum_a \left\{ \int_0^{u_a(t)} g_{1a}[x_a(t), \omega] d\omega + \int_0^{v_a(t)} g_{2a}[x_a(t), \omega] d\omega \right\} \\
&+ \frac{1}{\theta} \left[\sum_{rs} \sum_p \sum_a u_{ap}^{rs}(t) \ln u_{ap}^{rs}(t) - \sum_{rs} \sum_p \sum_{a \notin B(s)} v_{ap}^{rs}(t) \ln v_{ap}^{rs}(t) \right] \\
&+ \sum_{rs} \sum_{ap} \lambda_{ap}^{rs}(t)[u_{ap}^{rs}(t) - v_{ap}^{rs}(t)] + \sum_r \sum_{s \neq r} \sum_p \nu_p^{rs}(t) e_p^{rs}(t) \\
&+ \sum_s \sum_{r \neq s} \sigma_r^{rs}(t)[f^{rs}(t) - \sum_{a \in A(r)} \sum_p u_{ap}^{rs}(t)] \\
&+ \sum_{rs} \sum_{j \neq rs} \sum_p \sigma_{jp}^{rs}(t)[\sum_{a \in B(j)} v_{ap}^{rs}(t) - \sum_{a \in A(j)} u_{ap}^{rs}(t)] \\
&+ \sum_r \sum_{s \neq r} \sigma_s^{rs}(t)[\sum_{a \in B(s)} \sum_p v_{ap}^{rs}(t) - e^{rs}(t)] \\
&+ \sum_{rs} \sum_p \sum_{a \in p} \mu_{ap}^{rs}(t) \left\{ x_{ap}^{rs}(t) + \sum_{b \in \bar{p}} x_{bp}^{rs}(t) + E_p^{rs}(t) \right. \\
&- \left. \sum_{b \in \bar{p}} x_{bp}^{rs}[t + \bar{\tau}_a(t)] - E_p^{rs}[t + \bar{\tau}_a(t)] \right\}
\end{aligned}$$

where $\lambda_{ap}^{rs}(t)$ are Lagrange multipliers associated with the link state equations, $\nu_p^{rs}(t)$ are Lagrange multipliers associated with the destination node state equations, $\sigma_{jp}^{rs}(t)$ are Lagrange multipliers associated with the node flow conservation equations, and $\mu_{ap}^{rs}(t)$ are Lagrange multipliers associated with the flow propagation equations. For each link a which points from node l to node m, the first order necessary conditions of the instantaneous SDUO route choice program (10.69)-(10.82) include

$$\frac{\partial \mathcal{H}}{\partial u_{ap}^{rs}(t)} = g_{1a}[x_a(t), u_a(t)] + \frac{1}{\theta} \left[\ln u_{ap}^{rs}(t) + 1 \right] + \lambda_{ap}^{rs}(t) - \sigma_{lp}^{rs}(t) \geq 0,$$

$$\forall a \in A(l), p, r, s, \qquad (10.83)$$

$$\text{and} \qquad u_{ap}^{rs}(t)\frac{\partial \mathcal{H}}{\partial u_{ap}^{rs}(t)} = 0 \qquad \forall a \in A(l), p, r, s; \tag{10.84}$$

$$\frac{\partial \mathcal{H}}{\partial v_{ap}^{rs}(t)} = g_{2a}[x_a(t), v_a(t)] - \frac{1}{\theta}\left[\ln v_{ap}^{rs}(t) + 1\right] - \lambda_{ap}^{rs}(t) + \sigma_{mp}^{rs}(t) \geq 0,$$

$$\forall a \in B(m), m \neq s; p, r, s, \tag{10.85}$$

$$\frac{\partial \mathcal{H}}{\partial v_{ap}^{rs}(t)} = g_{2a}[x_a(t), v_a(t)] - \lambda_{ap}^{rs}(t) + \sigma_{mp}^{rs}(t) \geq 0,$$

$$\forall a \in B(m), m = s; p, r, s, \tag{10.86}$$

$$\text{and} \qquad v_{ap}^{rs}(t)\frac{\partial \mathcal{H}}{\partial v_{ap}^{rs}(t)} = 0 \qquad \forall a, p, r, s; \tag{10.87}$$

$$\frac{\partial \mathcal{H}}{\partial e_p^{rs}(t)} \geq 0, \qquad \forall p, r, s, \tag{10.88}$$

$$\text{and} \qquad e_p^{rs}(t)\frac{\partial \mathcal{H}}{\partial e_p^{rs}(t)} = 0 \qquad \forall p, r, s; \tag{10.89}$$

$$\frac{d\lambda_{ap}^{rs}(t)}{dt} = -\frac{\partial \mathcal{H}}{\partial x_{ap}^{rs}(t)} \qquad \forall a, p, r, s; \tag{10.90}$$

$$\frac{d\nu_p^{rs}(t)}{dt} = -\frac{\partial \mathcal{H}}{\partial E_p^{rs}(t)} \qquad \forall p, r, s; \tag{10.91}$$

$$x_{ap}^{rs}(t) \geq 0, \qquad u_{ap}^{rs}(t) \geq 0, \qquad v_{ap}^{rs}(t) \geq 0, \qquad \forall a, p, r, s; \tag{10.92}$$

$$e_p^{rs}(t) \geq 0, \qquad E_p^{rs}(t) \geq 0, \qquad \forall p, r, s. \tag{10.93}$$

Note that $\sigma_{rp}^{rs}(t) = \sigma_r^{rs}(t)$ when node l equals origin r.

Equivalence Analysis

We now show that the set of link states, inflows and exit flows that solves this program also satisfies the instantaneous stochastic dynamic user-optimal route choice conditions. This equivalence is demonstrated below by proving that the first order necessary conditions for optimal control program (10.69)-(10.82) are identical to the instantaneous stochastic dynamic user-optimal route choice conditions. The equivalence between the instantaneous SDUO route choice conditions and the first order necessary conditions of the optimal control program means that the instantaneous SDUO route choice conditions are satisfied at the optimal solution of this program.

Combining equations (10.83)-(10.84) with equations (10.85)-(10.87), the following equations can be derived for each link a pointing from node l to node m.

$$\frac{\partial \mathcal{H}}{\partial u_{ap}^{rs}(t)} + \frac{\partial \mathcal{H}}{\partial v_{ap}^{rs}(t)} = c_a(t) - \sigma_{lp}^{rs}(t) + \sigma_{mp}^{rs}(t) + \frac{1}{\theta}\left[\ln u_{ap}^{rs}(t) - \ln v_{ap}^{rs}(t)\right] \geq 0,$$

$$\forall a \in A(l) \cap B(m); p, r, s; \qquad (10.94)$$

$$\frac{\partial \mathcal{H}}{\partial u_{ap}^{rs}(t)} + \frac{\partial \mathcal{H}}{\partial v_{ap}^{rs}(t)} = c_a(t) - \sigma_{lp}^{rs}(t) + \sigma_s^{rs}(t) + \frac{1}{\theta}\left[\ln u_{ap}^{rs}(t) + 1\right] \geq 0,$$

$$\forall a \in A(l) \cap B(s); p, r, s. \qquad (10.95)$$

where node m equals destination s so that $\sigma_{mp}^{rs}(t) = \sigma_s^{rs}(t)$.

Furthermore, if $u_{ap}^{rs}(t) > 0$ and $v_{ap}^{rs}(t) > 0$, by (10.83)-(10.87),

$$c_a(t) = \sigma_{lp}^{rs}(t) - \sigma_{mp}^{rs}(t) - \frac{1}{\theta}\left[\ln u_{ap}^{rs}(t) - \ln v_{ap}^{rs}(t)\right]$$

$$\forall a \in A(l) \cap B(m); p, r, s; \qquad (10.96)$$

$$c_a(t) = \sigma_{lp}^{rs}(t) - \sigma_s^{rs}(t) - \frac{1}{\theta}\left[\ln u_{ap}^{rs}(t) + 1\right]$$

$$\forall a \in A(l) \cap B(s); p, r, s. \qquad (10.97)$$

For route p between origin node r and destination node s, let i denote node r or any intermediate node on this route. Denote route $\tilde{p}$ as a sequence of nodes $(i, 1, 2, \cdots, n, s)$ and also as a sequence of links $(1, 2, \cdots, k)$. The mean instantaneous travel time $\psi_{\tilde{p}}^{is}(t)$ for the remaining route $\tilde{p}$ between i and s is

$$\psi_{\tilde{p}}^{is}(t) = \sum_{a \in is\tilde{p}} c_a[x_a(t), u_a(t), v_a(t)] \qquad \forall i \in p, r, s. \qquad (10.98)$$

Consider a set of routes p which are from $r \to i \to s$ and the corresponding set of subroutes $\tilde{p}$. The flow conservation constraint at node i can be revised as

$$\sum_{a \in B(i)} v_{ap}^{rs}(t) = \sum_{a \in A(i)} u_{ap}^{rs}(t) \qquad \forall i, p, r, s; i \neq r, s. \qquad (10.99)$$

The fourth term in the Hamiltonian function should be revised as

$$\sum_{rs} \sum_{i \neq rs} \sigma_i^{rs}(t) \sum_p [\sum_{a \in B(j)} v_{ap}^{rs}(t) - \sum_{a \in A(j)} u_{ap}^{rs}(t)$$

so that $\sigma_{ip}^{rs}(t) = \sigma_i^{rs}(t)$ for the set of subroutes $\tilde{p}$. Note that all derivations from equation (10.83) to equation (10.97) follow for this set of subroutes $\tilde{p}$.

Now if $u_{ap}^{rs}(t) > 0$ and $v_{ap}^{rs}(t) > 0$, route $\tilde{p}$ is being used at time t. Thus, by equations (10.96)-(10.97),

$$\begin{aligned}
\psi_{\tilde{p}}^{is}(t) &= [\sigma_i^{rs}(t) - \sigma_1^{rs}(t)] + [\sigma_{1p}^{rs}(t) - \sigma_{2p}^{rs}(t)] + \cdots \\
&+ [\sigma_{n-1,p}^{rs}(t) - \sigma_{np}^{rs}(t)] + [\sigma_{np}^{rs}(t) - \sigma_s^{rs}(t)] \\
&- \frac{1}{\theta}[\ln u_{1\tilde{p}}^{rs}(t) - \ln v_{1\tilde{p}}^{rs}(t)] - \frac{1}{\theta}[\ln u_{2\tilde{p}}^{rs}(t) - \ln v_{2\tilde{p}}^{rs}(t)] - \cdots \\
&- \frac{1}{\theta}[\ln u_{(k-1)\tilde{p}}^{rs}(t) - \ln v_{(k-1)\tilde{p}}^{rs}(t)] - \frac{1}{\theta}[\ln u_{k\tilde{p}}^{rs}(t) + 1] \\
&= \sigma_i^{rs}(t) - \sigma_s^{rs}(t) - \frac{1}{\theta}[\ln u_{1\tilde{p}}^{rs}(t) + 1] \qquad (10.100)
\end{aligned}$$

for every route $\tilde{p}$ being used at time t. In equation (10.100), for any intermediate node j between upstream link a and downstream link b,

$$v_{a\tilde{p}}^{rs}(t) = u_{b\tilde{p}}^{rs}(t) \tag{10.101}$$

Thus, in the second line of equation (10.100), all terms except $\ln u_{1\tilde{p}}^{rs}(t)$ are canceled. Reorganizing equation (10.100), we have

$$u_{1\tilde{p}}^{rs}(t) = \exp[-\theta\psi_{\tilde{p}}^{is}(t)]\exp[\theta\sigma_i^{rs}(t) - \theta\sigma_s^{rs}(t) - 1] \tag{10.102}$$

Summing the above equation over all subroutes $\tilde{q}$,

$$\sum_{\tilde{q}} u_{1\tilde{q}}^{rs}(t) = \exp[\theta\sigma_i^{rs}(t) - \theta\sigma_s^{rs}(t) - 1]\sum_{\tilde{q}}\exp[-\theta\psi_{\tilde{q}}^{is}(t)] \tag{10.103}$$

so that

$$\exp[\theta\sigma_i^{rs}(t) - \theta\sigma_s^{rs}(t) - 1] = \frac{\sum_{\tilde{q}} u_{1\tilde{q}}^{rs}(t)}{\sum_{\tilde{q}}\exp[-\theta\psi_{\tilde{q}}^{is}(t)]} \tag{10.104}$$

Combining equations (10.102) and (10.104) for any subroute $\tilde{p}$ of the set $\{\tilde{q}\}$,

$$u_{1\tilde{p}}^{rs}(t) = \sum_{\tilde{q}} u_{1\tilde{q}}^{rs}(t)\frac{\exp[-\theta\psi_{\tilde{p}}^{is}(t)]}{\sum_{\tilde{q}}\exp[-\theta\psi_{\tilde{q}}^{is}(t)]} \tag{10.105}$$

where $\sum_{\tilde{q}} u_{1\tilde{q}}^{rs}(t)$ is the total flow entering the first links on subroutes $\tilde{q}$ from origin r arriving at node i and departing toward destination s at each time instant t; link 1 is the first link on subroute $\tilde{p}$ from i to s.

Equation (10.105) holds for each remaining route $\tilde{p}$ between i and s, where i is any intermediate node (including the origin) between each O-D pair rs in the network. For route $\tilde{p}$ connecting node i and destination s, flows for each O-D pair are assigned according to the above logit function of measured instantaneous route travel times.

When $\theta \to \infty$, the variance of the perceived instantaneous route travel times is zero. Thus, by equations (10.17) and (10.100), we have

$$\Psi_{\tilde{p}}^{is}(t) \to \psi_{\tilde{p}}^{is}(t) = \sigma_i^{rs}(t) - \sigma_s^{rs}(t) \tag{10.106}$$

Since we assume $u_a^{rs}(t) > 0$ and $v_a^{rs}(t) > 0$, route $\tilde{p}$ is being used by definition. Thus, the routes which are being used at time t have minimal instantaneous travel times equal to $[\sigma_i^{rs}(t) - \sigma_s^{rs}(t)]$. This is the case of deterministic instantaneous DUO route choice.

The deviations from the instantaneous DUO can be attributed to drivers' ignorance of (or non-compliance with) current travel time information provided. This is also a learning process. The value, $\theta = 0$, corresponds to total driver ignorance (no compliance) in route travel times in which case either route is equally probable. As drivers increase their compliance to the provided prevailing travel time information at each decision node, the flow on the shortest instantaneous travel time route will increase until an instantaneous DUO state is attained.

We note that the above analysis applies to any two decision nodes i and j between $r-s$. Since the intermediate node i could be the origin node r, the above results also hold for routes from r to s. It follows that

$$f_p^{rs}(t) = u_{1p}^{rs}(t) = f^{rs}(t)\frac{\exp[-\theta\psi_p^{rs}(t)]}{\sum_q \exp[-\theta\psi_q^{rs}(t)]} \tag{10.107}$$

With the above interpretation, it is now clear that equations (10.105) state the instantaneous stochastic dynamic user-optimal route choice conditions.

10.4 Numerical Example

We illustrate the solutions of instantaneous SDUO route choice model with the 4-link, 4-node test network shown in Figure 10.2. Time period $[0,T]$ is subdivided into $K = 8$ small time intervals. The algorithms were coded in FORTRAN and solved on a HP workstation.

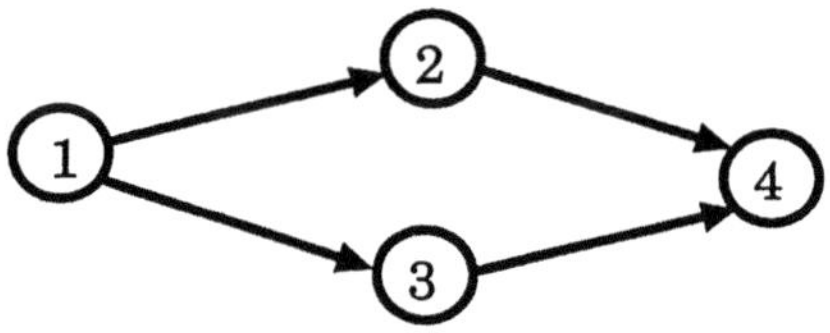

Figure 10.2: Test Network

The following link travel time functions were used in the computations:

$$c_a(k) = \tau_a(k) = g_{1a}(k) + g_{2a}(k)$$

$$g_{1a}(k) = \beta_{1a} + \beta_{2a}[u_a(k)]^2 + \beta_{3a}[x_a(k)]^2$$

$$g_{2a}(k) = \beta_{4a} + \beta_{5a}[v_a(k)]^2 + \beta_{6a}[x_a(k)]^2$$

where time interval $k = 1, 2, \cdots, 8$. The trip table is given in Table 10.1. The parameters for each link travel time function are given in Tables 10.2. The route choice dispersion parameter θ is given as 1. The optimal link flows and the corresponding optimal link travel times are given in Tables 10.3. The optimal route travel times are given in Tables 10.4. In this discrete time example, $x_a(k)$ represents vehicles on the link at the beginning of interval k; $u_a(k)$ and $v_a(k)$ represent inflow and exit flow during interval k.
In this example, note from Table 10.4 that the mean instantaneous travel times on routes 1–2–4 and 1–3–4 are not equal during each interval. These travel times are a property of the logit-type route flows in each interval.

Table 10.1: Required Flows from Origin 1 to Destination 4

Time Interval k	1	2	3	4	5	6	7	8
Flow/interval	40.	0.	0.	0.	0.	0.	0.	0.

Table 10.2: Parameters of Link Travel Time Functions for Instantaneous Model

link a	β_{1a}	β_{2a}	β_{3a}	β_{4a}	β_{5a}	β_{6a}
1—2	1.	0.001	0.	0.	0.015	0.002
1—3	2.	0.006	0.	0.	0.030	0.004
2—4	1.	0.001	0.	0.	0.015	0.002
3—4	1.	0.006	0.	0.	0.030	0.004

10.5 Notes

We have formulated an instantaneous stochastic dynamic user-optimal route choice problem as a general variational inequality. Furthermore, the problem is formulated as an optimal control model under the assumption of logit route choice probability. The solutions of the variational inequality and the optimal control model result in dispersed dynamic route choice governed by a logit distribution incorporating mean instantaneous route travel times. In stochastic dynamic route choice models, the O-D departure flows are dispersed across different travel time routes so that stochastic dynamic models represent more realistic travel choice behavior than deterministic models. This is one of the major merits of stochastic dynamic route choice models over deterministic dynamic route choice models.

In reality, drivers may rely on current information or predictive information to choose routes. Thus, drivers can be stratified into different classes. In each class, drivers may have different compliance to current information or future predictive information. In other words, a more realistic model could be a combined instantaneous and ideal SDUO route choice model with multiple classes of travelers.

It is expected that other SDUO route choice models based on different types of distributions of travel time perception errors will be developed. Figure 10.3 describes how realistic SDUO route choice models might be developed. Our future research also includes calibration of the route choice dispersion parameters and development of more general models incorporating stochastic mode and departure time choice.

In this chapter, we have also suggested algorithms for solving the instantaneous stochastic dynamic route choice problems. The solutions of the models

Table 10.3: Optimal Link Flows and Travel Times for Instantaneous Model

Interval k	Link a	Vehicles $x_a(k+1)$	Inflow $u_a(k)$	Exit Flow $v_a(k)$	Vehicles $x_a(k)$	Travel Time $c_a(k)\&\tau_a(k)$
1	1 — 2	28.8	28.8	0.0	0.0	1.83
2	1 — 2	25.1	0.0	3.6	10.8	2.85
3	1 — 2	14.3	0.0	10.8	25.1	4.02
4	1 — 2	0.0	0.0	14.3	14.3	4.49
1	1 — 3	11.2	11.2	0.0	0.0	2.76
2	1 — 3	11.2	0.0	0.0	11.2	2.50
3	1 — 3	2.3	0.0	8.9	2.3	2.18
4	1 — 3	0.0	0.0	2.3	0.0	2.00
1	2 — 4	0.0	0.0	0.0	0.0	1.00
2	2 — 4	3.6	3.6	0.0	0.0	1.01
3	2 — 4	10.8	10.8	3.6	3.6	1.34
4	2 — 4	21.5	14.3	3.6	10.8	1.64
5	2 — 4	14.3	0.0	7.2	21.5	2.70
6	2 — 4	14.3	0.0	0.0	14.3	1.41
7	2 — 4	5.9	0.0	8.4	14.3	2.47
8	2 — 4	0.0	0.0	5.9	5.9	1.60
1	3 — 4	0.0	0.0	0.0	0.0	1.00
2	3 — 4	0.0	0.0	0.0	0.0	1.00
3	3 — 4	8.9	8.9	0.0	0.0	1.48
4	3 — 4	8.5	2.3	2.7	8.9	1.57
5	3 — 4	2.3	0.0	6.2	8.5	2.46
6	3 — 4	2.1	0.0	0.2	2.3	1.02
7	3 — 4	0.0	0.0	2.1	2.1	1.15

Table 10.4: Instantaneous Route Travel Times

Interval	Mean Instantaneous Route Travel Times	
k	Route 1-2-4	Route 1-3-4
1	2.83	3.76
2	3.86	3.50
3	5.36	3.66
4	6.13	3.57
5	3.70	4.46
6	2.41	3.02
7	3.47	3.15
8	2.60	3.0

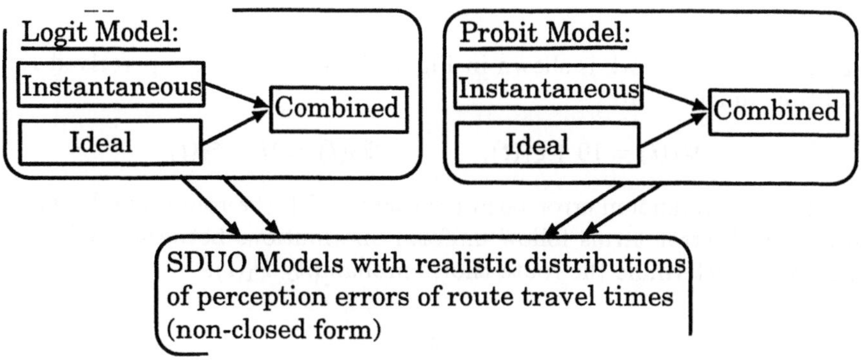

Figure 10.3: Toward Realistic SDUO Route Choice Models

result in dispersed dynamic route choice governed by a probit or logit distribution incorporating mean instantaneous route travel times. The MSA method and Monte Carlo sampling are used to solve the problem in combination with the relaxation method. Especially, the DYNASTOCH2 algorithm is proposed to solve the subproblems of the logit-based instantaneous SDUO route choice model. The DYNASTOCH2 algorithm are implemented using link flow variables so that explicit route enumeration can be avoided. The algorithms are implemented over an expanded time-space network which enables our previously developed solution techniques for dynamic network models to be fully used.

Problems

10.1 Discuss the relationship of instantaneous SDUO route choice conditions and instantaneous DUO route choice conditions. Explain how instantaneous SDUO route choice conditions reduce to instantaneous DUO route choice conditions under certain conditions.

10.2 Compare STOCH algorithm and DYNASTOCH2 algorithm. Discuss when DYNASTOCH2 algorithm reduces to the STOCH algorithm.

10.3 The problems of the IID property for logit functions are well known in static route choice. Discuss the impact of the IID property on dynamic route choice. Compare the impact of the IID property on both static and dynamic route choice.

10.4 The probit route choice function has a negative tail (one part of distribution function located in the area of negative travel times). Discuss how the effect of this negative tail can be eliminated in solving an instantaneous SDUO route choice problem.

10.5 A two-link network is shown in Figure 10.4. Assume the perceived route travel time functions are

$$\Psi_1(t) = 10 + \xi_1(t), \qquad \Psi_2(t) = 15 + \xi_2(t)$$

where $\xi_p^{rs}(t)$ is a random error term associated with the route p under consideration. Both error terms follow uniform distributions between [-5,5], [-8,8], respectively. Determine the route choice probability $P_1(t)$.

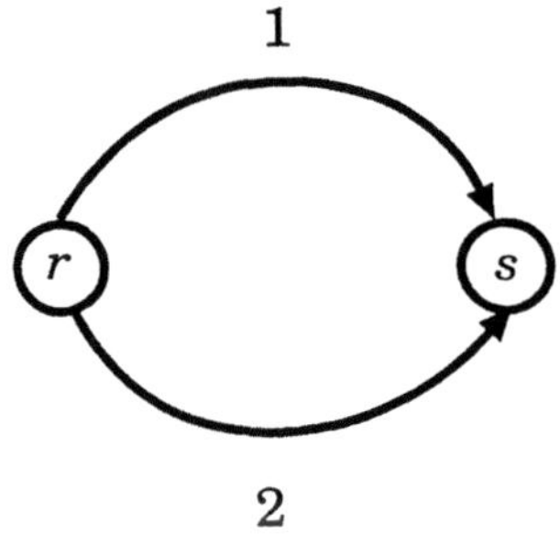

Figure 10.4: Network of Problem 10.5

10.6 For the network shown in Figure 10.2, the mean link travel time functions are

$$c_1(t) = u_1(t) + 7, \qquad c_2(t) = u_2(t) + 10$$

$$c_3(t) = 15, \qquad c_4(t) = u_4(t) + 10$$

Assume that the following logit-based stochastic route choice function is used to assign traffic between node 1 and node 4:

$$P_1(t) = \frac{\exp[-\psi_1^{14}(t)]}{\exp[-\psi_1^{14}(t)] + \exp[-\psi_2^{14}(t)]}$$

The O-D flow is $f^{14}(t) = 15t$ units. Determine mean inflows on links 1 and 2 for time period $[0, 10]$ and find the mean instantaneous route travel times. [Hint: assume no flow dispersion in the link flow propagation.]

10.7 When link travel time function is simplified as

$$c_a(t) = c_a[x_a(t)].$$

Formulate a route-based optimal control model of the instantaneous SDUO route choice problem.

10.8 When link travel time function is simplified as

$$c_a(t) = c_a[x_a(t)].$$

Formulate a link-based optimal control model of the instantaneous SDUO route choice problem.

10.9 Discuss when the optimal control model (10.69)-(10.82) reduces to Fisk's static logit route choice model.

10.10 Compare the solution algorithms for the instantaneous SDUO route choice problem and ideal SDUO route choice problem. Identify the major differences.

10.11 Explore a solution algorithm for instantaneous SDUO route choice problem, which does not require time-space network expansion.

Chapter 11

Combined Departure Time/ Route Choice Models

A dynamic route guidance system seeks to improve the utilization of transportation network capacity and reduce travel times, congestion and the effect of incidents. Provided with early detection of incidents and congestion, users of the system may be able to choose alternative routes, if there is excess capacity in the network, or shift their departure times to avoid congestion when little road capacity is available.

Journey-to-work travelers have especially important requirements for avoiding congested routes in order to arrive at work on time. Each departure time choice is based on minimal origin-destination travel times at each possible departure time. Of course, any change in departure times alters the traffic flow patterns in the network, so route and departure time decisions of other travelers will be affected.

The choice of departure time has been addressed by several researchers, including Abkowitz (1981) and Hendrickson and Plank (1984), who developed work trip scheduling models. De Palma et al (1983) and Ben-Akiva et al (1984) modeled departure time choice over a simple network with one bottleneck using the general continuous logit model. Mahmassani and Herman (1984) used a traffic flow model to derive the equilibrium joint departure time and route choice pattern over a parallel route network. Mahmassani and Chang (1987) further developed the concept of equilibrium departure time choice and presented the boundedly-rational user-equilibrium concept under which all drivers in the system are satisfied with their current travel choices, and thus feel no need to improve their outcome by changing decisions. More recently, several departure time choice models have been proposed by various researchers using different approaches on dynamic traffic networks. Janson (1993) formulated a dynamic user-equilibrium route choice model in which O-D flows have variable departure times and scheduled arrival times. Friesz et al (1993) presented a joint departure time and route choice model using the variational inequality approach. Ghali and Smith (1993) also considered this problem using microscopic representation of vehicle streams.

In this chapter, we present a dynamic, user-optimal departure time and route choice model for a general network with multiple origin-destination pairs. We model this choice problem by specifying that a given number of travelers are ready for departure between each origin-destination pair at time 0. However, their departure times may be delayed to reduce their overall travel costs. This model extends our initial DUO route choice model in one important respect: both departure time and route over a road network must be chosen. Additional network constraints are first formulated in Section 11.1. A route-based variational inequality model for joint departure time/route choice is presented in Section 11.2. In a parallel fashion, a link-based variational inequality model is proposed in Section 11.3. The properties of the models and the relationship between the variational inequality models and the optimization models is discussed in Section 11.4. Subsequently, a solution algorithm for link-based variational inequality model is proposed and then a numerical example is given to illustrate that total travel time can be decreased by choosing appropriate departure times.

11.1 Additional Network Constraints

We consider the following joint departure time and route choice situation: a given number of vehicles are scheduled to depart from each origin r to each destination s at an initial time 0. Denote the *cumulative* number of departing vehicles from origin r to destination s from time 0 to t as the state variable $F^{rs}(t)$. In this problem, the total number of departing vehicles $F^{rs}(T)$ for each O-D pair rs is assumed to be given. Also, $F_p^{rs}(t)$ denotes the cumulative number of departing vehicles from origin r toward destination s along route p by time t.

In addition, denote the instantaneous departure rate from origin node r toward destination node s at time t as $f^{rs}(t)$, which is a function of time; $f_p^{rs}(t)$ denotes the departure rate on route p and $f_p^{rs}(t)$ and $f^{rs}(t)$ are control variables to be determined according to the actual travel time between the origin and the destination. Then, we have an additional state equation for each origin r

$$\frac{dF_p^{rs}(t)}{dt} = f_p^{rs}(t) \qquad \forall p, r \neq s, s. \tag{11.1}$$

Also, at initial time $t = 0$,

$$F_p^{rs}(0) = 0, \qquad \forall p, r, s. \tag{11.2}$$

Assume that there are P routes from origin r to destination s (these can be generated as needed). Denote the indicator parameters δ_{ap}^{rs} as

$$\delta_{ap}^{rs} = \begin{cases} 1 & \text{if link } a \text{ is on route } p \text{ between O-D pair } rs \\ 0 & \text{otherwise.} \end{cases}$$

Flow conservation at origin node r relates the departure rates ($f^{rs}(t)$ and $f_p^{rs}(t)$) to the flow entering each link emanating from the origin. These flow conservation equations for origin r can be expressed as

$$f_p^{rs}(t) = \sum_{a \in A(r)} \delta_{ap}^{rs} u_{ap}^{rs}(t) \qquad \forall p, r, s; r \neq s; \tag{11.3}$$

$$\sum_p f_p^{rs}(t) = f^{rs}(t) \qquad \forall r, s; r \neq s. \tag{11.4}$$

Denote the *cumulative* number of vehicles arriving at destination s from origin r by time t as the state variable $E^{rs}(t)$; $E_p^{rs}(t)$ denotes the cumulative number of vehicles arriving at destination s from origin r along route p by time t. Denote the instantaneous flows *arriving at* destination node s from origin node r at time t as $e^{rs}(t)$, which is also a control variable. The control variable $e_p^{rs}(t)$ denotes the arrival rate on route p. Thus, we have an additional state equation for each destination s

$$\frac{dE_p^{rs}(t)}{dt} = e_p^{rs}(t) \qquad \forall p, r, s \neq r. \tag{11.5}$$

At the initial time $t = 0$,

$$E_p^{rs}(0) = 0 \qquad \forall p, r, s. \tag{11.6}$$

These variables must be nonnegative at all times:

$$E_p^{rs}(t) \geq 0, \quad F_p^{rs}(t) \geq 0, \quad e_p^{rs}(t) \geq 0, \quad f_p^{rs}(t) \geq 0, \quad \forall p, r, s. \tag{11.7}$$

Flow conservation at destination node s relates the arriving flow ($e^{rs}(t)$ and $e_p^{rs}(t)$) to the flow exiting each link leading to destination s at time t. Thus, the flow conservation equations for destination s can be expressed as

$$e_p^{rs}(t) = \sum_{a \in B(s)} \delta_{ap}^{rs} v_{ap}^{rs}(t) \qquad \forall p, r, s; s \neq r; \tag{11.8}$$

$$\sum_p e_p^{rs}(t) = e^{rs}(t) \qquad \forall r, s; s \neq r. \tag{11.9}$$

11.2 A Route-Based Model

A number of vehicles are ready to depart at an initial time 0, but these drivers may prefer to delay their departure times in order to reduce their driving times. Drivers are assumed to make their departure time choices so as to minimize their individual disutility functions defined on travel time and pre-trip delay. Of course, the change of departure flow rates will change the traffic in the network so that the travel times for other travelers could change.

11.2.1 Route-Based Conditions

We first consider the joint departure time/route choice conditions. A disutility function $\mathcal{U}_p^{rs}(t)$ based on departure times is defined for travelers departing from origin r to destination s on route p at time t. This disutility function is a weighted sum of:

1. waiting time at the origin node;
2. driving time during the trip;
3. a bonus for early arrival or a penalty for late arrival.

Consider the flow which originates at node r at time t and is destined for node s. There is a set of routes $\{p\}$ between O-D pair rs. Define $\eta_p^{rs}(t)$ as the travel time actually experienced over route p by vehicles departing origin r toward destination s at time t. We use a recursive formula to compute the route travel time $\eta_p^{rs}(t)$ for each allowable route. Assume route p consists of nodes $(r, 1, \cdots, i-1, i, \cdots, s)$. Denote $\eta_p^{ri}(t)$ as the travel time actually experienced over route p from origin r to node I by vehicles departing from origin r at time t. Then, a recursive formula for route travel time $\eta_p^{rs}(t)$ is:

$$\eta_p^{ri}(t) = \eta_p^{r(i-1)}(t) + \tau_a[t + \eta_p^{r(i-1)}(t)] \qquad \forall p, r, i; i = 1, 2, \cdots, s;$$

where link $a = (i-1, i)$.

We define one unit of disutility to equal one unit of in-vehicle driving time, and one unit of waiting time prior to departure to be equivalent to α units of disutility ($\alpha \leq 1$); α could be negative since staying at home may have positive utility. Since all travelers are able to depart at time 0, αt is the disutility for a departure at time t due to waiting. Furthermore, we assume there is a desired arrival time interval $[t_{rs}^* - \Delta_{rs}, t_{rs}^* + \Delta_{rs}]$ for travelers at each destination s, where t_{rs}^* is the center of the required arrival time interval (e.g. work starting time) associated with travelers departing from origin r toward destination s. Δ_{rs} represents the arrival time flexibility at destination s for travelers departing from origin r toward destination s.

The disutility function for the route-based joint departure time and route choice problem is constructed as

$$\mathcal{U}_p^{rs}(t) = \alpha t + \eta_p^{rs}(t) + \mathcal{V}_p^{rs}[t, \eta_p^{rs}(t); t_{rs}^*] \qquad \forall p, r, s, \tag{11.10}$$

where t is the departure time of travelers and $\mathcal{V}_p^{rs}[t, \eta_p^{rs}(t); t_{rs}^*]$ is the disutility for early or late arrival which is defined as follows

$$\mathcal{V}_p^{rs}[t, \eta_p^{rs}(t); t_{rs}^*] =$$

$$\begin{cases} \gamma_1[t + \eta_p^{rs}(t) - t_{rs}^* + \Delta_{rs}^*]^2 & \text{if } t + \eta_p^{rs}(t) < t_{rs}^* - \Delta_{rs}^* \quad \text{(early arrival)} \\ 0 & \text{if } |t + \eta_p^{rs}(t) - t_{rs}^*| \leq \Delta_{rs}^* \\ \gamma_2[t + \eta_p^{rs}(t) - t_{rs}^* - \Delta_{rs}^*]^2 & \text{if } t + \eta_p^{rs}(t) > t_{rs}^* + \Delta_{rs}^* \quad \text{(late arrival)} \end{cases}$$

where t is the departure time of travelers and γ_1 and γ_2 are parameters ($\gamma_2 \gg \alpha$). γ_1 is negative because early arrival should be encouraged rather than discouraged. This arrival time disutility function is shown in Figure 11.1.

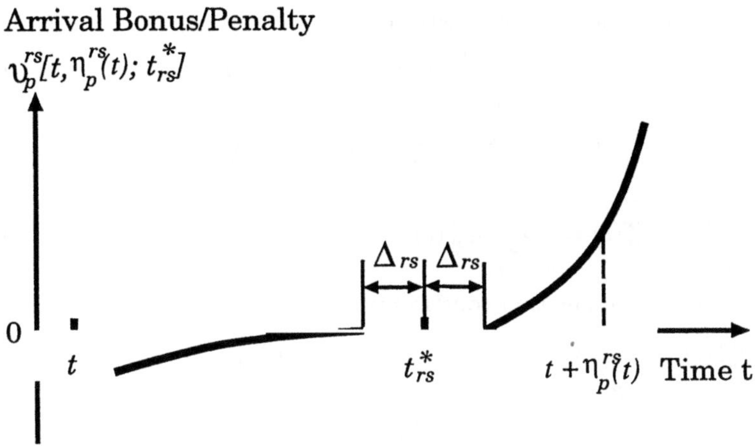

Figure 11.1: Arrival Time Disutility

The dynamic user-optimal departure time/route choice conditions require that for each O-D pair rs at any time t, if there is a positive departure flow $f_p^{rs^*}(t) > 0$ over route p, the disutility $\mathcal{U}_p^{rs^*}(t)$ for route p must equal the minimal rs disutility $\mathcal{U}_{min}^{rs^*}$ over time t. Furthermore, if the departure flow $f_p^{rs^*}(t)$ over route p equals 0 at time t, the disutility $\mathcal{U}_p^{rs^*}(t)$ over route p at time t must be greater than or equal to the minimal rs disutility $\mathcal{U}_{min}^{rs^*}$. The route-based DUO departure time/route choice conditions can be written as

$$\mathcal{U}_p^{rs^*}(t) - \mathcal{U}_{min}^{rs^*} \geq 0 \qquad \forall p, r, s; \tag{11.11}$$

$$f_p^{rs^*}(t)\ [\mathcal{U}_p^{rs^*}(t) - \mathcal{U}_{min}^{rs^*}] = 0 \qquad \forall p, r, s; \tag{11.12}$$

$$f_p^{rs}(t) \geq 0 \qquad \forall p, r, s. \tag{11.13}$$

where the asterisk denotes that the travel disutility is computed using DUO departure flows and route flows.

11.2.2 Dynamic Network Constraints

In this section, the constraint set for our dynamic user-optimal departure time/route choice problem is first summarized as follows.

Relationships between state and control variables:

$$\frac{dx_{ap}^{rs}}{dt} = u_{ap}^{rs}(t) - v_{ap}^{rs}(t) \qquad \forall a, p, r, s; \tag{11.14}$$

$$\frac{dE_p^{rs}(t)}{dt} = e_p^{rs}(t) \qquad \forall r,s,p; \tag{11.15}$$

$$\frac{dF_p^{rs}(t)}{dt} = f_p^{rs}(t) \qquad \forall r,s,p; \tag{11.16}$$

Flow conservation constraints:

$$f_p^{rs}(t) = \sum_{a\in A(r)} \delta_{ap}^{rs} u_{ap}^{rs}(t) \qquad \forall p,r,s; \tag{11.17}$$

$$e_p^{rs}(t) = \sum_{a\in B(s)} \delta_{ap}^{rs} v_{ap}^{rs}(t) \qquad \forall p,r,s; \tag{11.18}$$

$$\sum_{a\in B(j)} v_{ap}^{rs}(t) = \sum_{a\in A(j)} u_{ap}^{rs}(t) \qquad \forall j,p,r,s; j \neq r,s; \tag{11.19}$$

Flow propagation constraints:

$$x_{ap}^{rs}(t) = \sum_{b\in\tilde{p}} \{x_{bp}^{rs}[t+\tau_a(t)] - x_{bp}^{rs}(t)\} + \{E_p^{rs}[t+\tau_a(t)] - E_p^{rs}(t)\}$$

$$\forall r,s,p,j; a \in B(j); j \neq r; \tag{11.20}$$

Definitional constraints:

$$\sum_{rsp} u_{ap}^{rs}(t) = u_a(t), \qquad \sum_{rsp} v_{ap}^{rs}(t) = v_a(t), \qquad \forall a; \tag{11.21}$$

$$\sum_{rsp} x_{ap}^{rs}(t) = x_a(t), \qquad \sum_{rs} x_a^{rs}(t) = x_a(t), \qquad \forall a,r,s; \tag{11.22}$$

$$\sum_{p} E_p^{rs}(t) = E^{rs}(t), \qquad \sum_{p} F_p^{rs}(t) = F^{rs}(t), \qquad \forall r,s; \tag{11.23}$$

$$\sum_{p} f_p^{rs}(t) = f^{rs}(t), \qquad \sum_{p} e_p^{rs}(t) = e^{rs}(t), \qquad \forall r,s; \tag{11.24}$$

Nonnegativity conditions:

$$x_{ap}^{rs}(t) \geq 0, \quad u_{ap}^{rs}(t) \geq 0, \quad v_{ap}^{rs}(t) \geq 0 \quad \forall a,p,r,s; \tag{11.25}$$

$$e_p^{rs}(t) \geq 0, \; f_p^{rs}(t) \geq 0, \; E_p^{rs}(t) \geq 0, \; F_p^{rs}(t) \geq 0 \;\; \forall p,r,s; \tag{11.26}$$

$$f^{rs}(t) \geq 0, \qquad F^{rs}(t) \geq 0 \qquad \forall r,s; \tag{11.27}$$

Boundary conditions:

$$F^{rs}(T) \text{ given} \qquad \forall r,s; \tag{11.28}$$

$$E_p^{rs}(0) = 0, \quad F_p^{rs}(0) = 0 \quad \forall p,r,s; \quad x_{ap}^{rs}(0) = 0, \quad \forall a,p,r,s. \tag{11.29}$$

The first three constraints (11.14)-(11.16) are state equations for each link a and for the cumulative effects at origins and destinations. Equations (11.17)-(11.19) are flow conservation constraints at each node including origins and destinations. Other constraints include flow propagation constraints, definitional constraints, nonnegativity and boundary conditions. In summary, the control variables are $u_{ap}^{rs}(t)$, $v_{ap}^{rs}(t)$, $e_p^{rs}(t)$ and $f_p^{rs}(t)$; the state variables are $x_{ap}^{rs}(t)$, $E_p^{rs}(t)$ and $F_p^{rs}(t)$; the functionals are $\pi^{rs}(t)$.

11.2.3 The Variational Inequality Problem

The equivalent variational inequality formulation of route-based DUO departure time/route choice conditions (11.11)-(11.13) may be stated as follows.

Theorem 11.1. *The dynamic traffic flow satisfying constraints (11.14)-(11.29) is in a route-based DUO departure time/route choice state if and only if it satisfies the variational inequality problem:*

$$\int_0^T \left\{ \sum_{rs} \sum_{p} \mathcal{U}_p^{rs^*}(t) \left[f_p^{rs}(t) - f_p^{rs^*}(t) \right] \right\} dt \geq 0 \qquad (11.30)$$

Proof of Necessity. [1]

We need to prove that route-based DUO departure time/route choice conditions (11.11)-(11.13) imply variational inequality (11.30). Multiplying equations (11.11) and (11.13), we have

$$f_p^{rs}(t) \, \{\mathcal{U}_p^{rs^*}(t) - \mathcal{U}_{min}^{rs^*}\} \geq 0 \qquad \forall p, r, s. \qquad (11.31)$$

We subtract equation (11.12) from equation (11.31) and obtain

$$[\mathcal{U}^{rs^*}(t) - \mathcal{U}_{min}^{rs^*}] \, [f_p^{rs}(t) - f_p^{rs^*}(t)] \geq 0 \qquad \forall p, r, s. \qquad (11.32)$$

Summing equation (11.32) for all routes p and all O-D pairs rs, it follows that

$$\sum_{rs} \sum_{p} [\mathcal{U}_p^{rs^*}(t) - \mathcal{U}_{min}^{rs^*}] \, [f_p^{rs}(t) - f_p^{rs^*}(t)] \geq 0 \qquad (11.33)$$

Integrating the above equation (11.33) from time 0 to T, we have

$$\int_0^T \sum_{rs} \sum_{p} [\mathcal{U}_p^{rs^*}(t) - \mathcal{U}_{min}^{rs^*}] \, [f_p^{rs}(t) - f_p^{rs^*}(t)] \, dt \geq 0 \qquad (11.34)$$

or

$$\int_0^T \sum_{rs} \left\{ \sum_{p} \mathcal{U}_p^{rs^*}(t) \, [f_p^{rs}(t) - f_p^{rs^*}(t)] \right.$$

$$\left. - \sum_{rs} \sum_{p} \mathcal{U}_{min}^{rs^*} \, [f_p^{rs}(t) - f_p^{rs^*}(t)] \right\} dt \geq 0 \qquad (11.35)$$

By the definition of departure flows, we have

$$\int_0^T \sum_{p} f_p^{rs}(t) \, dt = F^{rs}(T) = \int_0^T \sum_{p} f_p^{rs^*}(t) \, dt$$

[1] The necessity and sufficiency proofs can be skipped without loss of continuity.

Thus, the second term of equation (11.35) is 0 and equation (11.35) becomes variational inequality (11.30).

Proof of Sufficiency.

We need to prove that any solutions $f_p^{rs^*}(t)$ to variational inequality (11.30) satisfy DUO departure time/route choice conditions (11.11)-(11.13). We know that the first and third DUO departure time/route choice conditions (11.11) and (11.13) hold by definition. Thus, we only need to prove that the second DUO departure time/route choice condition (11.12) also holds.

Assume that DUO departure time/route choice condition (11.12) does not hold only for route q between O-D pair kn during a short time interval $[d-\delta, d+\delta] \in [0,T]$, i.e.,

$$f_q^{kn^*}(t) > 0 \qquad \text{and} \qquad \mathcal{U}_q^{kn^*}(t) - \mathcal{U}_{min}^{kn^*} > 0 \tag{11.36}$$

or

$$f_q^{kn^*}(t)\ \{\mathcal{U}_q^{kn^*}(t) - \mathcal{U}_{min}^{kn^*}\} > 0 \tag{11.37}$$

Integrating equation (11.37) from $(d-\delta)$ to $(d+\delta)$, we have

$$\int_{d-\delta}^{d+\delta} f_q^{kn^*}(t)\ \{\mathcal{U}_q^{kn^*}(t) - \mathcal{U}_{min}^{kn^*}\}\ dt > 0 \tag{11.38}$$

or

$$\int_{d-\delta}^{d+\delta} f_q^{kn^*}(t)\ \mathcal{U}_q^{kn^*}(t)\ dt\ >\ \int_{d-\delta}^{d+\delta} f_q^{kn^*}(t)\ \mathcal{U}_{min}^{kn^*}\ dt \tag{11.39}$$

We now need to find a set of feasible departure route inflows $f_p^{rs}(t)$ which will contradict variational inequality (11.30).

For routes between O-D pairs $rs \neq kn$, we allow the feasible departure route inflows $f_p^{rs}(t)$ to equal the optimal departure route inflows $f_p^{rs^*}(t)$ at each instant of time. For O-D pair kn, we can always find a route l with the minimal disutility $\mathcal{U}_l^{kn^*}(t) = \mathcal{U}_{min}^{kn^*}$ for a short time interval $[b-\epsilon, b+\epsilon] \in [0,T]$. We note that route l is evaluated under the optimal departure route inflow pattern $\{f_p^{rs^*}(t)\}$. For other routes $p \neq q, l$ between O-D pair kn, we allow the feasible departure route inflows $f_p^{kn}(t)$ to equal the optimal departure route inflows $f_p^{kn^*}(t)$ at each instant of time. For routes q, l between O-D pair kn, we allow the feasible departure route inflows $f_q^{kn}(t)$, $f_l^{kn}(t)$ to equal optimal departure route inflows $f_q^{kn^*}(t)$ and $f_l^{kn^*}(t)$ for times outside the time interval $[d-\delta, d+\delta]$ and $[b-\epsilon, b+\epsilon]$, respectively. Furthermore, we shift all departure inflows over route q during time interval $[d-\delta, d+\delta]$ to route l over the time interval $[b-\epsilon, b+\epsilon]$, during which disutility $\mathcal{U}_l^{kn^*}(t) = \mathcal{U}_{min}^{kn^*}$. It follows that

$$\int_{b-\epsilon}^{b+\epsilon} f_l^{kn}(t)dt - \int_{b-\epsilon}^{b+\epsilon} f_l^{kn^*}(t)dt = \int_{d-\delta}^{d+\delta} f_q^{kn^*}(t)dt \tag{11.40}$$

Variational inequality (11.30) becomes

$$\int_0^T \sum_{rs}\sum_p \mathcal{U}_p^{rs^*}(t)\left[f_p^{rs}(t) - f_p^{rs^*}(t)\right] dt$$
$$= \int_0^T \left\{\mathcal{U}_l^{kn^*}(t)\left[f_l^{kn}(t) - f_l^{kn^*}(t)\right]\right.$$
$$\left. + \ \mathcal{U}_q^{kn^*}(t)\left[f_q^{kn}(t) - f_q^{kn^*}(t)\right]\right\} dt \tag{11.41}$$

The second term of equation (11.41) becomes

$$\int_0^T \mathcal{U}_q^{kn^*}(t)\left[f_q^{kn}(t) - f_q^{kn^*}(t)\right] dt$$
$$= \int_0^T \mathcal{U}_q^{kn^*}(t)\, f_q^{kn}(t)\, dt - \int_0^T \mathcal{U}_q^{kn^*}(t) f_q^{kn^*}(t)\, dt$$
$$= \int_0^T \mathcal{U}_q^{kn^*}(t)\, f_q^{kn}(t)\, dt - \left\{\int_0^{d-\delta} + \int_{d+\delta}^T\right\} \mathcal{U}_q^{kn^*}(t)\, f_q^{kn^*}(t)\, dt$$
$$- \int_{d-\delta}^{d+\delta} \mathcal{U}_q^{kn^*}(t) f_q^{kn^*}(t)\, dt \tag{11.42}$$

Substituting equation (11.39) into equation (11.42), we have

$$\int_0^T \mathcal{U}_q^{kn^*}(t)\left[f_q^{kn}(t) - f_q^{kn^*}(t)\right] dt$$
$$< \int_0^T \mathcal{U}_q^{kn^*}(t)\, f_q^{kn}(t)\, dt - \left\{\int_0^{d-\delta} + \int_{d+\delta}^T\right\} \mathcal{U}_q^{kn^*}(t)\, f_q^{kn^*}(t)\, dt$$
$$- \int_{d-\delta}^{d+\delta} \mathcal{U}_{min}^{kn^*} f_q^{kn^*}(t)\, dt \tag{11.43}$$

The first term of equation (11.41) becomes

$$\int_0^T \mathcal{U}_l^{kn^*}(t)\ \left[f_l^{kn}(t) - f_l^{kn^*}(t)\right] dt$$
$$= \int_{b-\epsilon}^{b+\epsilon} \mathcal{U}_l^{kn^*}(t)\ \left[f_l^{kn}(t) - f_l^{kn^*}(t)\right] dt$$
$$= \int_{b-\epsilon}^{b+\epsilon} \mathcal{U}_{min}^{kn^*}\ \left[f_l^{kn}(t) - f_l^{kn^*}(t)\right] dt$$
$$= \mathcal{U}_{min}^{kn^*} \int_{b-\epsilon}^{b+\epsilon} \left[f_l^{kn}(t) - f_l^{kn^*}(t)\right] dt \tag{11.44}$$

where $\mathcal{U}_l^{kn^*}(t) = \mathcal{U}_{min}^{kn^*}$ for time interval $[b-\epsilon, b+\epsilon]$. Substituting equation (11.40) into equation (11.44), we have

$$\int_0^T \mathcal{U}_l^{kn^*}(t) \left[f_l^{kn}(t) - f_l^{kn^*}(t)\right] dt = \mathcal{U}_{min}^{kn^*} \int_{d-\delta}^{d+\delta} f_q^{kn^*}(t)\, dt \qquad (11.45)$$

Summing equations (11.43) and (11.45) and substituting into equation (11.41), we obtain

$$\begin{aligned} &\int_0^T \sum_{rs}\sum_p \mathcal{U}_p^{rs^*}(t) \left[f_p^{rs}(t) - f_p^{rs^*}(t)\right] dt \\ < \quad &\int_0^T \mathcal{U}_q^{kn^*}(t)\, f_q^{kn}(t)\, dt - \left\{\int_0^{d-\delta} + \int_{d+\delta}^T\right\} \mathcal{U}_q^{kn^*}(t)\, f_q^{kn^*}(t)\, dt \quad (11.46) \end{aligned}$$

Furthermore, since $f_q^{kn}(t) = 0$ for time interval $[d-\delta, d+\delta]$ and $f_q^{kn}(t) = f_q^{kn^*}(t)$ for time intervals $[0, d-\delta]$, and $[d+\delta, T]$, we simplify equation (11.46) as

$$\begin{aligned} &\int_0^T \sum_{rs}\sum_p \mathcal{U}_p^{rs^*}(t) \left[f_p^{rs}(t) - f_p^{rs^*}(t)\right] dt \\ < \quad &\left\{\int_0^{d-\delta} + \int_{d+\delta}^T\right\} \mathcal{U}_q^{kn^*}(t)\, f_q^{kn}(t)\, dt - \left\{\int_0^{d-\delta} + \int_{d+\delta}^T\right\} \mathcal{U}_q^{kn^*}(t)\, f_q^{kn^*}(t)\, dt \\ = \quad &\left\{\int_0^{d-\delta} + \int_{d+\delta}^T\right\} \mathcal{U}_q^{kn^*}(t) \left[f_q^{kn}(t) - f_q^{kn^*}(t)\right] dt = 0 \qquad (11.47) \end{aligned}$$

The above equation contradicts variational inequality (11.30). Therefore, any optimal solutions $\{f_p^{rs^*}(t)\}$ to variational inequality (11.30) satisfy the second DUO departure time/route choice condition (11.12). Since we have proved the necessity and sufficiency of the equivalence of variational inequality (11.30) to the route-based DUO departure time/route choice conditions (11.11)-(11.13), the proof is complete.

11.3 A Link-Based Model

Because dynamic traffic flow does not have a constant flow rate during propagation over links and routes, the route-based VI can not be transformed into a link-based VI. Thus, it is very difficult to develop a solution algorithm for a route-based VI without explicit route enumeration. In Chapter 5, we presented a link-based VI model for the ideal DUO route choice problem so that route enumeration can be avoided in both the formulation and the solution procedure. This approach allows the dynamic VI route choice model to be applied to realistic transportation networks.

Using a similar approach, we extend the dynamic route choice model to include departure time choice as well. A link-based ideal dynamic user-optimal

(DUO) departure time/route choice model is presented for a network with multiple origin-destination pairs in this section. The set of dynamic network constraints for the link-based VI model is identical to constraint set (11.14)-(11.29) of the route-based VI model in Section 11.2. The basic difference between the two models is that the link-based VI model is formulated using link-based flow variables instead of route-based variables.

11.3.1 Link-Based Conditions

Departure Time Choice Conditions

We first consider the problem of departure time choice. A disutility function $\mathcal{U}^{rs}(t)$ based on departure times is defined for travelers departing from origin r to destination s at time t. This disutility function is a weighted sum of:

1. waiting time at the origin node;
2. driving time during the trip;
3. a bonus for early arrival or a penalty for late arrival.

A disutility function $\mathcal{U}^{rs}(t)$ based on departure times is defined for travelers departing from origin r to destination s at time t. Denote $\pi^{rs}(t)$ as the minimal travel time experienced by vehicles departing from origin r to destination s at time t. We also define the disutility for early or late arrival as follows

$$\mathcal{V}^{rs}[t, \pi^{rs}(t); t^*_{rs}] =$$

$$\begin{cases} \gamma_1[t + \pi^{rs}(t) - t^*_{rs} + \Delta^*_{rs}]^2 & \text{if } t + \pi^{rs}(t) < t^*_{rs} - \Delta^*_{rs} \quad \text{(early arrival)} \\ 0 & \text{if } |t + \pi^{rs}(t) - t^*_{rs}| \leq \Delta^*_{rs} \\ \gamma_2[t + \pi^{rs}(t) - t^*_{rs} - \Delta^*_{rs}]^2 & \text{if } t + \pi^{rs}(t) > t^*_{rs} + \Delta^*_{rs} \quad \text{(late arrival)} \end{cases}$$

where t is the departure time of travelers and γ_1 and γ_2 are parameters ($\gamma_1 \leq 0$, $\gamma_2 \gg \alpha$). Thus, the disutility function for the joint departure time and route choice problem is constructed as

$$\mathcal{U}^{rs}(t) = \alpha t + \pi^{rs}(t) + \mathcal{V}^{rs}[t, \pi^{rs}(t); t^*_{rs}] \qquad \forall r, s, \qquad (11.48)$$

where t is the departure time of travelers.

The dynamic user-optimal departure time choice conditions require that for each O-D pair rs at any time t, if there is a positive departure flow $f^{rs}(t) > 0$, the disutility $\mathcal{U}^{rs}(t)$ must equal the minimal rs disutility $\mathcal{U}^{rs}_{min}$ over time t. Furthermore, if the departure flow $f^{rs}(t)$ equals 0 at time t, the disutility $\mathcal{U}^{rs}(t)$ at time t must be greater than or equal to the minimal rs disutility $\mathcal{U}^{rs}_{min}$. The DUO departure time choice conditions can be written as

$$\mathcal{U}^{rs^*}(t) - \mathcal{U}^{rs}_{min} \geq 0 \qquad \forall r, s; \qquad (11.49)$$

$$f^{rs^*}(t)\ \{\mathcal{U}^{rs^*}(t) - \mathcal{U}^{rs}_{min}\} = 0 \qquad \forall r, s; \qquad (11.50)$$

$$f^{rs}(t) \geq 0 \qquad \forall r, s. \tag{11.51}$$

where the asterisk denotes that the travel disutility is computed using DUO departure flows.

Ideal DUO Route Choice Conditions

Next we discuss the problem of route choice. The dynamic user-optimal route choice problem is to find the dynamic trajectories of link states and inflow and exit flow control variables, given the network, the link travel time functions and the time-dependent O-D departure rate requirements. The O-D departure rates are specified by the departure time choice problem and are therefore exogenous to the route choice problem. In this joint DUO departure time/route choice problem, we use the ideal DUO route choice model of Chapter 5 as the basis.

Recall the travel-time-based ideal DUO route choice state is defined as:

> **Travel-Time-Based Ideal DUO State:** *If, for each O-D pair at each instant of time, the* actual *travel times experienced by travelers departing at the same time are equal and minimal, the dynamic traffic flow over the network is in a travel-time-based ideal dynamic user-optimal state.*

For vehicles departing from origin r at time t, denote $\Omega_a^{rj^*}(t)$ as the difference between the minimal travel time from r to j and the travel time from origin r to node j via the minimal travel time route from origin r to node i and link a. It follows that

$$\Omega_a^{rj^*}(t) = \pi^{ri^*}(t) + \tau_a[t + \pi^{ri^*}(t)] - \pi^{rj^*}(t). \qquad \forall a, r; a = (i, j). \tag{11.52}$$

Thus, the link-time-based ideal DUO route choice conditions are:

$$\Omega_a^{rj^*}(t) \geq 0 \qquad \forall a = (i, j), r; \tag{11.53}$$

$$u_a^{rs^*}[t + \pi^{ri^*}(t)]\, \Omega_a^{rj^*}(t) = 0 \qquad \forall a = (i, j), r, s; \tag{11.54}$$

$$u_a^{rs}[t + \pi^{ri^*}(t)] \geq 0 \qquad \forall a = (i, j), r, s. \tag{11.55}$$

Joint DUO Departure Time/Route Choice Conditions

In order to simplify the presentation, we rewrite the combined link-based DUO departure time/route choice conditions as follows:

$$\Omega_a^{rj^*}(t) \geq 0 \qquad \forall a = (i, j), r; \tag{11.56}$$

$$u_a^{rs^*}[t + \pi^{ri^*}(t)]\, \Omega_a^{rj^*}(t) = 0 \qquad \forall a = (i, j), r, s; \tag{11.57}$$

$$u_a^{rs}[t + \pi^{ri^*}(t)] \geq 0 \qquad \forall a = (i, j), r, s; \tag{11.58}$$

$$\mathcal{U}^{rs^*}(t) - \mathcal{U}_{min}^{rs^*} \geq 0 \qquad \forall r, s; \tag{11.59}$$

$$f^{rs^*}(t)\, \{\mathcal{U}^{rs^*}(t) - \mathcal{U}_{min}^{rs^*}\} = 0 \qquad \forall r, s; \tag{11.60}$$

$$f^{rs}(t) \geq 0 \qquad \forall r, s. \tag{11.61}$$

where $\mathcal{U}_{min}^{rs}$ is the minimal rs disutility over time t.

11.3.2 The Variational Inequality Problem

The equivalent variational inequality formulation of link-based DUO departure time/route choice conditions (11.56)-(11.61) may be stated as follows.

Theorem 11.2. *Dynamic traffic flow satisfying constraints (11.14)-(11.29) is in a DUO departure time/route choice state if and only if it satisfies the variational inequality problem:*

$$\int_0^T \left\{ \sum_{rs} \sum_a \Omega_a^{rj^*}(t) \left\{ u_a^{rs}[t + \pi^{ri^*}(t)] - u_a^{rs^*}[t + \pi^{ri^*}(t)] \right\} + \sum_{rs} \mathcal{U}^{rs^*}(t) \left\{ f^{rs}(t) - f^{rs^*}(t) \right\} \right\} dt \geq 0 \tag{11.62}$$

Proof of Necessity.[2]

We need to prove that DUO departure time/route choice conditions (11.56)-(11.61) imply variational inequality (11.62). We first discuss the ideal DUO route choice conditions (11.56)-(11.58). Multiplying equations (11.56) and (11.58), we have

$$u_a^{rs}[t + \pi^{ri^*}(t)]\, \Omega_a^{rj^*}(t) \geq 0 \qquad \forall a, r, s; a = (i, j). \tag{11.63}$$

Subtracting equation (11.57) from equation (11.63), we obtain

$$\left\{ u_a^{rs}[t + \pi^{ri^*}(t)] - u_a^{rs^*}[t + \pi^{ri^*}(t)] \right\} \Omega_a^{rj^*}(t) \geq 0 \quad \forall a, r, s; a = (i, j). \tag{11.64}$$

Summing equation (11.64) for all links a and all O-D pairs rs, it follows that

$$\sum_{rs} \sum_a \left\{ u_a^{rs}[t + \pi^{ri^*}(t)] - u_a^{rs^*}[t + \pi^{ri^*}(t)] \right\} \Omega_a^{rj^*}(t) \geq 0 \qquad \text{where } a = (i, j). \tag{11.65}$$

Integrating the above equation (11.65) from time 0 to T, we have

$$\int_0^T \sum_{rs} \sum_a \left\{ u_a^{rs}[t + \pi^{ri^*}(t)] - u_a^{rs^*}[t + \pi^{ri^*}(t)] \right\} \Omega_a^{rj^*}(t)\, dt \geq 0 \tag{11.66}$$

We next discuss DUO departure time choice conditions (11.59)-(11.61). Multiplying equations (11.59) and (11.61), we have

$$f^{rs}(t) \left\{ \mathcal{U}^{rs^*}(t) - \mathcal{U}_{min}^{rs} \right\} \geq 0 \qquad \forall r, s. \tag{11.67}$$

[2] The necessity and sufficiency proofs can be skipped without loss of continuity.

We subtract equation (11.60) from equation (11.67) to obtain

$$\mathcal{U}^{rs^*}(t)\left\{f^{rs}(t)-f^{rs^*}(t)\right\}-\mathcal{U}^{rs}_{min}\left\{f^{rs}(t)-f^{rs^*}(t)\right\}\geq 0 \qquad \forall r,s. \qquad (11.68)$$

Summing equation (11.68) for all O-D pairs rs, it follows that

$$\sum_{rs}\mathcal{U}^{rs^*}(t)\left\{f^{rs}(t)-f^{rs^*}(t)\right\}-\sum_{rs}\mathcal{U}^{rs}_{min}\left\{f^{rs}(t)-f^{rs^*}(t)\right\}\geq 0 \qquad (11.69)$$

Integrating the above equation from time 0 to T, we have

$$\int_0^T\sum_{rs}\mathcal{U}^{rs^*}(t)\left\{f^{rs}(t)-f^{rs^*}(t)\right\}dt - \int_0^T\sum_{rs}\mathcal{U}^{rs}_{min}\left\{f^{rs}(t)-f^{rs^*}(t)\right\}dt\geq 0 \qquad (11.70)$$

or

$$\int_0^T\sum_{rs}\mathcal{U}^{rs^*}(t)\left\{f^{rs}(t)-f^{rs^*}(t)\right\}dt - \sum_{rs}\mathcal{U}^{rs}_{min}\int_0^T\left\{f^{rs}(t)-f^{rs^*}(t)\right\}dt\geq 0 \qquad (11.71)$$

By the definition of departure flows, we have

$$\int_0^T f^{rs}(t)\,dt = F^{rs}(T) = \int_0^T f^{rs^*}(t)\,dt$$

Thus, the second term of equation (11.71) is 0 and equation (11.71) becomes

$$\int_0^T\sum_{rs}\mathcal{U}^{rs^*}(t)\left\{f^{rs}(t)-f^{rs^*}(t)\right\}dt\geq 0 \qquad (11.72)$$

Combining equations (11.66) and (11.72), we obtain variational inequality (11.62).

Proof of Sufficiency.

We need to prove that any solutions $u_a^{rs^*}[t+\pi^{ri^*}(t)]$ and $f^{rs^*}(t)$ to variational inequality (11.62) satisfy DUO departure time/route choice conditions (11.56)-(11.61). We know that the first and third ideal DUO route choice conditions (11.56) and (11.58) hold by definition. The fourth and sixth DUO departure time choice conditions (11.59) and (11.61) also hold by definition. Thus, we only need to prove that the second ideal DUO route choice condition (11.57) and the fifth DUO departure time choice condition (11.60) also hold.

In order to prove the above statement, we need to prove that the following three cases are not true.

1. The second ideal DUO route choice condition (11.57) does not hold, but the fifth DUO departure time choice condition (11.60) does hold.

2. The fifth DUO departure time choice condition (11.60) does not hold, but the second ideal DUO route choice condition (11.57) does hold.

3. Both the second ideal DUO route choice condition (11.57) and the fifth DUO departure time choice condition (11.60) do not hold.

Case 1

The proof for Case 1 is similar to that for a pure ideal DUO route choice problem. Assume that the second ideal DUO route choice condition (11.57) does not hold only for a link $b = (l, m)$ for O-D pair pq during a short time interval $[d - \delta, d + \delta]$ where $[d - \delta, d + \delta] \in [0, T]$, i.e.,

$$u_b^{pq^*}[t + \tau^{kl^*}(t)] > 0 \qquad \text{and}$$
$$\Omega_b^{km^*}(t) = \tau^{kl^*}(t) + \tau_b[t + \tau^{kl^*}(t)] - \tau^{km^*}(t) > 0 \tag{11.73}$$

Multiplying the above two equations, we have

$$u_b^{pq^*}[t + \tau^{kl^*}(t)]\, \Omega_b^{km^*}(t) > 0 \tag{11.74}$$

The first term in the variational inequality (11.62) becomes

$$\int_0^T \sum_{rs} \sum_a \Omega_a^{rj^*}(t)\; u_a^{rs^*}[t + \pi^{ri^*}(t)]\; dt$$
$$= \int_{d-\delta}^{d+\delta} u_b^{pq^*}[t + \tau^{kl^*}(t)]\;\; \Omega_b^{km^*}(t)\; dt > 0 \tag{11.75}$$

We note that all other terms in the above equation vanish because the ideal DUO route choice condition (11.57) holds for other links and O-D pairs at each time instant and for link $b = (l, m)$ for O-D pair pq at time instants which are not within time interval $[d - \delta, d + \delta]$.

For each O-D pair rs, we can always find one minimal travel time route k for vehicles departing origin r at time t, which was evaluated under the optimal flow pattern $\{u_a^{rs^*}[t + \pi^{ri^*}(t)]\}$. For this route k, the first ideal DUO route choice condition (11.56) becomes equality by definition. It follows that

$$\Omega_a^{rj^*}(t) = \pi^{ri^*}(t) + \tau_a[t + \pi^{ri^*}(t)] - \pi^{rj^*}(t) = 0$$

$$\forall a, r, s; a = (i, j); a \in k. \tag{11.76}$$

Next, we need to find a set of feasible inflows $u_a^{rs}[t + \pi^{ri^*}(t)]$ so that the equation

$$u_a^{rs}[t + \pi^{ri^*}(t)]\; \Omega_a^{rj^*}(t) = 0 \qquad \forall a, r, s; a = (i, j) \tag{11.77}$$

always holds. We choose the feasible departure flows $f^{rs}(t)$ to equal the optimal departure flows $f^{rs^*}(t)$ for all O-D pairs rs at each instant of time. Thus, the second term in (11.62) will vanish. We also need to re-route all feasible departure flows $f^{rs}(t)$ for all O-D pairs at each instant of time. For each O-D pair rs, we assign the feasible O-D departure flow $f^{rs}(t)$ to the minimal travel time route k, which was evaluated under the optimal flow patterns $\{u_a^{rs^*}[t + \pi^{ri^*}(t)]\}$. This generates a set of feasible inflow patterns $\{u_a^{rs}[t+\pi^{ri^*}(t)]\}$ which always satisfies equation (11.77) (because either $\Omega_a^{rj^*}(t) = 0$ for links on route k or $u_a^{rs}[t + \pi^{ri^*}(t)] = 0$ since no flow is routed onto those links which are not on route k). Summing equations (11.77) for all links a and all O-D pairs rs, it follows that

$$\sum_{rs}\sum_{a} u_a^{rs}[t + \pi^{ri^*}(t)]\ \Omega_a^{rj^*}(t) = 0 \qquad \text{where } a = (i, j). \tag{11.78}$$

Integrating the above equation, we have

$$\int_0^T \sum_{rs}\sum_{a} \Omega_a^{rj^*}(t)\ u_a^{rs}[t + \pi^{ri^*}(t)]\ dt = 0 \tag{11.79}$$

We subtract equation (11.75) from equation (11.79) and obtain

$$\int_0^T \sum_{rs}\sum_{a} \Omega_a^{rj^*}(t)\ \left\{u_a^{rs}[t + \pi^{ri^*}(t)] - u_a^{rs^*}[t + \pi^{ri^*}(t)]\right\}\ dt < 0 \tag{11.80}$$

Note that the second term in (11.62) equals 0. It follows that

$$\int_0^T \left\{\sum_{rs}\sum_{a} \Omega_a^{rj^*}(t)\ \left\{u_a^{rs}[t + \pi^{ri^*}(t)] - u_a^{rs^*}[t + \pi^{ri^*}(t)]\right\}\right.$$
$$\left. + \sum_{rs} \mathcal{U}^{rs^*}(t)\left\{f^{rs}(t) - f^{rs^*}(t)\right\}\right\}\ dt < 0 \tag{11.81}$$

The above equation contradicts variational inequality (11.62). Therefore, any optimal solutions $\{u_a^{rs^*}[t + \pi^{ri^*}(t)]\}$ and $\{f^{rs^*}(t)\}$ to variational inequality (11.62) that satisfy the fifth DUO departure time choice condition (11.60) also satisfy the second ideal DUO route choice condition (11.57).

Case 2

Assume that the fifth DUO departure time choice condition (11.60) does not hold only for O-D pair pq during a short time interval $[d - \delta, d + \delta]$ where $[d - \delta, d + \delta] \in [0, T]$, i.e.,

$$f^{pq^*}(t) > 0 \qquad \text{and} \qquad \mathcal{U}^{pq^*}(t) - \mathcal{U}_{min}^{pq} > 0 \tag{11.82}$$

or

$$f^{pq^*}(t)\ \{\mathcal{U}^{pq^*}(t) - \mathcal{U}_{min}^{pq}\} > 0 \tag{11.83}$$

We now need to find a set of feasible departure flows $f^{rs}(t)$ and link inflows $u_a^{rs}[t + \pi^{ri^*}(t)]$ which will contradict variational inequality (11.62).

For O-D pairs $rs \neq pq$, we allow the feasible departure flows $f^{rs}(t)$ to equal the optimal departure flows $f^{rs^*}(t)$ at each instant of time. For O-D pair pq, we shift all departure flows during time interval $[d-\delta, d+\delta]$ to the time interval $[d+\delta, T]$, during which disutility $\mathcal{U}^{pq^*}(t) = \mathcal{U}^{pq}_{min}$, (if $d+\delta = T$, we shift all departure flows during time interval $[d-\delta, d+\delta]$ to the time interval $[0, d-\delta]$ and the proof will follow.) Thus, the second term of variational inequality (11.62) becomes

$$\int_0^T \sum_{rs} \mathcal{U}^{rs^*}(t) \left\{ f^{rs}(t) - f^{rs^*}(t) \right\} dt$$

$$= \int_0^T \mathcal{U}^{pq^*}(t) \left\{ f^{pq}(t) - f^{pq^*}(t) \right\} dt$$

$$= \left\{ \int_0^{d-\delta} + \int_{d+\delta}^T \right\} \mathcal{U}^{pq^*}(t) \left\{ f^{pq}(t) - f^{pq^*}(t) \right\} dt$$

$$+ \int_{d-\delta}^{d+\delta} \mathcal{U}^{pq^*}(t) \left\{ f^{pq}(t) - f^{pq^*}(t) \right\} dt$$

$$= \mathcal{U}^{pq}_{min} \left\{ \int_0^{d-\delta} + \int_{d+\delta}^T \right\} \left\{ f^{pq}(t) - f^{pq^*}(t) \right\} dt$$

$$- \int_{d-\delta}^{d+\delta} \mathcal{U}^{pq^*}(t)\, f^{pq^*}(t)\, dt \tag{11.84}$$

Note that in the above equation, $\mathcal{U}^{pq^*}(t) = \mathcal{U}^{pq}_{min}$ for time instants which do not lie within time interval $[d-\delta, d+\delta]$. For any time instant t during time interval $[d-\delta, d+\delta]$, the adjusted feasible departure flow $f^{pq}(t) = 0$. By definition of departure flows, we have

$$\int_0^T f^{pq}(t)\, dt = F^{pq}(T) = \int_0^T f^{pq^*}(t)\, dt$$

Substituting into equation (11.84), it follows that

$$\left\{ \int_0^{d-\delta} + \int_{d+\delta}^T \right\} \left\{ f^{pq}(t) - f^{pq^*}(t) \right\} dt$$

$$= \int_0^T \left\{ f^{pq}(t) - f^{pq^*}(t) \right\} dt - \int_{d-\delta}^{d+\delta} \left\{ f^{pq}(t)dt - f^{pq^*}(t) \right\} dt$$

$$= \int_{d-\delta}^{d+\delta} f^{pq^*}(t)\, dt \tag{11.85}$$

Integrating equation (11.83) from $(d-\delta)$ to $(d+\delta)$, we have

$$\int_{d-\delta}^{d+\delta} f^{pq^*}(t)\, \{\mathcal{U}^{pq^*}(t) - \mathcal{U}^{pq}_{min}\}\, dt > 0 \tag{11.86}$$

or

$$\int_{d-\delta}^{d+\delta} f^{pq^*}(t)\,\mathcal{U}^{pq^*}(t)\,dt > \int_{d-\delta}^{d+\delta} f^{pq^*}(t)\,\mathcal{U}^{pq}_{min}\,dt \tag{11.87}$$

Substituting equations (11.85) and (11.87) into equation (11.84), we have

$$\mathcal{U}^{pq}_{min}\left\{\int_0^{d-\delta} + \int_{d+\delta}^{T}\right\}\left\{f^{pq}(t) - f^{pq^*}(t)\right\}\,dt - \int_{d-\delta}^{d+\delta} \mathcal{U}^{pq^*}(t)\,f^{pq^*}(t)\,dt$$

$$= \mathcal{U}^{pq}_{min}\int_{d-\delta}^{d+\delta} f^{pq^*}(t)\,dt - \int_{d-\delta}^{d+\delta} \mathcal{U}^{pq^*}(t)\,f^{pq^*}(t)\,dt$$

$$< \mathcal{U}^{pq}_{min}\int_{d-\delta}^{d+\delta} f^{pq^*}(t)\,dt - \int_{d-\delta}^{d+\delta} \mathcal{U}^{pq}_{min}\,f^{pq^*}(t)\,dt = 0 \tag{11.88}$$

Following the above adjustment of feasible departure flows, the link inflows should be adjusted accordingly so as to be feasible. As illustrated in Case 1, for each O-D pair rs, we can always find one minimal travel time route k for vehicles departing origin r at time t, which was evaluated under the optimal flow pattern $\{u_a^{rs^*}[t + \pi^{ri^*}(t)]\}$. For this route k, the first ideal DUO route choice condition (11.56) becomes an equality by definition. It follows that

$$\Omega_a^{rj^*}(t) = \pi^{ri^*}(t) + \tau_a[t + \pi^{ri^*}(t)] - \pi^{rj^*}(t) = 0$$

$$\forall a, r, s; a = (i, j); a \in k. \tag{11.89}$$

Next, we need to find a set of feasible inflows $u_a^{rs}[t+\pi^{ri^*}(t)]$ so that the equation

$$u_a^{rs}[t + \pi^{ri^*}(t)]\,\Omega_a^{rj^*}(t) = 0 \qquad \forall a, r, s; a = (i, j) \tag{11.90}$$

always holds. For each O-D pair rs at each time instant t, we assign the feasible O-D departure flow $f^{rs}(t)$ to the minimal travel time route k only, which was evaluated under the optimal flow pattern $\{u_a^{rs^*}[t+\pi^{ri^*}(t)]\}$. This will generate a feasible inflow pattern $\{u_a^{rs}[t+\pi^{ri^*}(t)]\}$ which always satisfy equation (11.90). Summing equations (11.90) for all links a and all O-D pairs rs, it follows that

$$\sum_{rs}\sum_{a} u_a^{rs}[t + \pi^{ri^*}(t)]\,\Omega_a^{rj^*}(t) = 0 \qquad \text{where } a = (i, j). \tag{11.91}$$

Subtracting equation (11.57) from equation (11.91) and integrating the resulting equation, we have

$$\int_0^T \sum_{rs}\sum_{a} \Omega_a^{rj^*}(t)\,\left\{u_a^{rs}[t + \pi^{ri^*}(t)] - u_a^{rs^*}[t + \pi^{ri^*}(t)]\right\}dt = 0 \tag{11.92}$$

Combining equations (11.84), (11.88) and (11.92), it follows that

$$\int_0^T \left\{\sum_{rs}\sum_{a} \Omega_a^{rj^*}(t)\,\left\{u_a^{rs}[t + \pi^{ri^*}(t)] - u_a^{rs^*}[t + \pi^{ri^*}(t)]\right\}\right.$$

$$\left. + \sum_{rs} \mathcal{U}^{rs^*}(t)\left\{f^{rs}(t) - f^{rs^*}(t)\right\}\right\}\,dt < 0 \tag{11.93}$$

The above equation contradicts variational inequality (11.62). Therefore, any optimal solutions $\{u_a^{rs^*}[t+\pi^{ri^*}(t)]\}$ and $\{f^{rs^*}(t)\}$ to variational inequality (11.62) that satisfy the second ideal DUO route choice condition (11.57) also satisfy the fifth DUO departure time choice condition (11.60).

Case 3

Case 3 includes the following two sub-cases: 3a) conditions (11.57) and (11.60) do not hold for different O-D pairs; 3b) conditions (11.57) and (11.60) do not hold for the same O-D pair.

Case 3a

Assume that the second ideal DUO route choice condition (11.57) does not hold for O-D pair kn for time interval $[d_1-\delta_1, d_1+\delta_1]$ and the fifth DUO departure time choice condition (11.60) does not hold for O-D pair pq for time interval $[d_2-\delta_2, d_2+\delta_2]$. Note that the two O-D pairs are different, but the two time intervals may or may not be different. For O-D pair kn, we assume that the second ideal DUO route choice condition (11.57) does not hold only for a link $b=(l,m)$ during time interval $[d_1-\delta_1, d_1+\delta_1]$, i.e.,

$$\begin{aligned} &u_b^{kn^*}[t+\tau^{kl^*}(t)]>0 \qquad \text{and} \\ &\Omega_b^{km^*}(t)=\tau^{kl^*}(t)+\tau_a[t+\tau^{kl}(t)]-\tau^{km^*}(t)>0 \end{aligned} \tag{11.94}$$

Following the derivation from (11.73) to (11.81) in Case 1, we can find a set of feasible inflows $u_a^{kn}[t+\pi^{ri^*}(t)]$ for O-D pair kn so that the following equation holds:

$$\begin{aligned} \int_0^T &\left\{\sum_a \Omega_a^{kj^*}(t)\left\{u_a^{kn}[t+\tau^{ki^*}(t)]-u_a^{kn^*}[t+\tau^{ki^*}(t)]\right\}\right. \\ &+\left.\mathcal{U}^{kn^*}(t)\left\{f^{kn}(t)-f^{kn^*}(t)\right\}\right\}dt<0 \end{aligned} \tag{11.95}$$

Note that in the derivation of the above equation, we follow the process in Case 1 by assuming there is only one O-D pair kn.

For O-D pair pq, we assume that the fifth DUO departure time choice condition (11.60) does not hold during time interval $[d_2-\delta_2, d_2+\delta_2]$.

$$f^{pq^*}(t)>0 \qquad \text{and} \qquad \mathcal{U}^{pq^*}(t)-\mathcal{U}_{min}^{pq}>0 \tag{11.96}$$

Following the derivation from (11.82) to (11.93) in Case 2, we can find a set of feasible departure flows $f^{pq}(t)$ and inflows $u_a^{kn}[t+\tau^{ki^*}(t)]$ so that the following equation holds:

$$\begin{aligned} \int_0^T &\left\{\sum_a \Omega_a^{pj^*}(t)\left\{u_a^{pq}[t+\tau^{pi^*}(t)]-u_a^{pq^*}[t+\tau^{pi^*}(t)]\right\}\right. \\ &+\left.\mathcal{U}^{pq^*}(t)\left\{f^{pq}(t)-f^{pq^*}(t)\right\}\right\}dt<0 \end{aligned} \tag{11.97}$$

Note that in the derivation of the above equation, we follow the process in Case 2 by assuming there is only one O-D pair pq.

For other O-D pairs rs, we choose the feasible departure flows $f^{rs}(t)$ to equal the optimal departure flows $f^{rs^*}(t)$ at each instant of time t. Thus, the second terms in (11.62) for those O-D pairs will vanish. We also need to re-route all feasible departure flows $f^{rs}(t)$ for those O-D pairs at each instant of time. For each O-D pair rs at each instant of time, we assign the feasible O-D departure flow $f^{rs}(t)$ to the minimal travel time route h, which was evaluated under the optimal flow pattern $\{u_a^{rs^*}[t+\pi^{ri^*}(t)]\}$. This will generate a set of feasible inflow patterns $\{u_a^{rs}[t+\pi^{ri^*}(t)]\}$ which always allows the first terms in (11.62) to equal 0 for those O-D pairs rs, as illustrated in Case 1.

Combining equations (11.95) and (11.97) and the above analysis,

$$\int_0^T \left\{ \sum_{rs}\sum_a \Omega_a^{rj^*}(t) \left\{ u_a^{rs}[t+\pi^{ri^*}(t)] - u_a^{rs^*}[t+\pi^{ri^*}(t)] \right\} \right.$$

$$\left. + \sum_{rs} \mathcal{U}^{rs^*}(t) \left\{ f^{rs}(t) - f^{rs^*}(t) \right\} \right\} dt < 0 \tag{11.98}$$

The above equation contradicts variational inequality (11.62). Therefore, any optimal solutions $\{u_a^{rs^*}[t+\pi^{ri^*}(t)]\}$ and $\{f^{rs^*}(t)\}$ to variational inequality (11.62) will satisfy both the second ideal DUO route choice condition (11.57) and the fifth DUO departure time choice condition (11.60).

Case 3b

Assume that the second ideal DUO route choice condition (11.57) and the fifth DUO departure time choice condition (11.60) do not hold for the same O-D pair pq, but for time intervals $[d_1-\delta_1, d_1+\delta_1]$ and $[d_2-\delta_2, d_2+\delta_2]$, respectively. Note that the two time intervals can be either identical or different. Since the fifth DUO departure time choice condition (11.60) does not hold during time interval $[d_2-\delta_2, d_2+\delta_2]$, it follows that

$$f^{pq^*}(t) > 0 \quad \text{and} \quad \mathcal{U}^{pq^*}(t) - \mathcal{U}^{pq}_{min} > 0 \tag{11.99}$$

Following the derivation from (11.82) to (11.88) in Case 2, we can find a set of feasible departure flows $f^{pq}(t)$ so that the following equation holds for O-D pair pq:

$$\int_0^T \mathcal{U}^{pq^*}(t) \left\{ f^{pq}(t) - f^{pq^*}(t) \right\} dt < 0 \tag{11.100}$$

Note that in the derivation of the above equation, we follow the process in Case 2 by assuming there is only one O-D pair pq.

For O-D pair pq, we also need to adjust the link inflow pattern accordingly so that they will be feasible. We assume that the second ideal DUO route choice condition (11.57) does not hold only for a link $b = (l, m)$ during time interval $[d_1 - \delta_1, d_1 + \delta_1]$, i.e.,

$$u_b^{pq^*}[t + \tau^{pl^*}(t)] > 0 \qquad \text{and}$$

$$\Omega_b^{pm^*}(t) = \tau^{pl^*}(t) + \tau_a[t + \tau^{pl}(t)] - \tau^{pm^*}(t) > 0 \tag{11.101}$$

Following the derivation from equations (11.73) to (11.80) in Case 1, we can find a set of feasible inflows $u_a^{pq}[t+\tau^{pi^*}(t)]$ so that the following equation always hold for O-D pair pq:

$$\int_0^T \sum_a \Omega_a^{pj^*}(t) \left\{u_a^{pq}[t + \tau^{pi^*}(t)] - u_a^{pq^*}[t + \tau^{pi^*}(t)]\right\} dt < 0 \tag{11.102}$$

Note that in the derivation of the above equation, we follow the process in Case 1 by assuming there is only one O-D pair pq.

Combining equations (11.100) and (11.102), we have

$$\int_0^T \left\{\sum_a \Omega_a^{pj^*}(t) \left\{u_a^{pq}[t + \tau^{pi^*}(t)] - u_a^{pq^*}[t + \tau^{pi^*}(t)]\right\}\right.$$
$$\left. + \ \mathcal{U}^{pq^*}(t)\left\{f^{pq}(t) - f^{pq^*}(t)\right\}\right\} dt < 0 \tag{11.103}$$

Thus, we found a set of feasible departure flows $f^{pq}(t)$ and inflows $u_a^{pq}[t+\tau^{pi^*}(t)]$ so that the above equation holds for O-D pair pq.

For other O-D pairs rs, we choose the feasible departure flows $f^{rs}(t)$ to equal the optimal departure flows $f^{rs^*}(t)$ at each instant of time. Thus, the second terms in (11.62) for those O-D pairs vanish. We also reroute the feasible departure flows $f^{rs}(t)$ for those O-D pairs at each time instant. For each O-D pair rs, we assign the feasible O-D departure flow $f^{rs}(t)$ to the minimal travel time route h, which was evaluated under the optimal flow pattern $\{u_a^{rs^*}[t + \pi^{ri^*}(t)]\}$. This generates a set of feasible inflow patterns $\{u_a^{rs}[t+\pi^{ri^*}(t)]\}$ which always allow the first terms in VI (11.62) for those O-D pairs to equal 0, as illustrated in Case 1.

Combining equations (11.103) and the above analysis, we have

$$\int_0^T \left\{\sum_{rs}\sum_a \Omega_a^{rj^*}(t) \left\{u_a^{rs}[t + \pi^{ri^*}(t)] - u_a^{rs^*}[t + \pi^{ri^*}(t)]\right\}\right.$$
$$\left. + \sum_{rs} \mathcal{U}^{rs^*}(t)\left\{f^{rs}(t) - f^{rs^*}(t)\right\}\right\} dt < 0 \tag{11.104}$$

The above equation contradicts variational inequality (11.62). Therefore, any optimal solutions $\{u_a^{rs^*}[t + \pi^{ri^*}(t)]\}$ and $\{f^{rs^*}(t)\}$ to variational inequality (11.62) will satisfy both the second ideal DUO route choice condition (11.57) and the fifth DUO departure time choice condition (11.60). Since we have proved the necessity and sufficiency of the equivalence of variational inequality (11.62) to DUO departure time/route choice conditions (11.56)-(11.61), the proof is complete.

11.4 Solution Algorithm and An Example

11.4.1 Discrete Variational Inequality Problem

To convert our continuous VI model into a discrete VI model, the time period $[0, T]$ is subdivided into N small time increments. Each time increment is a unit of time. Then, $u_a(n)$ represents the inflow into link a during interval n and $v_a(n)$ represents the exit flow from link a during interval n. To simplify the formulation, we modify the estimated actual travel time on each link in the following way so that each estimated travel time is equal to a multiple of the time increment.

$$\bar{\tau}_a(n) = i \quad \text{if} \quad i - 0.5 \leq \bar{\tau}_a(n) < i + 0.5,$$

where i is an integer and $0 \leq i \leq K$. We note that this round-off method is used only in the flow propagation constraints. More accurate flow propagation constraints can be obtained by making the time intervals smaller. We also note that evaluation of the link travel time function and objective function does not have this round-off error so that the subsequent minimal-cost route search does not have this round-off error.

We now consider the relationship between VI models and relaxation subproblems. We do not discuss each VI model in this chapter. Instead, we focus our analysis on the link-time-based VI model for the ideal DUO departure time/route choice problem. We show in the following that the VI model can be reformulated as a discrete optimization problem under relaxation. Similar analyses can be performed for the route-based VI model for the ideal DUO departure time/route choice problem.

In this section, we discuss the following VI problem:

$$\int_0^T \left\{ \sum_{rs} \sum_a \Omega_a^{rj^*}(t) \left\{ u_a^{rs}[t + \pi^{ri^*}(t)] - u_a^{rs^*}[t + \pi^{ri^*}(t)] \right\} \right.$$
$$\left. + \sum_{rs} \mathcal{U}^{rs^*}(t) \left\{ f^{rs}(t) - f^{rs^*}(t) \right\} \right\} dt \geq 0 \tag{11.105}$$

The above continuous VI problem is converted into the following discrete VI problem:

$$\sum_{n=1}^{N} \left\{ \sum_{rs} \sum_a \Omega_a^{rj^*}(n) \left\{ u_a^{rs}[n + \pi^{ri^*}(n)] - u_a^{rs^*}[n + \pi^{ri^*}(n)] \right\} \right.$$
$$\left. + \sum_{rs} \mathcal{U}^{rs^*}(n) \left\{ f^{rs}(n) - f^{rs^*}(n) \right\} \right\} \geq 0 \tag{11.106}$$

To simplify our analysis, we assume the time period $[0, T]$ is long enough so that all departure flows can be cleared at the final time T. In other words, any positive departure from origin r at time t will arrive at destination s at time $t + \pi^{rs^*}(t) \leq T$.

11.4.2 Relaxation Procedure and Optimization Problem

For each relaxation iteration, we temporarily fix: 1) actual travel time $\tau_a(n)$ in link flow propogation constraints as $\bar{\tau}_a(n)$; 2) minimal travel times $\pi^{ri^*}(n)$ as $\bar{\pi}^{ri^*}(n)$ and $\pi^{rj^*}(n)$ as $\bar{\pi}^{rj^*}(n)$ for each link and each origin node. Thus, the disutility function for the joint departure time and route choice problem becomes

$$\bar{\mathcal{U}}^{rs}(n) = \alpha n + \bar{\pi}^{rs}(n) + \mathcal{V}^{rs}[n, \bar{\pi}^{rs}(n); t^*_{rs}] \qquad \forall r, s, \tag{11.107}$$

where n is the departure time interval of travelers.

Then, the objective function of an optimization problem (NLP) can be formulated, which is equivalent to the discrete VI under relaxation. It follows that

$$\begin{aligned} \min_{u,v,x,E,f} \quad Z &= \sum_{k=1}^{K}\Bigg\{\sum_a \Bigg\{\int_0^{u_a(k)} \{\tau_a[x_a(k), \omega, v_a(k)]d\omega \\ &+ \sum_r u_a^r(k)[\bar{\pi}^{ri}(\xi_a^r) - \bar{\pi}^{rj}(\xi_a^r)]\Bigg\} \\ &+ \sum_{rs}\int_0^{f^{rs}(k)} \mathcal{U}^{rs}(k)d\omega\Bigg\} \end{aligned} \tag{11.108}$$

where the new time interval index k is defined as

$$k = \xi_a^r + \bar{\pi}^{ri}(\xi_a^r) \tag{11.109}$$

so that departure time interval ξ_a^r can be calculated by using the above equation once interval k, origin r, and arrival node i are given. Note that both minimal travel times $\bar{\pi}^{ri}(\xi_a^r)$ and $\bar{\pi}^{rj}(\xi_a^r)$ are fixed temporarily at each relaxation iteration.

To ensure that the optimization program with the above objective function is indeed equivalent to the discrete VI, we should demonstrate that the gradient of the above discrete objective function (11.108) is equivalent to the cost term of discrete variational inequality (11.106) under relaxation, i.e.,

$$\frac{\partial Z}{\partial u_a^{rs}[n + \bar{\pi}^{ri}(n)]} = \tau_a[n + \bar{\pi}^{ri}(n)] + \bar{\pi}^{ri}(n) - \bar{\pi}^{rj}(n) = \bar{\Omega}_a^{rj}(n) \qquad \forall a, r, s; n. \tag{11.110}$$

$$\frac{\partial Z}{\partial f^{rs}(n)} = \mathcal{U}^{rs}(n) \qquad \forall a, r, s; n. \tag{11.111}$$

It is obvious that equation (11.113) holds. In the following, we prove that equation (11.110) holds. Taking partial derivative of objective function (11.108) with respect to each link inflow, it follows that

$$\frac{\partial Z}{\partial u_a^{rs}(k)} = \tau_a(k) + \bar{\pi}^{ri}(\xi_a^r) - \bar{\pi}^{rj}(\xi_a^r) \qquad \forall a, r, s; k. \tag{11.112}$$

Since the above equation applies for each time interval k, it applies for each time interval ξ_a^r. Replacing the departure time interval ξ_a^r by n in the above equation, we have

$$\frac{\partial Z}{\partial u_a^{rs}(k)} = \tau_a(k) + \bar{\pi}^{ri}(n) - \bar{\pi}^{rj}(n) \qquad \forall a, r, s; k. \quad (11.113)$$

Substituting equation (11.109) (replacing departure time interval ξ_a^r by n) into the above equation, we obtain equation (11.110). Therefore, we have demonstrated that the gradient of discrete objective function (11.108) is equivalent to the cost term of discrete variaitonal inequality (11.106) under relaxation. In other words, this objective function (11.108) could be used in the relaxation optimization problem.

Note that the network constraints apply for each time interval, regardless whether the index of time interval is n or k. Thus, for simplicity, we use the time interval index k in the network constraint set.

The discrete network flow constraints are now summarized as follows:

$$x_{ap}^{rs}(k+1) = x_{ap}^{rs}(k) + u_{ap}^{rs}(k) - v_{ap}^{rs}(k) \qquad \forall a, p, r, s; k = 1, \cdots, K; \quad (11.114)$$

$$F^{rs}(k+1) = F^{rs}(k) + \sum_{a \in A(r)} \sum_p u_{ap}^{rs}(k) \qquad \forall r; s \neq r; k = 1, \cdots, K; \quad (11.115)$$

$$E^{rs}(k+1) = E^{rs}(k) + \sum_{a \in B(s)} \sum_p v_{ap}^{rs}(k) \qquad \forall r; s \neq r; k = 1, \cdots, K; \quad (11.116)$$

$$\sum_{a \in A(r)} \sum_p u_{ap}^{rs}(k) = f^{rs}(k) \qquad \forall r \neq s; k = 1, \cdots, K; \quad (11.117)$$

$$\sum_{a \in B(j)} v_{ap}^{rs}(k) - \sum_{a \in A(j)} u_{ap}^{rs}(k) = 0 \qquad \forall j, p, r, s; j \neq r, s; k = 1, \cdots, K; \quad (11.118)$$

$$x_{ap}^{rs}(k) = \sum_{b \in \tilde{p}} \{x_{bp}^{rs}[k + \bar{\tau}_a(k)] - x_{bp}^{rs}(k)\} + \{E_p^{rs}[k + \bar{\tau}_a(k)] - E_p^{rs}(k)\}$$

$$\forall a \in B(j); j \neq r; p, r, s; k = 1, \cdots, K+1; \quad (11.119)$$

$$u_{ap}^{rs}(k) \geq 0, \quad v_{ap}^{rs}(k) \geq 0, \quad x_{ap}^{rs}(k+1) \geq 0, \qquad \forall a, p, r, s; k = 1, \cdots, K; \quad (11.120)$$

$$E_p^{rs}(k+1) \geq 0, \qquad F_p^{rs}(k+1) \geq 0, \qquad \forall p, r, s; k = 1, \cdots, K; \quad (11.121)$$

$$E_p^{rs}(1) = 0, \qquad F_p^{rs}(1) = 0, \qquad \forall p, r, s; \quad (11.122)$$

$$x_{ap}^{rs}(1) = 0, \qquad \forall a, p, r, s. \quad (11.123)$$

In summary, in each relaxation step, we need to solve a discrete NLP problem consisting of objective function (11.108) and constraints (11.114)-(11.123).

In the resulting discrete time problem, $x_a(k)$ represents vehicles on the link at the beginning of interval k; $u_a(k)$ and $v_a(k)$ represent inflow and exit flow during interval k. Let $\tau_a(k)$ denote the travel time for vehicles entering link a at the beginning of interval $k = [k, k+1]$, and let $\pi^{rs}(k)$ be the average minimal $r-s$ travel time for vehicles departing origin r during interval k. Let $f^{rs}(k)$ denote the O-D departure flow during interval k.

By discretizing the time period, the variational inequality can be converted into a discrete time NLP under relaxation. In our algorithm, at each iteration, the departure time choice model is solved together with the route choice problem by the Frank-Wolfe technique (Frank and Wolfe, 1956). The flowchart of the solution procedure is shown in Figure 11.2.

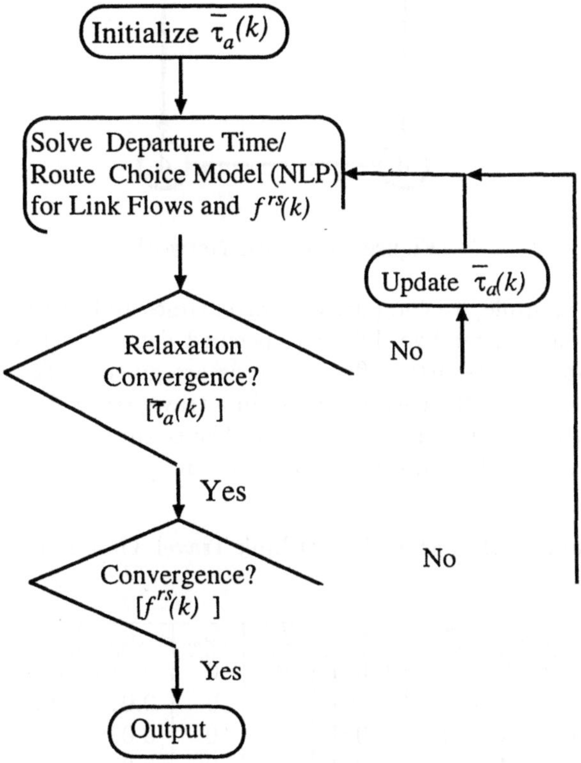

Figure 11.2: Flowchart of the Solution Algorithm

11.4.3 Numerical Example

We illustrate the solution of our departure time/route choice model with the 4-link, 4-node test network shown in Figure 11.3. The assignment time period $[0, T]$ is subdivided into $K = 8$ small time intervals. The algorithm was coded in FORTRAN and solved on a HP-715 workstation. As proposed in Chapter 5, the following link travel time functions were used in the computations:

$$\tau_a(k) = c_a(k) = g_{1a}(k) + g_{2a}(k)$$
$$g_{1a}(k) = \beta_{1a} + \beta_{2a}[u_a(k)]^2 + \beta_{3a}[x_a(k)]^2$$
$$g_{2a}(k) = \beta_{4a} + \beta_{5a}[v_a(k)]^2 + \beta_{6a}[x_a(k)]^2$$

where the time interval $k = 1, 2, \cdots, 8$. The parameters for each link travel time function are given in Table 11.1.

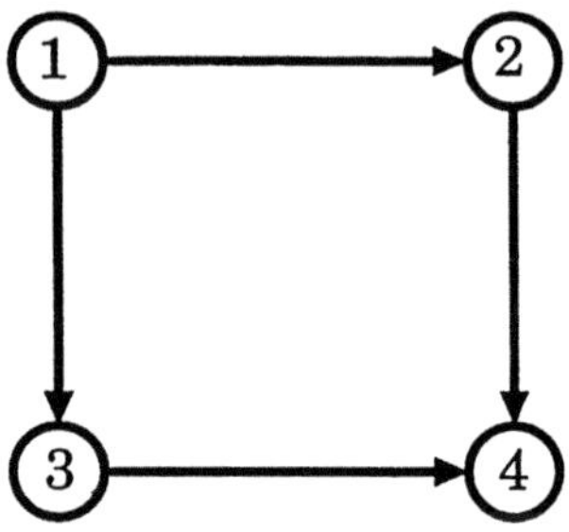

Figure 11.3: Test Network

In this example, we compare the travel times and disutilities from origin 1 to destination 4 under two different specified departure flow patterns. The initial and improved departure flow patterns are given in Table 11.2; the user optimal link flows and the corresponding link travel times for the initial departure flow pattern are shown in Table 11.3. The optimal link flows and optimal link travel times for the improved departure flow pattern are given in 11.4.

Table 11.1: Parameters of Link Travel Time Functions

link a	β_{1a}	β_{2a}	β_{3a}	β_{4a}	β_{5a}	β_{6a}
1—2	1.	0.001	0.	0.	0.015	0.002
1—3	1.	0.001	0.	0.	0.015	0.002
2—4	1.	0.001	0.	0.	0.015	0.002
3—4	1.	0.001	0.	0.	0.015	0.002

We first discuss the flow propagation in the network, using link 2–4 in Table 11.4 as an example (see Figure 11.5).

Interval 2: Vehicles from link 1–2 enter link 2–4 during interval 2: $u_{24}(2) = 6.5$. Since the travel time on link 2–4, $\tau_{24}(2)$, is 1.0 interval for the first vehicles entering at the beginning of interval 2, these vehicles exit link 2–4 during interval 3: $v_{24}(3) = 6.5$.

Table 11.2: Departure Flow Patterns from Origin 1 to Destination 4

Initial Pattern

Interval k	1	2	3	4	5	6	7	8
O-D Flow	60.	0.	0.	0.	0.	0.	0.	0.

Improved Pattern

Interval k	1	2	3	4	5	6	7	8
O-D Flow	30.	0.	30.	0.	0.	0.	0.	0.

Interval 3: There are $u_{24}(3) = 8.5$ vehicles entering link 2–4 during interval 3. Since travel time $\tau_{24}(3)$ is 1.8 intervals for the first vehicles entering at the beginning of interval 3, only $v_{24}(4) = 7.4$ of the 8.5 vehicles exit during interval 4, and the remaining 1.1 vehicles exit during interval 5.

Interval 4: The travel time $\tau_{24}(4)$ is 2.0 intervals for the first vehicles entering at the beginning of interval 4. There are $u_{24}(4) = 6.2$ additional vehicles entering during interval 4. The remaining 1.1 vehicles already on the link (see the paragraph above) exit during interval 5, allowing 5.5 vehicles also to exit, for a total of 6.6 exiting vehicles; $6.2 - 5.5 = 0.7$ vehicles of the 6.2 vehicles entering during interval 4 exit during interval 6.

Interval 5: $u_{24}(5) = 8.8$ vehicles enter during interval 5. Those vehicles begin to exit link 2–4 during interval 6 (6.2 vehicles) and finish exiting during interval 7 (2.6 vehicles).

We now consider the improvement of travel times and disutilities from origin 1 to destination 4 as a result of the changed departure flow pattern. Assume that parameter α for waiting time equals 0.5; there is no arrival penalty in this example.

With the *initial* departure flow pattern shown in Table 11.2, the O-D travel time for the first vehicle departing at the beginning of interval 1 is 3 time intervals. (In Table 11.3, vehicles start to exit links 2–4 and 3–4 during interval 3; the O-D travel times are rounded integers.) Also, the O-D travel time for the last vehicle departing at the end of interval 1 is 6 time intervals. (In Table 11.3, vehicles finish exiting link 2–4 and 3–4 during interval 7). The *average* O-D travel time for these vehicles is $(3 + 6)/2 = 4.5$ time intervals. Since the average waiting time is zero for the departure flow pattern shown in Table 11.2, total disutility averages $(\alpha \cdot 0 + 4.5) \cdot 60 = 270$ units.

By way of comparison, with the *improved* departure flow pattern shown in Table 11.2, the O-D travel time for the first vehicle departing at the beginning of interval 1 is also 3.0 intervals. (Vehicles start to exit links 2–4 and 3–4 during interval 3; see footnote $*$ in Figure 11.4.) Also the O-D travel time for the last vehicle departing at the end of interval 1 is 4.0 intervals; see footnote $*$ in Figure 11.4. The O-D travel time for the first vehicle departing at the beginning of interval 3 is 3.0 intervals. Also, the O-D travel time for the last vehicle departing at the end of interval 3 is $7.0 - 3.0 = 4.0$ intervals; see footnote $**$ in Figure 11.5. The average O-D travel time for vehicles departing during interval 1 is $(3+4)/2 = 3.5$ time intervals, and the waiting time is 0. The average O-D travel time for vehicles departing during interval 3 is $(3+4)/2 = 3.5$ intervals, and the waiting time is 2 intervals. Total disutility averages $[(\alpha \cdot 0 + 3.5) \cdot 30 + (\alpha \cdot 2 + 3.5) \cdot 30] = 240$ units. (Recall that $\alpha = 0.5$.) Thus, by delaying some departures, total disutility due to waiting and traveling is reduced.

Table 11.3: Optimal Link Flows and Travel Times for Initial O-D Flows

Interval k	Link a	Vehicles $x_a(k+1)$	Inflow $u_a(k)$	Exit Flow $v_a(k)$	Vehicles $x_a(k)$	Travel Time $\tau_a(k)$
1	1 — 2	30.0	30.0	0.0	0.0	1.9
2	1 — 2	21.7	0.0	8.3	30.0	3.8
3	1 — 2	21.7	0.0	0.0	21.7	1.9
4	1 — 2	12.4	0.0	9.2	21.7	3.2
5	1 — 2	0.0	0.0	12.4	12.4	3.6
1	1 — 3	30.0	30.0	0.0	0.0	1.9
2	1 — 3	21.7	0.0	8.3	30.0	3.8
3	1 — 3	21.7	0.0	0.0	21.7	1.9
4	1 — 3	12.4	0.0	9.2	21.7	3.2
5	1 — 3	0.0	0.0	12.4	12.4	3.6
1	2 — 4	0.0	0.0	0.0	0.0	1.0
2	2 — 4	8.3	8.3	0.0	0.0	1.1
3	2 — 4	0.0	0.0	8.3	8.3	2.2
4	2 — 4	9.2	9.2	0.0	0.0	1.1
5	2 — 4	13.4	12.4	8.3	9.2	2.4
6	2 — 4	3.8	0.0	9.6	13.4	2.7
7	2 — 4	0.0	0.0	3.8	3.8	1.2
1	3 — 4	0.0	0.0	0.0	0.0	1.0
2	3 — 4	8.3	8.3	0.0	0.0	1.1
3	3 — 4	0.0	0.0	8.3	8.3	2.2
4	3 — 4	9.2	9.2	0.0	0.0	1.1
5	3 — 4	13.4	12.4	8.3	9.2	2.4
6	3 — 4	3.8	0.0	9.6	13.4	2.7
7	3 — 4	0.0	0.0	3.8	3.8	1.2

Table 11.4: Optimal Link Flows and Travel Times for Improved O-D Flows

Interval k	Link a	Vehicles $x_a(k+1)$	Inflow $u_a(k)$	Exit Flow $v_a(k)$	Vehicles $x_a(k)$	Travel Time $\tau_a(k)$
1	1 — 2	15.0	15.0	0.0	0.0	1.2
2	1 — 2	8.5	0.0	6.5	15.0	2.1
3	1 — 2	15.0	15.0	8.5	8.5	2.5
4	1 — 2	8.8	0.0	6.2	15.0	2.0
5	1 — 2	0.0	0.0	8.8	8.8	2.3
1	1 — 3	15.0	15.0	0.0	0.0	1.2
2	1 — 3	8.5	0.0	6.5	15.0	2.1
3	1 — 3	15.0	15.0	8.5	8.5	2.5
4	1 — 3	8.8	0.0	6.2	15.0	2.0
5	1 — 3	0.0	0.0	8.8	8.8	2.3
1	2 — 4	0.0	0.0	0.0	0.0	1.0
2	2 — 4	6.5	6.5	0.0	0.0	1.0
3	2 — 4	8.5	8.5	6.5	6.5	1.8
4	2 — 4	7.4	6.2	7.4	8.5	2.0
5	2 — 4	9.5	8.8	6.6	7.4	1.9
6	2 — 4	2.6	0.0	6.9	9.5	1.9
7	2 — 4	0.0	0.0	2.6	2.6	1.1
1	3 — 4	0.0	0.0	0.0	0.0	1.0
2	3 — 4	6.5	6.5	0.0	0.0	1.0
3	3 — 4	8.5	8.5	6.5	6.5	1.8
4	3 — 4	7.4	6.2	7.4	8.5	2.0
5	3 — 4	9.5	8.8	6.6	7.4	1.9
6	3 — 4	2.6	0.0	6.9	9.5	1.9
7	3 — 4	0.0	0.0	2.6	2.6	1.1

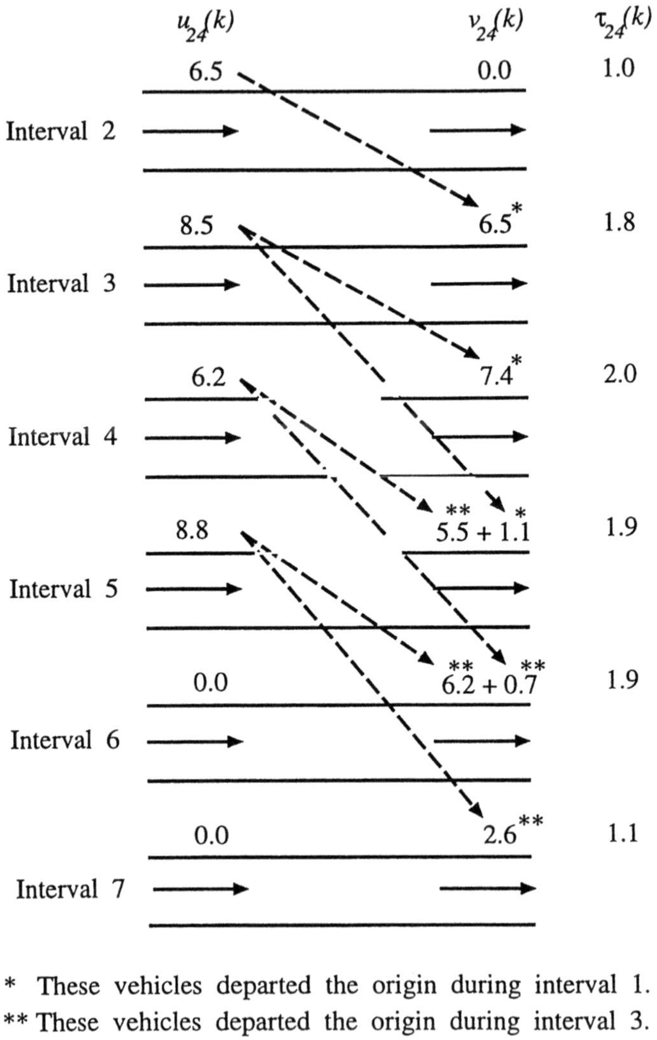

Figure 11.4: Flow Propagation on Link 2–4

11.5 Notes

In contrast to departure time choice, arrival time choice may be more significant for drivers. One example is the home-to-work trip in which a latest arrival time needs to be strictly guaranteed. Our model uses desired arrival time to determine driver departure times. The penalty for late arrival must be very large. On the other hand, the disutility due to waiting at the origin is not important. In Chapter 12, we explored how the desired arrival time can be specified for different groups of travelers in a joint mode/departure time/route choice problem. In future research, we will also address work-to-home trips whose arrival time requirements seem to be more elastic since drivers are more concerned with avoiding congestion. The utility due to waiting at the origin is important because drivers prefer to stay at workplaces until traffic congestion decreases. Consequently, the arrival penalty is not significant.

Several departure time choice models have been proposed by various researchers using different approaches on dynamic traffic networks. Janson (1993) formulated a dynamic user-optimal route choice model in which trips have variable departure times and scheduled arrival times. Friesz et al (1993) presented a joint departure time and route choice model using the variational inequality approach. Ghali and Smith (1993) also considered this problem using a microscopic representation of vehicle streams.

In this chapter, a route-based VI model and a link-based VI model were presented for DUO departure time/route choice. The route choice program and the departure time choice program are integrated in one framework and solved simultaneously. The necessity and sufficiency proofs of the VI model demonstrate that this model is consistent with the link-based DUO departure time/route choice conditions. Using a link-based VI formulation, explicit route enumeration can be avoided in computation. This feature allows our model to be applied to large-scale dynamic transportation networks with general link travel time functions.

Two major constraints prevent us from applying existing dynamic transportation network models to ATIS systems. The first concerns the accurate representation of travelers' choice behavior. In future extensions, utility functions instead of pure travel times should be used in route choice problems. Different perceptions and compliance with information must be investigated by stratifying travelers into multiple groups. The second concerns the accurate representation of traffic dynamics on each street link. Link traffic dynamics may be very complicated; as pointed out by Newell (1990) and Daganzo (1993), a set of appropriate closed-form link travel time functions might involve the interactions of neighboring link flows. This feature prevents formulating an appropriate optimization model for a realistic departure time/route choice problem. Thus, the general VI formulation approach was proposed for such applications. However, VI models require more computational capability than optimization models.

The proposed link-based VI model for DUO departure time/route choice can be extended to include arrival time choice, destination choice and mode choice as well. Our next step is to investigate other efficient solution algorithms for the DUO departure time/route choice VI model. Solution algorithms, such as the projection algorithm, implemented by Nagurney (1986) for static network equilibrium VI models, are also extendible to our dynamic VI problem. We note that the solution algorithm for our DUO departure time/route choice VI model has to be implemented on an expanded time-space network as proposed in Boyce et al (1991). Other important problems, such as incident related dynamic route choice problems and dynamic congestion pricing problems, will be studied as extensions of this VI model.

Problems

11.1 Compare the relative importance of departure time choice and arrival time choice under various circumstances. Explain your answer.

11.2 Discuss when the early arrival at destinations can be considered as a utility and when it can be considered as a disutility for travelers.

11.3 Identify the possible desired arrival time intervals for the following trips:

1. home to work;
2. work to home;
3. home to school;
4. home to shopping.

11.4 Leaving home earlier is a typical way to avoid peak hour congestion in the morning commute trip. Formulate a disutility function for a joint departure time/route choice problem. Discuss how to modify the formulation in this chapter to handle this departure time choice problem.

11.5 The disutility function is defined as follows:

$$\mathcal{U}_p^{rs}(t) = \alpha t + \pi_p^{rs}(t) \qquad \forall r, s, p$$

where t is the departure time of travelers and $\pi_p^{rs}(t)$ is the actual route travel time. The network has only one link and the link travel time function (minutes):

$$\tau_1(t) = 15 + u_1(t)$$

Determine ideal DUO departure flows for time period [8:00 am, 8:10 am]. Assume parameter $\alpha = 0.3$ and the total number of O-D departures at time 8:00 am is $F^{13}(0) = 30$ units and they must depart within time period [8:00 am, 8:05 am].

11.6 In problem 11.5, assume that all travelers should arrive at the destination at time 8:15 am. The penalty for late arrival is computed by the equation

$$\mathcal{V}^{rs}[t, \pi^{rs}(t); t^*_{rs}] = [t + \pi^{rs}(t) - t^*_{rs} - \Delta^*_{rs}]^2$$

Determine ideal DUO departure flows for time period [8:00 am, 8:10 am].

11.7 When is arrival time choice more important than departure time choice?

11.8 Travel time reliability is sometimes more important than travel time itself. The basic reason is that travel time variance could be very high in congested urban area. The most common solution for travelers is to leave their origins much earlier than necessary in order to guarantee arrival on time at destination. Discuss how the departure time choice model can be modified to handle this type of problem.

11.9 Compare the solution algorithms for combined ideal DUO departure time/route choice problem and ideal DUO route choice problem. Identify the major differences.

11.10 The disutility function is defined as follows:

$$\mathcal{U}^{rs}_p(t) = \alpha t + \pi^{rs}_p(t) \qquad \forall r, s, p$$

where t is the departure time of travelers and $\pi^{rs}_p(t)$ is the actual route travel time. The network is shown in Figure 11.5. The travel time functions are as follows:

$$\tau_1(t) = 2 + u_1(t), \qquad \tau_2(t) = 5 + 2t, \qquad \tau_3(t) = 6.5$$

Determine ideal DUO departure flows for time period $[0, 20]$. Assume parameter $\alpha = 0.5$ and the total number of O-D departures at time 0 is $F^{13}(0) = 100$ units and they must depart within time period $[0, 20]$.

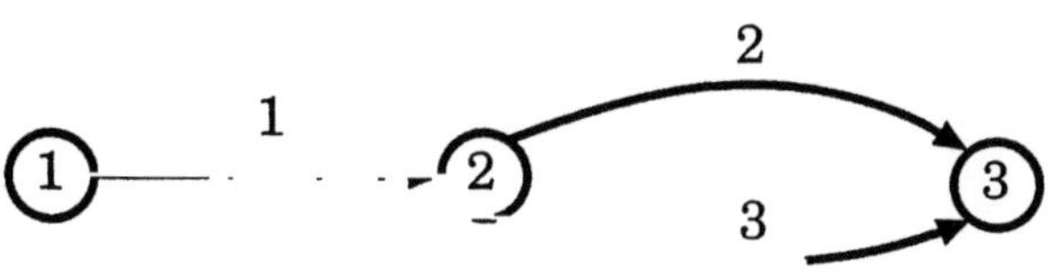

Figure 11.5: Network of Problem 11.10

Chapter 12

Combined Mode/Departure Time/Route Choice Models

We consider the efficient operation of an integrated transportation system within an ITS environment. A dynamic route guidance system would improve utilization of the overall capacity of the transportation system so as to reduce travel times, congestion and incidents. By providing early detection of incidents and congestion in the transportation network, the route guidance system would redistribute traffic among the available modes and routes when there is excess capacity in some parts of the road network or shift the departure times of travelers to avoid peak-hour congestion when no additional road capacity is available. Furthermore, the route guidance system would provide travelers with accurate, current information on both transit and road networks so that some motorists could make their own time-cost tradeoffs and shift to transit, if appropriate.

In this chapter, we address the dynamic mode/departure time/route choice problem with multiple stratifications of users, each with a different propensity to use transit or high occupancy vehicles. This stratification could be made by income level, automobile ownership or vehicle occupancy regulation. As in conventional planning models, socio-economic factors can be considered in mode choice models in time-dependent circumstances. The advantage of formally considering multiple classes of users, who value their time and convenience differently, is to model travelers' mode choices more accurately.

Furthermore, different people have different propensities to use different travel modes. For example, senior citizens may prefer local streets to freeways; therefore, freeways may constitute a specific mode for this particular class of travelers. In principle, by considering many classes of travelers with a specified time-cost tradeoff in each class, it is possible to predict mode choice deterministically.

A shift of travelers from cars to transit or from low-occupancy cars to high-occupancy cars may significantly decrease road congestion and increase the efficiency of the overall transportation system. Moreover, in the road network, journey-to-work trips have especially important requirements for avoiding congested routes so as to arrive at work on time. Since each departure time choice is based on prevailing origin-destination travel times, any change in departure times will alter the traffic flow patterns in the network so that route choice decisions of other travelers will be modified. Therefore, in order to achieve this balanced allocation to various departure times and different modes, an integrated model including all elements (mode, departure time and route choice) should be constructed.

There have been extensive studies in mode choice analysis. Wilson (1969) studied the trip distribution, modal split and trip assignment problem using entropy maximizing methods. Florian and Nguyen (1978) presented a combined trip distribution, modal split and trip assignment model. Route choice in their model is based on the user-optimal principle, and the mode choice is given by a logit model. Boyce (1978) also discussed the equilibrium solutions for combined location, mode choice and trip assignment models.

Studies into multiple classes of travelers in travel choice models were begun by Dafermos (1972). She presented a multiclass user equilibrium traffic assignment model. Later, LeBlanc and Abdulaal (1982) presented combined mode split/assignment and distribution/mode split/assign- ment models with multiple classes of travelers.

In this chapter, we present a dynamic user-optimal (DUO) mode, departure time and route choice model for a transportation network with multiple origin-destination pairs. The model developed in this chapter extends the joint departure time/route choice model in Chapter 11 to the case in which the combined mode, departure time and route choice should be considered with multiple classes of travelers. We model this choice problem by specifying that a given number of travelers are ready for departure between each origin-destination pair at the beginning of each of several short time periods. However, motorists may shift to transit or delay their departure times to reduce their overall travel costs. The model extends our previous dynamic user-optimal departure time and route choice model in two important respects:

1. alternative mode choices are available;
2. travelers are stratified into different classes according to travelers' socio-economic characteristics.

The problem is formulated as a variational inequality model. The dynamic logit-type modal choice is considered together with the joint DUO departure time and route choice problem for motorists.

The combined travel choice problem is first simply described in Section 12.1. The stratification of travelers into multiple classes is presented for accurate mode choice analysis. Then, major components of the combined travel choice, including mode choice, departure time choice and route choice problems, are discussed in detail in Section 12.2. The formulation of the variational inequality model is presented in Section 12.3.

12.1 The Combined Travel Choice Problem

A multiple origin-destination transportation network is considered. For simplicity, the transportation network is defined to consist of a transit network and a road network. Consider a fixed time period $[0, T]$. The length of the time period is sufficient to allow all travelers in the peak period to complete their trips. We consider the following mode, departure time, and route choice situation with $(K+1)$ points $T_1, T_2, \cdots, T_{K+1}$ on the time horizon for the fixed time period $[0, T]$. These points divide the time period $[0, T]$ into K intervals, where $T_1 = 0$ and $T_{K+1} = T$. Any interval k is denoted as $[T_k, T_{k+1}]$, $k = 1, 2, \cdots, K$, and these intervals may or may not be equal in length. The length of each interval would typically be 15 to 30 minutes for non-peak periods. For peak periods, this interval should be set to be consistent with the time headway of transit operations, such as 5 or 10 minutes. We also assume no departure time choice option for transit users, implying that headways are uniform over the time period $[0, T]$.

For each O-D pair rs, the class of travelers departing during period k can be further stratified into K smaller sub-classes according to the socio-economic characteristics of each traveler. There are several approaches to stratifying travelers. A typical one for mode choice is to classify travelers based on income and age (see Table 12.1). There are 9 combinations in this approach. Other approaches of stratification are discussed in the multi-class route choice problems in Chapter 5.

Table 12.1: Stratification of Travelers Based on Income and Age

	Degree of Change		
Income	High	Middle	Low
Age	Old	Middle	Young

For any interval k, travelers in class m are ready to depart by transit or car at an initial time T_k. Thus, based on the total disutilities of using transit and using car, travelers are split into 2 classes. For each O-D pair rs and each traveler class m, the total number of travelers departing during interval k is denoted as $R_m^{rs}(k)$ and is given exogenously. Similarly, let $Q_m^{rs}(k)$ and $G_m^{rs}(k)$ denote the total numbers of travelers of class m departing by transit and automobile, respectively, from origin r to destination s during interval k. It follows that

$$G_m^{rs}(k) + Q_m^{rs}(k) = R_m^{rs}(k) \qquad \text{given} \qquad \forall m, r, s, k. \tag{12.1}$$

Moreover, motorists may prefer to delay their departure times within period $[T_k, T_{k+1}]$ in order to reduce their driving times. Note that motorists departing during interval k can only shift their departure times within interval k. In future studies, the above assumption can be relaxed so that motorists departing during interval k can shift their departure times within intervals $l \geq k$.

Drivers are assumed to make their departure time choices so as to minimize their individual disutility functions defined on travel time and pre-trip delay. The criteria for choosing each departure time consist of:

1. the waiting time before departure;

2. the actual travel time between the origin and destination;

3. a bonus for early arrival or a penalty for late arrival.

Of course, a change of departure flow rate will change the traffic on the road network so that the actual origin-destination travel times will change.

Travelers' choices of mode, departure time and route are interrelated decisions. Given a desired arrival time, say at the workplace, choices of mode and departure time depend on the traveler's estimate of the en route travel time on each mode. Thus, choice of mode depends on the travel disutilities of alternative modes, which also may vary by time of day. Because there is a partition of the decision variables between two ordered stages; mode choice and departure time/route choice for motorists, we use a two-stage method to formulate the variational inequality model, in which mode choices are first-stage decision variables and departure time/route choices for motorists are second-stage decision variables. See Figure 12.1. Note that in the variational inequality, all three travel choice variables are determined simultaneously. The above formulation process serves to decompose the combined travel choice problem.

Furthermore, in the road network, choice of route depends on the travel times of alternative routes, which also may vary by time of day. In the second-stage travel choice (Figure 12.1), these choices are represented as a hierarchical structure. The dynamic route choice problem for motorists is considered to be the fundamendal element in the formulation; it corresponds to a single *controller* allocating fixed departure flows at each time t to user-optimal routes, given the departure frequencies. For each O-D pair, a *departure time coordinator* for motorists determines the departing flows at each time t, for which the user-optimal travel time at time t is exogenous. Since these problems are independent by O-D pair, $n(n-1)$ separate problems for each interval k can be used to represent all O-D pairs.

The hierarchical consideration for motorists assumes that the route choice controller, who is responsible for allocating all O-D flows to routes for each time interval k, knows the objective of each departure coordinator. Through this knowledge, the route choice coordinator is able to achieve a lower value of his/her objective function than if the departure time objectives were unknown. In contrast, the departure time coordinators know only the O-D travel times at time t provided to them. Knowledge of the route choice objective is not needed for their coordination task.

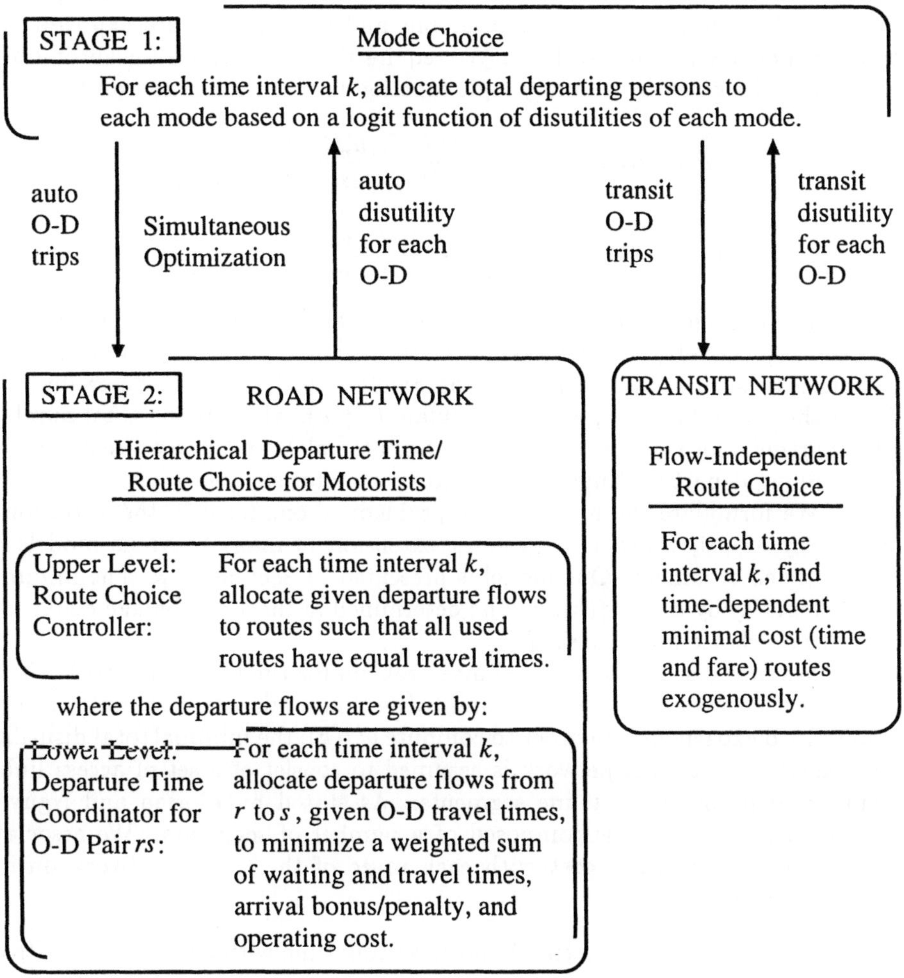

Figure 12.1: Two-Stage Simultaneous Travel Choices

12.2 Individual Travel Choice Problems

12.2.1 Mode Choice Problem

We consider the mode-choice problem for travelers in class m during any interval $[T_k, T_{k+1}]$. For simplicity, it is assumed that the automobile occupancy factor is 1; however, an occupancy factor could be used to convert the person flow to vehicle flow since both automobile and transit flows are expressed in terms of persons per unit of time. We require the modal share for automobile and transit be given by a binary logit function for each time interval k:

$$P(\text{auto}) = \frac{\exp(-\theta_m \mu_m^{rs}(k))}{\exp(-\theta_m \nu_m^{rs}(k)) + \exp(-\theta_m \mu_m^{rs}(k))} \tag{12.2}$$

$$P(\text{transit}) = \frac{\exp(-\theta_m \nu_m^{rs}(k))}{\exp(-\theta_m \nu_m^{rs}(k)) + \exp(-\theta_m \mu_m^{rs}(k))} \tag{12.3}$$

where $\mu_m^{rs}(k)$ and $\nu_m^{rs}(k)$ are the minimal total disutilities by auto and by transit, respectively, for travelers in class m departing during time interval k from r to s, and θ_m is a positive parameter which needs to be calibrated. Using the above function, we can calculate $G_m^{rs}(k)$, which is the total number of departing motorists in class m during interval k and is also required for determining departure time/route choices for motorists.

We formulate the mode choice problem as one term in the variational inequality model, which is a dynamic extension of many previous models in the static environment. Our model is presented in Section 12.3. The minimal total disutility by car, $\mu_m^{rs}(k)$, will be determined in the second-stage departure time/route choice for motorists.

We first discuss the route choice problem for the transit network. Since the transit network has fixed fares and travel times, minimal cost routes can be determined exogenously in order to compute $\nu_m^{rs}(k)$, the minimal total disutility by transit. The transit network is assumed to consist of a set of access links, transfer links and transit line segments. As stated by Florian and Nguyen (1978), a transit route is composed of a number of segments. We associate a time-dependent travel cost with each route of the transit network during interval k as follows:

1. a walking (or driving time) and a waiting time with access links to/from stations;

2. a walking time and a waiting time with a transfer link;

3. an in-vehicle time with a line segment;

4. origin-destination fare.

For any interval k, it is assumed that the link travel cost is independent of the transit link flows. Thus, the time-dependent travel cost $C_p(k)$ over route

p for travelers departing during interval k can be easily computed. We suppose that a dynamic user-optimal state occurs on the road network, as discussed in Section 12.2.2, where we consider departure time choice and route choice. Similarly, we suppose that a dynamic user-optimal state occurs on the transit network. That is

$$C_p(k) = \nu_m^{rs}(k)/\eta_m \qquad \text{if } Q_{mp}^{rs}(k) > 0; \tag{12.4}$$

$$C_p(k) \geq \nu_m^{rs}(k)/\eta_m \qquad \text{if } Q_{mp}^{rs}(k) = 0 \tag{12.5}$$

where η_m is the disutility scaling parameter associated with the social-economic characteristics of class m travelers. Since $C_p(k)$ is fixed, $\nu_m^{rs}(k)/\eta_m$ is the minimal travel cost from r to s for travelers departing during interval k. Thus, the route choice problem in a transit network is simply a time-dependent minimal route cost problem for each interval k.

12.2.2 Departure Time/Route Choice for Motorists

Next we consider the second-stage problem of departure time/route choice for the road network. The second-stage problem can be formulated as the major terms in the variational inequality model.

Departure Time Choice

A disutility function $\mathcal{U}_m^{rs}(t)$ based on departure times is defined for class m drivers departing from origin r to destination s at time t. This function represents a weighted sum of total elapsed time during the journey, including the waiting time at the origin node, the driving time during the trip, any bonus/penalty for early/late arrival, and auto operating cost. We focus the following discussion on the disutility for any class m motorists. We define one unit of in-vehicle driving time to equal γ_m units of disutility, and one unit of waiting time prior to departure to be equivalent to α_m units of disutility ($\alpha_m \leq 1$).

For each class of travelers departing during interval k, there is a required arrival time interval $[t_{rs}^*(k) - \Delta_{rs}(k), t_{rs}^*(k) + \Delta_{rs}(k)]$ at each destination s, where $t_{rs}^*(k)$ is the center of the required arrival time interval associated with travelers departing from origin r during interval k toward destination s. $\Delta_{rs}(k)$ represents the arrival time flexibility at destination s for travelers departing from origin r during interval k toward destination s. We also define the disutility for early or late arrival as follows:

$$\mathcal{V}_m^{rs}[t, \pi^{rs}(t); t_{rs}^*(k)] =$$

$$\begin{cases} \beta_{1i}[t + \pi^{rs}(t) - t_{rs}^*(k) + \Delta_{rs}^*(k)]^2 & \text{if } t + \pi^{rs}(t) < t_{rs}^*(k) - \Delta_{rs}^*(k) \\ 0 & \text{if } |t + \pi^{rs}(t) - t_{rs}^*(k)| \leq \Delta_{rs}^*(k) \\ \beta_{2i}[t + \pi^{rs}(t) - t_{rs}^*(k) - \Delta_{rs}^*(k)]^2 & \text{if } t + \pi^{rs}(t) > t_{rs}^*(k) + \Delta_{rs}^*(k) \end{cases}$$

where time t is the departure time of travelers and β_{1i}, β_{2i} are parameters ($\beta_{1i} \leq 0, \beta_{2i} \gg \alpha$). This relationship is shown in Figure 12.2. This disutility of late arrival is continuous and differentiable with respect to t and $\pi^{rs}(t)$.

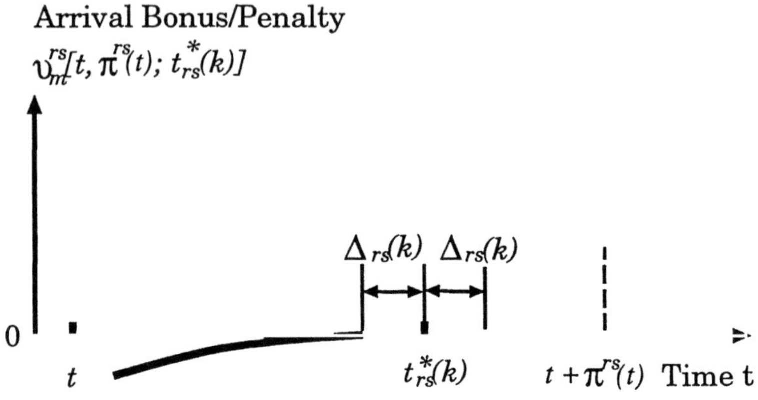

Figure 12.2: Arrival Time Disutility

Let C^{rs} denote the auto operating cost from origin r to destination s; for simplicity, we assume this cost is fixed for each O-D pair. We define one unit of operating cost to equal ξ_m units of disutility. Even though the operating cost is not directly associated with the departure time choice of motorists, it is an important decision element when choosing between transit and car and is used in the mode choice model. Thus, the disutility function for class m motorists departing during interval k is constructed as

$$\mathcal{U}_m^{rs}(t;k) = \alpha_m(t - T_k) + \gamma_m \pi^{rs}(t) + \mathcal{V}_m^{rs}[t, \pi^{rs}(t); t_{rs}^*(k)] + \xi_m C^{rs}$$

$$\forall m, r, s, k, t \in (T_k, T_{k+1}) \qquad (12.6)$$

where time t is the departure time of travelers of class m departing during interval k and T_k is the starting time of interval k. Note that since travelers of class m desire to depart at time T_k, $\alpha_m(t - T_k)$ is the disutility of waiting.

For O-D pair rs and any time t, denote $f_m^{rs}(t)$ and $F_m^{rs}(t)$ as the departure rate and the cumulative number of departures of class m motorists, respectively. Then for motorists, we construct the dynamic user-optimal departure time choice conditions as

$$\mathcal{U}_m^{rs}(t;k) = \mu_m^{rs}(k) \qquad \text{if } f_m^{rs}(t) > 0 \qquad \forall m, r, s, k, t \in (T_k, T_{k+1}); \qquad (12.7)$$

$$\mathcal{U}_m^{rs}(t;k) \geq \mu_m^{rs}(k) \qquad \text{if } f_m^{rs}(t) = 0 \qquad \forall m, r, s, k, t \in (T_k, T_{k+1}); \qquad (12.8)$$

where $\mu_m^{rs}(k)$ is the minimal rs disutility for class m motorists departing during interval k. By definition of $f_m^{rs}(t)$ (the O-D departure rate) and $F_m^{rs}(t)$ (the cumulative number of departing vehicles), it follows that

$$\int_0^t f_m^{rs}(t)dt = F_m^{rs}(t) \qquad \forall m, r, s, k, t \in (T_k, T_{k+1}); \qquad (12.9)$$

or

$$\frac{dF_m^{rs}(t)}{dt} = f_m^{rs}(t) \qquad \forall m, r, s, k, t \in (T_k, T_{k+1}); \qquad (12.10)$$

where $F_m^{rs}(t)$ and $f_m^{rs}(t)$ are state and control variables, respectively. Also, we have corner point or boundary conditions as follows:

$$F_m^{rs}(T_{k+1}) = \sum_{j=1}^{k} G_m^{rs}(j) \qquad \forall m, r, s, k; \qquad (12.11)$$

$$F_m^{rs}(T_1) = F_m^{rs}(0) = 0 \qquad \forall m, r, s; \qquad (12.12)$$

where the total number of departing class m motorists during interval j ($1 \leq j \leq k$), $G_m^{rs}(j)$, is given by the first-stage mode choice program. When solving the second-stage departure time/route choice program for motorists, the departure time choice model decides the instantaneous departure rate $f_m^{rs}(t)$ for each class m and each O-D pair rs. Then, the departure rates sum up as follows:

$$f^{rs}(t) = \sum_m f_m^{rs}(t) \qquad \forall r, s$$

where $f^{rs}(t)$ is the input for the route choice model for motorists.

Route Choice

In this section, we discuss the route choice in the hierarchical departure time/ route choice program for the road network. We assume that all motorists in all classes choose minimal travel time routes. The stratification of travelers is used for mode and departure time choices. Thus, the dynamic route choice problem is the same as in Chapter 5. The stratification of travelers for the route choice problem is addressed in Chapter 5.

The dynamic route choice problem is to find the dynamic trajectories of link states and inflow and exit flow control variables, given the road network, the link travel time functions and the time-dependent O-D flow requirements. The O-D requirements are specified by the departure time choice problem for motorists and are therefore temporarily fixed in the route choice problem for motorists. The formulation of the problem is based on the underlying choice criterion that each traveler uses the route that minimizes his/her actual travel time when departing from the origin to his/her destination.

Denote $\Omega_a^{rj^*}(t)$ as the difference in the travel time from r to j via the minimal travel time route from r to i plus link a and the minimal travel time from r to j for vehicles departing from origin r at time t. It follows that

$$\Omega_a^{rj^*}(t) = \pi^{ri^*}(t) + \tau_a[t + \pi^{ri^*}(t)] - \pi^{rj^*}(t) \qquad \forall a, r; a = (i, j). \quad (12.13)$$

As formulated in Chapter 5, we can rewrite the link-time-based ideal DUO route choice conditions as:

$$\Omega_a^{rj^*}(t) \geq 0 \qquad \forall a=(i,j), r; \tag{12.14}$$

$$u_a^{rs^*}[t+\pi^{ri^*}(t)]\,\Omega_a^{rj^*}(t) = 0 \qquad \forall a=(i,j), r, s; \tag{12.15}$$

$$u_a^{rs}[t+\pi^{ri^*}(t)] \geq 0 \qquad \forall a=(i,j), r, s. \tag{12.16}$$

The detailed analysis is given in Chapter 5. Note that in the above equation, the actual travel time $\pi^{rs}(t)$ may be greater than any interval $[T_k, T_{k+1}]$. Thus, we implicitly assume that $\pi^{rs}(t)$ is differentiable over the time period $[0,T]$.

12.3 The Link-Time-Based Model

12.3.1 Network Constraints

The constraints for dynamic user-optimal mode, departure time and route choice model are formulated as follows.

Mode Choice

$$G_m^{rs}(k) + Q_m^{rs}(k) = R_m^{rs}(k) \quad \text{given} \quad \forall m, r, s, k; \tag{12.17}$$

$$G_m^{rs}(k) \geq 0, \qquad Q_m^{rs}(k) \geq 0 \qquad \forall m, r, s, k; \tag{12.18}$$

Departure Time/Route Choice for Motorists

Departure Time Choice

$$\frac{dF_m^{rs}(t)}{dt} = f_m^{rs}(t) \qquad \forall m, r, s, k, t \in (T_k, T_{k+1}); \tag{12.19}$$

$$F_m^{rs}(T_{k+1}) = \sum_{j=1}^{k} G_m^{rs}(j) \qquad \forall m, r, s, k; \tag{12.20}$$

$$F_m^{rs}(T_1) = F_m^{rs}(0) = 0 \qquad \forall m, r, s; \tag{12.21}$$

$$f^{rs}(t) = \sum_m f_m^{rs}(t) \qquad \forall r, s, t \in (T_k, T_{k+1}); \tag{12.22}$$

$$f_m^{rs}(t) \geq 0, \qquad F_m^{rs}(t) \geq 0 \qquad \forall m, r, s, k, t \in (T_k, T_{k+1}); \tag{12.23}$$

where $G_m^{rs}(k)$ is given by mode choice problem.

Route Choice

$$\frac{dx_{ap}^{rs}}{dt} = u_{ap}^{rs}(t) - v_{ap}^{rs}(t) \qquad \forall a, p, r, s; \tag{12.24}$$

$$\frac{dE_p^{rs}(t)}{dt} = e_p^{rs}(t) \qquad \forall p, r, s; \tag{12.25}$$

$$\frac{dF_p^{rs}(t)}{dt} = f_p^{rs}(t) \qquad \forall p, r, s; \tag{12.26}$$

$$f_p^{rs}(t) = \sum_{a \in A(r)} \delta_{ap}^{rs} u_{ap}^{rs}(t) \qquad \forall p, r, s; \tag{12.27}$$

$$e_p^{rs}(t) = \sum_{a \in B(s)} \delta_{ap}^{rs} v_{ap}^{rs}(t) \qquad \forall p, r, s; \tag{12.28}$$

$$\sum_{a \in B(j)} v_{ap}^{rs}(t) = \sum_{a \in A(j)} u_{ap}^{rs}(t) \qquad \forall j, p, r, s; j \neq r, s; \tag{12.29}$$

$$x_{ap}^{rs}(t) = \sum_{b \in \bar{p}} \{x_{bp}^{rs}[t + \tau_a(t)] - x_{bp}^{rs}(t)\} + \{E_p^{rs}[t + \tau_a(t)] - E_p^{rs}(t)\}$$

$$\forall r, s, p, j; a \in B(j); j \neq r; \tag{12.30}$$

$$\sum_{rsp} u_{ap}^{rs}(t) = u_a(t), \qquad \sum_{rsp} v_{ap}^{rs}(t) = v_a(t), \qquad \forall a; \tag{12.31}$$

$$\sum_{rsp} x_{ap}^{rs}(t) = x_a(t), \qquad \sum_{rs} x_a^{rs}(t) = x_a(t), \qquad \forall a; \tag{12.32}$$

$$\sum_p E_p^{rs}(t) = E^{rs}(t), \qquad \sum_p F_p^{rs}(t) = F^{rs}(t), \qquad \forall r, s; \tag{12.33}$$

$$\sum_p f_p^{rs}(t) = f^{rs}(t), \qquad \sum_p e_p^{rs}(t) = e^{rs}(t), \qquad \forall r, s; \tag{12.34}$$

$$x_{ap}^{rs}(t) \geq 0, \quad u_{ap}^{rs}(t) \geq 0, \quad v_{ap}^{rs}(t) \geq 0 \quad \forall a, p, r, s; \tag{12.35}$$

$$e_p^{rs}(t) \geq 0, \ f_p^{rs}(t) \geq 0, \ E_p^{rs}(t) \geq 0, \ F_p^{rs}(t) \geq 0 \qquad \forall p, r, s; \tag{12.36}$$

$$E_p^{rs}(0) = 0, \quad F_p^{rs}(0) = 0 \quad \forall p, r, s; \quad x_{ap}^{rs}(0) = 0, \quad \forall a, p, r, s; \tag{12.37}$$

The mode choice model constraints are flow conservation (12.17) and nonnegativity (12.18). The decision variables are $G_m^{rs}(k)$ and $Q_m^{rs}(k)$, which represent the total numbers of travelers by car and by transit, respectively. The disutility function $\mu_m^{rs}(k)$ for motorists is determined by the departure time choice program. The disutility function $\nu_m^{rs}(k)$ for transit travelers is calculated exogenously.

The second-stage model constraints (12.19)-(12.37) are defined for motorists. In the departure time choice model for motorists, we have only state equations (12.19), boundary conditions (12.20) and (12.21), flow conservation equations (12.22) and nonnegativity conditions (12.23). The control variables are $f_m^{rs}(t)$ and $f^{rs}(t)$, and the state variables are $F_m^{rs}(t)$, which represent cumulative class m motorist departures from origin r to destination s. The inputs for this departure time choice model include $G_m^{rs}(k)$ and $\pi^{rs}(t)$. The total number of motorists $G_m^{rs}(k)$ in class m during interval k is determined by the mode choice program. The actual O-D travel time function $\pi^{rs}(t)$ for motorists is determined by the route choice model for motorists. As shown in the optimality conditions, the minimal disutility $\mu_m^{rs}(k)$ for motorists can be computed after solving the departure time model. Note that $\mu_m^{rs}(k)$ is an input to the mode choice model.

The ideal DUO route choice model for motorists was presented in Chapter 5. In this model, the route specific departure variables $f_p^{rs}(t)$ and $F_p^{rs}(t)$ must be found. The first three constraints (12.24)-(12.26) are state equations for each link a and cumulative effects at origins and destinations. Equations (12.27)-(12.29) are flow conservation constraints at each node including origins and destinations. The other constraints include flow propagation, definitional, nonnegativity, and boundary conditions. In summary, in this model the control variables are $u_{ap}^{rs}(t)$, $v_{ap}^{rs}(t)$, $e_p^{rs}(t)$ and $f_p^{rs}(t)$; the state variables are $x_{ap}^{rs}(t)$, $E_p^{rs}(t)$ and $F_p^{rs}(t)$; and the functionals are $\pi^{rs}(t)$. The inputs for this model include the instantaneous O-D flow rate $f^{rs}(t)$, which is a decision variable in the departure time choice model.

The relationship of decision variables in the simultaneous variational inequality model is shown in Figure 12.3, which can be used to help design the solution algorithm for the model. The departure time choice model provides a set of corner point (boundary) conditions at intervals $T_1, T_2, \cdots, T_{K+1}$ for the departure time/route choice program for motorists. Also, in the two-stage simultaneous program, the first-stage mode choice program provides boundary conditions for a cluster of hierarchical departure time/route choice models for each interval k in the second-stage. We prove in the next section that the optimal solution to the two-stage simultaneous problem is equivalent to a variational inequality model.

12.3.2 The Variational Inequality Problem

To summarize the above analysis, the combined dynamic user-optimal mode/ departure time/route choice conditions are presented as follows:

Mode Choice

$$G_m^{rs}(k) - R_m^{rs}(k)\frac{\exp(-\theta_m \mu_m^{rs}(k))}{\exp(-\theta_m \mu_m^{rs}(k)) + \exp(-\theta_m \nu_m^{rs}(k))} = 0, \qquad \forall m, r, s, k. \tag{12.38}$$

Departure Time Choice for Motorists

$$\mathcal{U}^{rs^*}(t) - \mathcal{U}_{min}^{rs^*} \geq 0 \qquad \forall r, s; \tag{12.39}$$

$$f^{rs^*}(t)\,\{\mathcal{U}^{rs^*}(t) - \mathcal{U}_{min}^{rs^*}\} = 0 \qquad \forall r, s; \tag{12.40}$$

$$f^{rs}(t) \geq 0 \qquad \forall r, s. \tag{12.41}$$

Route Choice for Motorists

$$\Omega_a^{rj^*}(t) \geq 0 \qquad \forall a = (i, j), r; \tag{12.42}$$

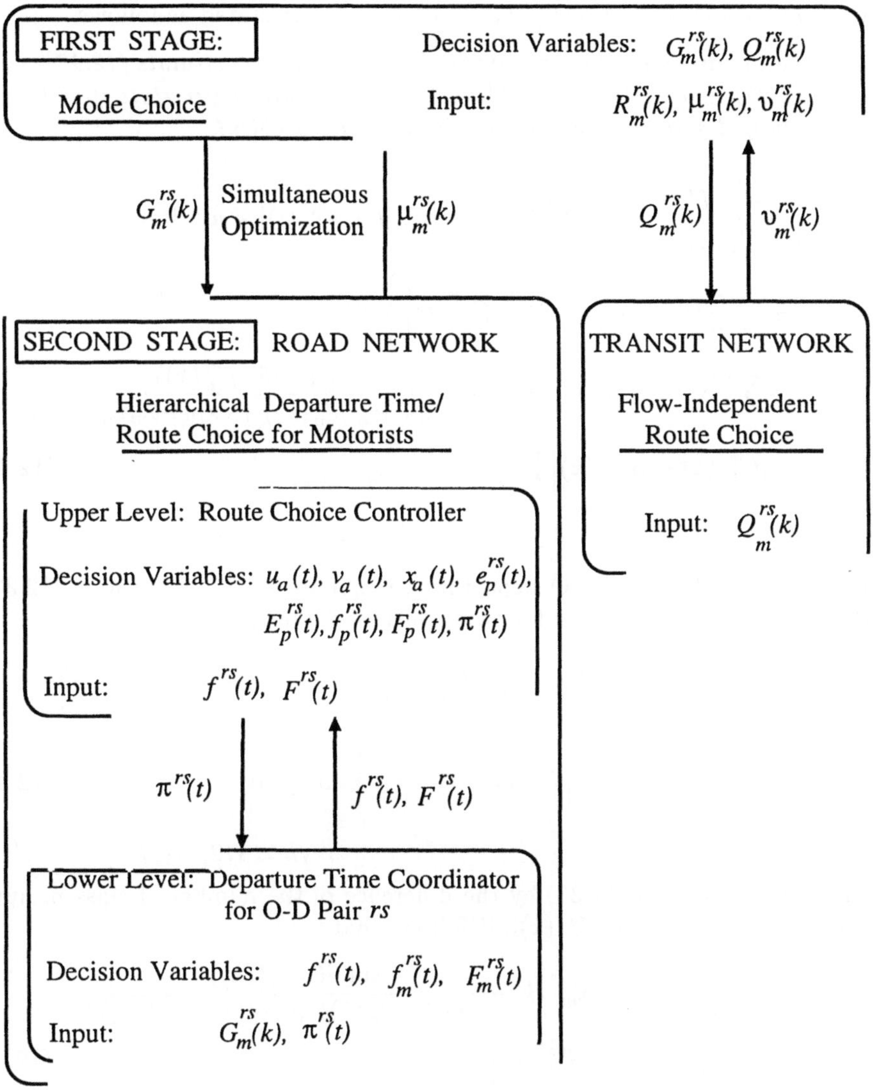

Figure 12.3: Decision Variables in the Combined Travel Choice Problem

$$u_a^{rs^*}[t+\pi^{ri^*}(t)]\,\Omega_a^{rj^*}(t)=0 \qquad \forall a=(i,j),r,s; \qquad (12.43)$$

$$u_a^{rs}[t+\pi^{ri^*}(t)]\geq 0 \qquad \forall a=(i,j),r,s; \qquad (12.44)$$

where $\mathcal{U}_{min}^{rs}$ is the minimal rs disutility over time t.

The equivalent variational inequality formulation of the DUO mode/departure time/route choice conditions (12.38)-(12.44) may be stated as follows.

Theorem 12.1. *Dynamic traffic flow satisfying constraints (12.17)-(12.37) is in a DUO mode/departure time/route choice state if and only if it satisfies the variational inequality problem:*

$$\sum_{k=1}^{K}\left\{\int_{T_k}^{T_{k+1}}\left\{\sum_{rs}\sum_{a}\Omega_a^{rj^*}(t)\ \left[u_a^{rs}[t+\pi^{ri^*}(t)]-u_a^{rs^*}[t+\pi^{ri^*}(t)]\right]\right.\right.$$
$$\left.+\sum_{rs}\mathcal{U}^{rs^*}(t)\ \left[f^{rs}(t)-f^{rs^*}(t)\right]\right\}dt$$
$$+\sum_{rs}\sum_{m}\left[G_m^{rs}(k)-R_m^{rs^*}(k)\frac{\exp(-\theta_m\mu_m^{rs}(k))}{\exp(-\theta_m\mu_m^{rs}(k))+\exp(-\theta_m\nu_m^{rs}(k))}\right]$$
$$\left.\left[G_m^{rs}(k)-G_m^{rs^*}(k)\right]\right\}\geq 0 \qquad (12.45)$$

Proof .

From Chapter 11, the DUO departure time/route choice conditions (12.39)-(12.44) are equivalent to the following variational inequality:

$$\int_{T_k}^{T_{k+1}}\left\{\sum_{rs}\sum_{a}\Omega_a^{rj^*}(t)\ \left\{u_a^{rs}[t+\pi^{ri^*}(t)]-u_a^{rs^*}[t+\pi^{ri^*}(t)]\right\}\right.$$
$$\left.+\sum_{rs}\mathcal{U}^{rs^*}(t)\ \left[f^{rs}(t)-f^{rs^*}(t)\right]\right\}dt\geq 0 \qquad (12.46)$$

$$\forall k=1,2,\cdots,K.$$

Multiplying equation (12.38) by the difference of the number of class m automobile trips, $[G_m^{rs}(k)-G_m^{rs^*}(k)]$, it follows that

$$\left[G_m^{rs}(k)-R_m^{rs^*}(k)\frac{\exp(-\theta_m\mu_m^{rs}(k))}{\exp(-\theta_m\mu_m^{rs}(k))+\exp(-\theta_m\nu_m^{rs}(k))}\right]$$
$$\cdot\left[G_m^{rs}(k)-G_m^{rs^*}(k)\right]=0 \qquad (12.47)$$

Summing the above equation for all O-D pairs and all traveler classes, it follows that

$$\sum_{rs}\sum_{m}\left[G_m^{rs}(k)-R_m^{rs^*}(k)\frac{\exp(-\theta_m\mu_m^{rs}(k))}{\exp(-\theta_m\mu_m^{rs}(k))+\exp(-\theta_m\nu_m^{rs}(k))}\right]$$
$$\cdot\left[G_m^{rs}(k)-G_m^{rs^*}(k)\right]=0 \qquad (12.48)$$

Adding inequality (12.46) and equation (12.48) and summing over all time periods k, we obtain variational inequality (12.45). Now, we conclude that DUO mode/departure time/route choice conditions (12.38)-(12.44) are equivalent to variational inequality (12.45).

12.4 Notes

We have presented a combined model in which dynamic user-optimal mode, departure time and route choice occurs. It is recognized that different travelers perceive the time-cost tradeoff differently, and thus distinct classes of travelers are included in the model. The advantage over sequential dynamic models is that the time-dependent interaction among the mode choice, departure time choice and route choice is inherently recognized.

The most likely application of the model which we outlined is in real-time ATIS in conjunction with APTS. Since time-dependent road pricing policies can be readily adopted in our model, an extended combined model incorporating time-dependent road pricing (especially congestion pricing) will find important applications in ATMS. The calibration of the mode choice model requires the determination of the parameter θ_m which plays a key role in determining the modal shares on the road and transit networks. In contrast to the static mode choice model, the disutilities for both motorists and transit users are time-dependent. Thus, the estimation procedure for θ_m is more complicated and remains a major task in the future.

The model presented in this chapter can be extended to include more than two modes. In particular, bus lanes and HOV (High Occupancy Vehicle) lanes can be specified as a special mode in road networks. The modal choice model assumes no interactions exist between the transit links and the auto links. However, this assumption is not generally applicable to bus transit. Buses move with road traffic and experience the same time-dependent congestion and delays as automobiles. The interaction of the two interdependent modes can be studied using the multi-class variational inequality models for route choice problems in Chapter 5.

Generally, a variational inequality model is more difficult to solve than the combined departure time/route choice model. For this combined model, since the mode choice program will result in a logit type mode choice function for transit users and motorists, the main difficulty lies in how to solve the departure time choice and route choice problems efficiently in response to a large set of feasible mode choice solutions.

Problems

12.1 Identify all major transportation modes for commuting in your area. Discuss which mode choices should be considered as dynamic.

12.2 In considering mode choice, security and comfort are two other major factors affecting travelers' choices in addition to travel cost consideration. Discuss how these factors affect the dynamic mode choice and how they can be considered in the dynamic mode choice model.

12.3 Discuss why a combined mode/departure time/route choice model has advantages over a sequential model, which implements each dynamic travel choice sequentially.

12.4 When adding dynamic mode choice to dynamic route models, formulation and computational difficulties have been significantly increased. Identify these difficulties.

12.5 Consider the transit network as one part of a generalized transportation network. The combined mode/departure time/route choice problem can then be transformed into a dynamic route choice problem on a generalized transportation network. Write the corresponding ideal DUO route choice conditions and possible additional constraints for the part of transit network.

12.6 Discuss the time-dependence of parameters in a logit-type dynamic mode choice model. Propose several possible approaches to calibrate these parameters.

12.7 When traveling from Berkeley to San Francisco, there are two modes to choose from: BART (train) and freeway. Suppose the travel time on BART is 30 minutes. On the other hand, because of congestion, driving by car on freeway has the following travel time function

$$\tau(t) = -4(t-10)^2/5 + 100$$

Assume that dispersion parameter $\theta = 2$ and O-D flow is $f^{rs}(t) = 100t$ units. Determine departure flows on BART and freeway for time period $[0, 20]$ by using a logit-type dynamic mode choice model. Draw a diagram showing the relationship of departure flows vs. time for BART and freeway. Also indicate when they have equal departure flows.

Chapter 13

Link Travel Time Functions for Dynamic Network Models

Extensive research has occurred in recent years on dynamic transportation network models, and especially on dynamic route choice models; these models have important applications in future ATIS and ATMS systems. However, most of the existing models lack a basis in traffic engineering. A significant problem for dynamic route choice is that the traditional BPR (Bureau of Public Roads, the predecessor of the Federal Highway Administration, U.S. DOT) volume-delay function is not applicable to a time-dependent traffic network. Meanwhile, since no proper dynamic link travel time functions exist, current dynamic route choice models assume various functional forms which are either too abstract or cannot provide realistic travel time estimates, even for a small network. Thus, it is becoming increasingly urgent to develop a set of time-dependent link travel time functions for dynamic route choice problems.

In this chapter, the independent variables necessary to describe the dynamic traffic flow and estimate the corresponding time-dependent travel time over a highway link are discussed. In order to standardize the dynamic route choice formulation to be used in ATIS and ATMS applications, this chapter seeks to provide a solid foundation based on the principles of traffic engineering. For the purpose of short-term travel time forecasting, dynamic link travel time functions are also necessary to transform traffic flow data from probe vehicles or roadway detectors into travel times. The application of those functions in IVHS projects can also be expected.

Link travel time or delay functions have been extensively studied in traffic flow theory and traffic engineering research. These functions can be classified based on road types. In general, the following types of roadway links have different link travel time functions:

1. Arterial Streets

    (a) Links with Signalized Intersections

- Fixed Signal Control
- Actuated Signal Control

(b) Links with Unsignalized Intersections

- Major/Minor Priority Intersections
- All-Way-Stop Intersections

2. Freeways

(a) Freeway Segment

(b) Ramps

(c) Weaving Sections

3. Local Streets

(a) Stop/Yield Control

(b) No Control

The objective of this chapter is to review currently available delay models, identify suitable functions, and develop them into dynamic link travel time functions which would be applicable to dynamic route choice models. The focus is on exploring dynamic travel time functions for signalized arterial network links and freeway segments. In Section 13.1, we discuss the classification of dynamic link travel times for various applications. In Sections 13.2 and 13.3, travel time functions for arterials with long and short time horizons are discussed separately, and two sets of functions are recommended for dynamic route choice models. The implications of those functional forms are analyzed in Section 13.4 and some modifications for dynamic models are suggested. In Section 13.5, we propose dynamic travel time functions for freeway segments.

13.1 Functions for Various Purposes

For an arterial link, travel time is considered to consist of two main components. The first is the travel time (or cruise time) over the uncongested portion of the link; the second is the congested travel time or queuing delay at the intersection, plus the travel time through the intersection to the downstream link. For a freeway segment, link travel time is also considered to consist of two main components. The first is the uncongested cruise time over the link; the second is the congested travel time or queuing delay on the link. Nevertheless, we use similar formulae for link travel time functions for both arterial and freeway segments.

Recall that the *instantaneous* travel time $c_a(t)$ at time t is the travel time that is experienced by vehicles traversing link a when prevailing traffic conditions remain unchanged. It is the sum of two components:

1. an instantaneous flow-dependent cruise time $D_{a1}(t)$ over the first part of link a for an arterial segment; for a freeway segment, the uncongested cruise time over the link;

2. an instantaneous queuing delay $D_{a2}(t)$.

It follows that

$$c_a(t) = D_{a1}(t) + D_{a2}(t). \tag{13.1}$$

The instantaneous route travel time function $\psi_p^{rs}(t)$ for each route p between O-D pair rs is defined as the sum of the instantaneous link travel times over all links in route p:

$$\psi_p^{rs}(t) = \sum_{a \in rsp} c_a(t) \qquad \forall p, r, s. \tag{13.2}$$

Thus, the instantaneous route travel time is that time experienced by a vehicle, if prevailing traffic conditions do not vary until the vehicle reaches its destination. This instantaneous route travel time provides a first approximation to the time-dependent vehicle travel time.

Also recall that $\tau_a(t)$ is the actual travel time over link a for vehicles entering link a at time t. Similarly, $\eta_p^{rs}(t)$ is the actual travel time experienced over route p by vehicles departing from origin r toward destination s at time t. Once the actual link travel time $\tau_a(t)$ is determined, the actual route travel time $\eta_p^{rs}(t)$ can be computed using the recursive formula given in Chapter 4.

Since network traffic conditions change over time, the actual route time may be significantly different from the instantaneous route travel time, especially when the route or travel time is long. Otherwise, the instantaneous route travel time provides a good estimate of the actual route travel time. The instantaneous route travel time is easily obtained or estimated compared to the actual route travel time since the prevailing traffic flow data can be obtained in real-time from a probe vehicle or a roadway detector.

The difference between the instantaneous link travel time and the actual link travel time may be insignificant since the length of links is generally short (0–1 miles), as is the travel time (several seconds to a few minutes). In conclusion, we would like to develop a temporal link travel time function which is a good representation of both the instantaneous link travel time and the actual link travel time. In the following, the link travel time function refers to the *actual* link travel time function. However, we note that when a link is extremely congested or oversaturated, these two kinds of link travel times may be quite different.

We would like to suggest a temporal link travel time function which would be a good representation of actual link travel time. The functions proposed in this chapter might be calibrated for arterial links with vehicle-actuated and fixed-time signals separately. We note that turning flows at an intersection are not considered and will be addressed in future research.

In the link travel time, vehicle delay is a major time-dependent component. Accurate determination of this component requires the experience and insights of traffic engineering. Vehicle delay at the exit from the link comprises both *deterministic* and *stochastic* components. When one considers a discrete time problem and the analysis time interval is shortened, the stochastic delay becomes less significant. We note that the stochastic delay also depends on the exit capacity of a link. On this basis, the application of queuing and delay models may be divided into two categories discussed below.

In the first category, the flow and capacity information is required to predict queues and delays defined on a longer time-scale (e.g. successive 5 to 30 minute intervals). It is only possible to predict overall quantities such as the average queue length, or average delay per vehicle evaluated over a complete traffic peak. In this case, the stochastic delay constitutes a significant part of the total delay and cannot be neglected in any travel time estimation. The travel time function in this category is suitable for off-line evaluation of ATMS and ATIS systems as well as transportation planning applications.

In the second category, the flow and capacity information is required to estimate queues and delays defined on a short time-scale (e.g. successive one to five minute intervals or shorter). It is possible to determine in detail the time variation of the average queue length and vehicular delay. The stochastic delay is therefore negligible, and only deterministic oversaturation delay must be considered in this case. Travel time functions in this category are suitable for real-time on-line evaluation of ATMS and ATIS systems.

For freeway segments, we consider the flow and capacity variation which is required to estimate queues and delays defined only on a short time scale (e.g., less than 5 minutes). Thus, it is possible to determine in detail the time variation of the average density and vehicular delay. The stochastic delay is therefore negligible, and only deterministic delay must be considered in this case. These travel time functions for freeway segments are suitable for real-time on-line evaluation of ATMS and ATIS systems.

13.2 Stochastic Link Travel Time Functions

In this section, we consider stochastic link travel time functions. We seek to apply existing delay formulae to derive corresponding temporal link travel time functions. These functions could be used for an analysis interval of 5–30 minutes or longer. Note that other factors could contribute to stochastic delays in travel time functions, although stochastic variation over the long horizon is our major concern in this chapter.

13.2.1 Moving Queue Concept

In the delay formula, the input is the average flow rate of arrival at the downstream intersection. In our dynamic route choice model (Chapter 5, 7), we use three variables ($x_a(t)$, $u_a(t)$, $v_a(t)$) to represent the dynamics of traffic on a link; however, the arrival flow rate at the downstream intersection of the link does not correspond to any of these three variables. To overcome this difficulty, we divide a physical link $a = (A, B)$ into two dummy links: link $a_1 = (A, D)$ and $a_2 = (D, B)$ (see Figure 1). The location of dummy node D indicates the end of the vehicle queue and is movable. We assume that link $a_1 = (A, D)$ contains an uncongested vehicle stream and $a_2 = (D, B)$ contains a traffic queue. Thus, the length of a_2 is the length of the physical queue on link a. When there is no queue on link a, the length of dummy link a_2 equals zero and dummy link a_1 has the same length as link a.

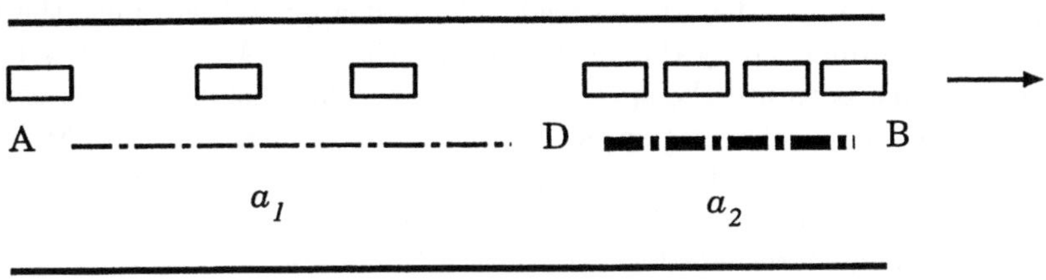

Figure 13.1: Traffic on an Arterial Road Link

The state equations for the two dummy links are:

$$\frac{dx_{a1}(t)}{dt} = u_{a1}(t) - v_{a1}(t) \qquad \forall a_1 \in a; a; \qquad (13.3)$$

$$\frac{dx_{a2}(t)}{dt} = u_{a2}(t) - v_{a2}(t) \qquad \forall a_2 \in a; a. \qquad (13.4)$$

The flow conservation for the two dummy links requires:

$$v_{a1}(t) = u_{a2}(t) \qquad \forall a_1, a_2 \in a; a. \qquad (13.5)$$

Now, we use six variables $x_{a1}(t)$,$u_{a1}(t)$,$v_{a1}(t)$,$x_{a2}(t)$,$u_{a2}(t)$,$v_{a2}(t)$ to describe the dynamic traffic for each link a. The link flow propagation constraints will be discussed below.

Next we consider a discrete time problem. Denote $\Delta k = [k, k+1]$ as the length of time interval in hours. This interval could be as long as 0.25-1 hours in the stochastic link travel time functions, or as short as 0.5-5 minutes in the deterministic link travel time functions. In this discrete time formulation, $x_a(k)$ represents vehicles on the link at the beginning of interval k; $u_a(k)$ and $v_a(k)$ represent inflow and exit flow during interval k. Writing the above equations in a discrete-time form, it follows that

$$x_{a1}(k+1) = x_{a1}(k) + u_{a1}(k)\Delta k - v_{a1}(k)\Delta k \qquad \forall a_1 \in a; a, k; \qquad (13.6)$$

$$x_{a2}(k+1) = x_{a2}(k) + u_{a2}(k)\Delta k - v_{a2}(k)\Delta k \qquad \forall a_2 \in a; a, k; \qquad (13.7)$$

$$v_{a1}(k) = u_{a2}(k) \qquad \forall a_1, a_2 \in a; a, k. \qquad (13.8)$$

Thus, in general the average link travel time $\tau_a(k)$ per vehicle during time interval k can be expressed as the sum of two components: 1) a cruise time $D_{a1}(k)$ over sublink a_1; and 2) a queuing delay $D_{a2}(k)$ on sublink a_2. It follows that

$$\tau_a(k) = D_{a1}(k) + D_{a2}(k) \qquad \forall a \tag{13.9}$$

13.2.2 Cruise Time

Next we consider the cruise time $D_{a1}(k)$ (in seconds) over sublink a_1. The cruise time $D_{a1}(k)$ can be computed by using the cruise speeds. There are two approaches to compute the cruise speed. The first approach is to use the free-flow speed recommended by the 1994 Highway Capacity Manual (HCM). The second approach is to use the speed-density-flow relationship. For an arterial link, the dependency of cruise speed on flow is not very significant (McShane and Roess, 1990). The cruise speed on an arterial link is mainly dependent on the classification of the arterial and the geometry of the link. Thus, the second approach of using speed-density-flow relationship is not the major one used in traffic engineering and is not strongly recommended here.

The arterial classifications are defined in the 1994 Highway Capacity Manual based on arterial function and design. A range of free-flow speeds within each classification are recommended by HCM as the cruise speeds $w_{a1}(k)$. Those speeds are the ones observed during very low traffic volume conditions. The arterial should be classified by functional category and design category. The functional category is either principal or minor arterial. The design category have three major examples: 1) typical suburban design; 2) typical intermediate design; 3) typical urban design. For example, for a typical suburban principal arterial, the typical free-flow speed is 40 miles/hour, while for a typical urban minor arterial, the free-flow speed is 27 miles/hour. Within an arterial classification, there are several unstated factors affecting free-flow speed, such as presence of parking, opportunities for side friction, intersection spacing, local development and street use. Thus, the cruise speeds should be adjusted accordingly for each arterial link based on actual road conditions.

Denote l_a as the length of link a (miles) and e_{am} (vehicles/mile) as the maximal density (jam density) of traffic on link a. The length of queuing vehicles at the initial instant and final time instant of interval k are $x_{a2}(k)/e_{am}$, $x_{a2}(k+1)/e_{am}$, respectively. The average queue length is $[x_{a2}(k) + x_{a2}(k+1)]/2e_{am}$. Thus, the actual length of cruise from inflow to reaching the queue is

$$l_a - [x_{a2}(k) + x_{a2}(k+1)]/2e_{am}$$

Therefore, the actual cruise time (in seconds) over link a for inflow entering link a during time interval k is

$$D_{a1}(k) \quad = \quad 3600\frac{l_a - [x_{a2}(k) + x_{a2}(k+1)]/2e_{am}}{w_{a1}(k)}$$

$$= \quad 1800 \frac{2l_a \, e_{am} - [x_{a2}(k) + x_{a2}(k+1)]}{w_{a1}(k) \, e_{am}} \tag{13.10}$$

13.2.3 Delay and Link Travel Time Functions

Before we analyze the intersection delay, the concept of link capacity needs to be clarified. In a time-varying traffic network, there are different maximal discharge rates of traffic on a link. Generally, there are three roadway components that have capacity constraints: entry, midblock road section and exit of a link. Therefore, when we use the word *capacity* in a temporal traffic network, it is necessary to indicate whether it is an inflow capacity, a midblock flow capacity or an exit flow capacity. Because both capacities at entry and midblock are usually higher than at the exit point, we will refer only to the exit capacity when considering the flow capacity of a link. Note that when capacity at entry or midblock is lower than at the exit point, a filtering factor is used to adjust the link delay. This will be discussed below.

It is assumed that the capacity at the exit of a link is a function of time, $\mu_a = \mu_a(k)$, for each link a (in vehicles/hour). In practice, it is necessary to evaluate the exit capacity for each time interval given the exit flow $v_{a2}(k)$ such that $v_{a2}(k) \leq v_{a2}(k)$ in the relevant interacting link flows. Capacity calculation methods also depend on the type of intersection; traffic signals; major/minor priority; and roundabouts. We only discuss signal capacity in this chapter.

Since we are considering a rather long time interval, the delay at the exit of the link involves deterministic and stochastic components. The mathematical models used to estimate intersection delay are queuing models. Deterministic queuing theory, in which the delay is obtained as a simple integral of demand minus capacity, ignores the random nature of traffic arrivals and departures. This may lead to serious underestimates in the delay, particularly when demand approaches capacity. On the other hand, steady state queuing theory predicts infinite queues and delays when the demand reaches the capacity. Thus, the most difficult region for delay estimation is where demand (inflow) and capacity are approximately equal; it is this region which is inadequately represented both by the deterministic approach and steady state queuing approach. Methods are needed that adequately treat the entire range of demand and capacity, and properly account for the random nature of traffic.

Here, we deal solely with queues and the corresponding approach delays. Other types of delay, for example geometric delay experienced for negotiating the junction in the absence of queues, are not treated. They may be considered in a practical implementation.

The average delay per vehicle $D_{a2}(k)$ for vehicles arriving at the downstream intersection of link a during time interval k can be expressed as the sum of two delay terms:

$$D_{a2}(k) = \alpha_a \, d_{a1}(k) + \beta_a \, d_{a2}(k) \qquad \forall a \tag{13.11}$$

where $d_{a1}(k)$ is the non-random delay (delay due to signal cycle effects calculated assuming non-random arrivals at the average inflow rate); and $d_{a2}(k)$ is the overflow delay including effects of random arrivals as well as any oversaturation delays experienced by vehicles arriving during the specified flow period; α_a is a delay adjustment factor for link a to account for the impact of signal progression and control type on delay ($0 \leq \alpha_a \leq 1$); and β_a is a delay adjustment factor to account for filtering effect on cycle-to-cycle variations of arrivals ($0 \leq \beta_a \leq 1$).

The progression factor α_a is recommended by the 1994 HCM. The two effects, signal progression and control type, are mutually exclusive. The quality of progression is of critical importance to determine the delay adjustment. The parameter that best describes this characteristic is the arrival type for each lane group. Six arrival types are defined for the arrival flow $u_{a2}(k)$, based on the density of arrival platoon and the arrival time instant in the red phase. Detailed studies, including microscopic simulation modeling, could be used to estimate the appropriate values of α_a for each link.

The filtering factor β_a incorporates the metering effect of the upstream node. In general, there are several factors affecting the stochastic delay term in the stochastic link travel time function for long time horizon. In the case of long time horizon, the stochastic delay term is traditionally derived by assuming that the downstream intersection is an isolated intersection subject to Poisson arrivals, independent of the stochastic delay at nearby intersections. Newell (1990) noted that under certain conditions, this assumption may grossly overestimate the stochastic delay on an arterial. Since an arterial link in a network can no longer be considered as isolated, the assumption of independent link will overestimate its delay for dynamic assignment. Newell (1990) further pointed out that the cumulative stochastic delay on links over an arterial is mainly dependent on the stochastic delay at the critical intersection. Also, Van As (1991) noted that the randomness of arrivals in a traffic network is significantly reduced when the arrivals at upstream intersections are highly congested. Thus, delay is significantly overestimated in networks if the proposed formulae do not allow for the less variable distribution of arrivals, especially if the network is near to saturation. Tarko and Rouphail (1994) implemented a simulation study to determine how filtering factor β_a depend on the critical intersection. Recently, Li (1994) presented a detailed model for estimating β_a. Although no recommendations have been reached so far, it is necessary to use an appropriate β_a in the delay estimation for dynamic assignment.

Next, denote $\rho_a(k)$ as the degree of saturation at the exit from link a during time interval k. It follows that

$$\rho_a(k) = \frac{u_{a2}(k)}{\mu_a(k)} \tag{13.12}$$

Non-random delay at intersection is estimated by assuming that the number of vehicles which arrive during each signal cycle is fixed and equivalent to the average flow (demand) rate per cycle. Different expressions are used for the non-random delay term according to the arrival characteristics (uniform or platooned) and the signal characteristics (one or two green periods). The uniform delay formula valid for single green periods with uniform arrivals was proposed by Webster (1958):

$$d_{a1}(k) = \frac{0.5c[1 - g(k)/c(k)]^2}{1 - \rho_a(k)\, g(k)/c(k)} \tag{13.13}$$

where $c(k)$ is the signal cycle time in seconds during time interval k, and $g(k)$ is the effective green time in seconds during time interval k. To include the effects of traffic progression on delays at traffic signals, Fambro et al (1991) and Messer (1990) suggested converting the above uniform delay formula into a non-uniform arrival term to account for the progression effects as follows:

$$\begin{aligned} d_{a1}(k) &= \frac{0.5c[1 - g(k)/c(k)]^2}{1 - \rho_a(k)\, g(k)/c(k)} \cdot \frac{1 - P(k)}{1 - g(k)/c(k)} \\ &= \frac{0.5c[1 - g(k)/c(k)] \cdot [1 - P(k)]}{1 - \rho_a(k)\, g(k)/c(k)} \end{aligned} \tag{13.14}$$

where $P(k)$ is the proportion of traffic arriving in the green phase during time interval k.

Recently, a more general delay formula for this uniform delay term was proposed as follows (Akcelik and Rouphail, 1991):

$$d_{a1}(k) = \begin{cases} 0.5[c(k) - g(k)] & \text{for } \rho_a(k) > 1.0 \\ \frac{0.5c(k)[1-g(k)/c(k)]^2}{1-\rho_a(k)\, g(k)/c(k)} & \text{for } \rho_a(k) \le 1.0 \end{cases}$$

Note that the above formula is not smooth (no continuous first-order derivative with respect to inflow $u_{a2}(k)$). If we want to combine those functions into our framework, we need to smooth those functions, as a smooth function is necessary for the solution of dynamic route choice problems.

Overflow delay estimation has attracted extensive research. Its development was initially reported by Kimber and Hollis (1979). Later on, Hurdle (1984) further discussed the assumptions and limitations of those delay models. Several countries proposed time-dependent delay formulae in their capacity guides, including the U.S. (TRB, 1994), Canada (Teply, 1984) and Australia (Akcelik, 1981). More recently, Tarko, Rouphail and Akcelik (1993) discussed the reduction in overflow delay on a given link caused by a constraint on the inflow rate.

Both the 1985 HCM and 1994 HCM suggested a delay formula to account for both short-term (random or Poisson) and long-term overflows of queues to subsequent cycles due to continuous oversaturation; however, it has been widely criticized in recent years. Akcelik (1988) noted that the HCM equation predicted higher delays for oversaturated conditions than the Australian and Canadian formulas. He recommended a general formula for the overflow delay as follows:

$$\begin{aligned} d_{a2}(k) &= 900\Delta k\, [\rho_a(k)]^n \{[\rho_a(k)-1] \\ &+ \left[[\rho_a(k)-1]^2 + \frac{m[\rho_a(k)-\rho_{ao}(k)]}{\mu_a(k)\,\Delta k}\right]^{\frac{1}{2}}\right\} \end{aligned} \tag{13.15}$$

where $\rho_{ao}(k)$ is the degree of saturation below which the overflow delay $d_{a2}(k)$ is negligible. This term can be expressed as

$$\rho_{ao}(k) = a + b\, s(k)\, g(k)$$

where $s(k)$ is the saturation flow rate in vehicles per second during time interval k, $s(k)\,g(k)$ is the capacity per cycle during time interval k, and a, b, m, and n are calibration parameters.

However, Akcelik's formula is not smooth at the point $\rho_{ao}(k)$ (no continuous first-order derivative with respect to inflow $u_{a2}(k)$). Thus, we suggest dropping this term in the above delay equation because its effect is negligible, and use the revised $d_{a2}(k)$ function in our link travel time functions for dynamic modeling. On the other hand, it is worthwhile to note that there is an alternative function for $d_{a2}(k)$ suggested by Burrow (1989). Burrow's function is a generalized version of Akcelik's model for overflow delay and is a smooth function:

$$\begin{aligned} d_{a2}(k) &= 900\Delta k\, [\rho_a(k)]^n \{[\rho_a(k)-1] + \alpha \\ &+ \left[[\rho_a(k)-1]^2 + \frac{m[\rho_a(k)+\beta]}{\mu_a(k)\,\Delta k}\right]^{\frac{1}{2}}\right\} \end{aligned} \tag{13.16}$$

where α is an additional term used to encompass the more general form above and β is a term related to the $\rho_a(k)$ term in Akcelik's model. Note that parameters α and β in the above equation are different from the progression and filtering factors in the total delay estimation.

In the above models, the initial queue $x_{a2}(k)$ is assumed to be zero when the overflow period Δk begins; however, this initial queue should be counted in our dynamic delay model, and the additional queuing delay (in seconds) caused by an initial queue $x_{a2}(k)$ can be assessed as (Akcelik and Rouphail, 1991):

$$3600\frac{x_{a2}(k)}{\mu_a(k)} \tag{13.17}$$

We note that the above formula is suitable for the case when the approaching flow $u_{a2}(k)$ is greater than or equal to the exiting flow $v_{a2}(k)$. For other cases, it may overestimate the delays caused by the initial queue. Those cases need further study in the future. In the meantime, we suggest the following overflow delay equation for the dynamic problem:

$$d_{a2}(k) = 3600\frac{x_{a2}(k)}{\mu_a(k)} \tag{13.18}$$

$$+ \quad 900\Delta k \, [\rho_a(k)]^n \left\{ [\rho_a(k) - 1] + \left[[\rho_a(k) - 1]^2 + \frac{m\rho_a(k)}{\mu_a(k)\,\Delta k} \right]^{\frac{1}{2}} \right\}$$

In summary, we propose the following stochastic link travel time function for the case of a long time interval ($\Delta k \geq 5$ minutes):

$$\tau_a(k) = D_{a1}(k) + \alpha_a d_{a1}(k) + \beta_a d_{a2}(k) \qquad \forall a \qquad (13.19)$$

where $D_{a1}(k)$ is given by (13.10), $d_{a1}(k)$ is given by (13.13) and $d_{a2}(k)$ is given by (13.18).

13.3 Deterministic Link Travel Time Functions

Now we consider a rather short time horizon of 1–5 minutes in length. In delay analysis, the major mathematical models are queuing models. In general, the delay at the exit of the link involves deterministic and stochastic delays. The stochastic delay is caused by the short-term variations of demand (arrivals) and capacity. If one considers a relatively short interval (1–5 minutes), these variations are already incorporated in the arrival pattern and capacity pattern. Thus, the arrival and exit flow rates can be assumed steady within short intervals with negligible error of delay estimation. In other words, it is reasonable to assume that the impact of randomness of traffic arrivals at the traffic signal is negligible during such a short time. The stochastic delay can be neglected and only deterministic uniform and oversaturation delays need to be considered. Nevertheless, we note that no explicit distinction between long-time and short-time horizons is attempted in this paper. The main criterion for selecting the corresponding travel time function is the variations of demand (arrivals) and supply (capacity) on the specific arterial network. If the variation is highly random, the stochastic travel time function should be used. Otherwise, the deterministic travel time function could be used.

There have been some recent studies in the deterministic travel time function. The model presented by Takaba (1991) is in this category. Furthermore, if one wants to consider a short time interval less than one minute, it is necessary to know the offset for each intersection signal and determine the non-random delays.

For the deterministic travel time function problem, the delay at the exit from the link is considered deterministic in contrast with that occurring in the stochastic travel time function problem. Nevertheless, the cruise time formula (13.10) is still applicable, since it is based on the average speed of the inflow. Thus, the average link travel time per vehicle during time interval k is still expressed as

$$\tau_a(k) = D_{a1}(k) + D_{a2}(k) \qquad \forall a.$$

The main difference between a stochastic travel time function and a deterministic travel time function is the second term $D_{a2}(k)$. The delay term

$D_{a2}(k)$ can be developed in a deterministic manner for the deterministic travel time function over a short time period such as 1–5 minutes. The average delay per vehicle $D_{a2}(k)$ for vehicles arriving at the exit from link a during time interval $[k, k+1]$ can also be expressed as the sum of two delay terms:

$$D_{a2}(k) = \alpha_a d_{a1}(k) + \beta_a d_{a2}(k) \qquad \forall a$$

where $d_{a1}(k)$ is the cyclic delay (delay due to signal cycle effects calculated assuming non-random arrivals at the average inflow rate in each cycle); $d_{a2}(k)$ is the delay due to oversaturation experienced by vehicles arriving during the specified flow period; α_a is a delay adjustment factor for link a to account for the impact of signal progression and control type on delay $(0 \leq \alpha_a \leq 1)$. Factor $\beta_a = 1$ in this case, since the filtering effect is fully incorporated through the capacity constraints on inflows and there are no additional inflow variations within the time intervals. The first delay term $d_{a1}(k)$ is given by Webster's formula. For the formula proposed by Akcelik and Rouphail (1991), smoothing is necessary for our purpose. Therefore, we will consider it in future research.

We next discuss the second delay term $d_{a2}(k)$. The decision variable is the queue length on physical link a. As discussed before, link $a = (A, B)$ is decomposed into two sublinks, $a_1 = (A, D)$ and $a_2 = (D, B)$. After the cruise time $D_{a1}(k)$, vehicles entering link a during time interval $[k, k+1]$ should either reach the queue on link (D, B) or proceed to downstream links. As before, the location of dummy node D is movable. Thus, the length of link $a_2 = (D, B)$ is the length of queuing flow.

The deterministic (initial) queue encountered by the first vehicle arriving at the beginning of time interval $[k, k+1]$ is $x_{a2}(k)$. The deterministic queue encountered by the last vehicle arriving at the end of time interval $[k, k+1]$ is $\{[u_{a2}(k) - v_{a2}(k)]\Delta k + x_{a2}(k)\}$. The average queue for vehicles arriving during time interval $[k, k+1]$ is $\{[u_{a2}(k) - v_{a2}(k)]\Delta k/2 + x_{a2}(k)\}$. Thus, the average deterministic queuing discharge time or queuing delay $d_{a2}(k)$ per vehicle arriving during time interval $[k, k+1]$ can be expressed as

$$\begin{aligned} d_{a2}(k) &= 3600 \frac{[u_{a2}(k) - v_{a2}(k)]\Delta k/2 + x_{a2}(k)}{\mu_a(k)} \\ &= 1800 \frac{[u_{a2}(k) - v_{a2}(k)]\Delta k + 2x_{a2}(k)}{\mu_a(k)} \end{aligned} \qquad (13.20)$$

where $d_{a2}(k)$ is in seconds.

In summary, we propose the following deterministic link travel time function for a short time interval problem.

$$\tau_a(k) = D_{a1}(k) + \alpha_a d_{a1}(k) + \beta_a d_{a2}(k) \qquad \forall a \qquad (13.21)$$

where $D_{a1}(k)$ is given by (13.10), $d_{a1}(k)$ is given by (13.13) and $d_{a2}(k)$ is given by (13.20).

13.4 Implications of the Proposed Functions

13.4.1 Number of Link Flow Variables

The selection of suitable link travel time functions for discrete-time dynamic route choice largely depends on the length of the analysis time interval and the randomness of arrival and capacity changes during the interval. Travel times for longer time intervals or large random variation of arrival/capacity must account for stochastic delays at intersections. The intersection delays in travel time functions for shorter time intervals or steady arrival/capacity problem are predominantly deterministic.

Various dynamic choice models have been proposed by many researchers. These models use some variations of time-dependent link travel time functions. A number of models still use the BPR travel time function for a time-dependent traffic network problem; however, the BPR function was based on steady state traffic flow. This assumption may not be valid in a time-dependent and stochastic traffic network. Furthermore, in using time-dependent delay formulae, traffic engineering practice implies that the BPR function cannot predict intersection delays properly (HCM, 1994). Therefore, using the BPR function as the basis to formulate dynamic traffic assignment might generate results that are too approximate to be realistic in a time-varying environment.

Some dynamic route choice models use a single flow variable (i.e. the number of vehicles on a link or average flow rate over a link for each time interval) to describe the dynamics of traffic flow on a link. From the above analysis, it is evident that one flow variable for a link is insufficient to capture the dynamic characteristics of traffic flow on a link and cannot be used to estimate the time-dependent delays at the intersection properly. Therefore, the six link flow variables $x_{a1}(t)$, $u_{a1}(t)$, $v_{a1}(t)$, $x_{a2}(t)$, $u_{a2}(t)$, $v_{a2}(t)$ and link state equations (see equations (13.6)-(13.8) and Table 13.1) are proposed as the basis to formulate dynamic route choice models.

Furthermore, link flow propagation constraints are used in dynamic assignment to ensure the progression of flows consistent with link travel time. To simplify our notation, a continuous time constraint is considered. For link a, if inflow $u_a(t)$ experiences link travel time $\tau_a(t)$, the summation of the cumulative number of vehicles entering link a during time period $[0, t]$ and the number of existing vehicles at initial time 0 must equal to the cumulative number of vehicles exiting link a during time period $[0, t + \tau_a(t)]$. It follows that

$$\int_0^t u_a(\omega)d\omega + x_a(0) = \int_0^{t+\tau_a(t)} v_a(\omega)d\omega \qquad \forall a. \tag{13.22}$$

Taking the derivative of this equation with respect to time t, it follows that

$$u_a(t) = \left[1 + \frac{\partial \tau_a(t)}{\partial t}\right] v_a[t + \tau_a(t)] \qquad \forall a. \tag{13.23}$$

If the above flow propagation constraint is formulated for two sublinks a_1 and a_2, there are five equations for link flow variables $u_{a1}(t)$, $v_{a1}(t)$, $x_{a1}(t)$ and

$u_{a2}(t)$, $v_{a2}(t)$, $x_{a2}(t)$. Thus, if an inflow $u_a(t)$ or $u_{a1}(t)$ is given, the flow conservation constraint and flow propagation constraint will first determine $x_{a1}(t)$ and $v_{a1}(t)$ for sublink a_1. The flow conservation on dummy node D will give $v_{a1}(t) = u_{a2}(t)$. Then, by using the flow conservation constraint and flow propagation constraint, one can determine $x_{a2}(t)$ and $v_{a2}(t)$ for sublink a_2.

Another type of flow propagation constraint can be formulated as well. The number of vehicles on link a at time t must equal the cumulative number of vehicles exiting link a during time period $[t, t+\tau_a(t)]$. It follows that

$$x_a(t) = \int_t^{t+\tau_a(t)} v_a(\omega)d\omega \qquad \forall a. \tag{13.24}$$

Note that equations (13.23) and (13.24) are two equivalent link flow propagation constraints. In summary, there are two flow conservation and two flow propagation equations for sublinks a_1 and a_2 as well as one flow conservation equation for dummy node d. Thus, only one out of the six link flow variables is independent.

Associated with these proposed dynamic link time functions, we need to modify our dynamic network models. If other formulation approaches are used, they may reduce the number of variables. However, the detailed delay calculation may not be possible due to the lack of queue length determination.

13.4.2 Physical Constraints for Link Traffic Flow

The maximal number of vehicles that link a can accommodate is $l_a e_{am}$. Then, we can write the following constraint for any dynamic network model:

$$x_{2a}(k) \leq l_a e_{am} \qquad \forall a. \tag{13.25}$$

Since this constraint is related to the state variable only, it will not cause any problem to our desired optimality conditions for DUO route choice models. When solving an assignment problem including the above constraint, we can put it as a penalty term in the objective function and solve an ordinary dynamic route choice model.

An extreme case is when the queue on link a approaches the maximum, i.e. $x_{2a}(k) = l_a e_{am}$. In this situation, the queue may extend into the upstream links, which causes a spillback problem; however, if the above constraint and upper bound constraints for cumulative O-D departures and associated access delays at origin nodes are added in the assignment problem, the physical spillback can be described and formulated in our dynamic models. The time-dependent link capacities have to be adjusted accordingly. For example, in Figure 14.2, downstream link BC has 2 lanes and upstream link AB has 3 lanes. When downstream link BC is filled with vehicles and vehicles spill back to upstream link AB, the capacity of upstream link AB will be reduced to the same level of downstream link BC. Thus, its capacity has to be adjusted dynamically according to the variation in the number of vehicles on downstream link BC. This changes the travel time of upstream link AB accordingly.

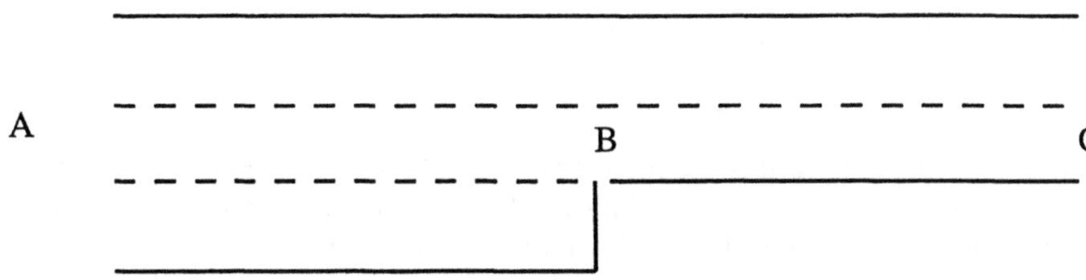

Figure 13.2: Link Capacity Changes on an Arterial

Another constraint is the flow capacity at the exit of a link. Since we consider this capacity in the travel time functions, it is not necessary to put it as an explicit constraint in the route choice models. The objective function of the dynamic models should also be revised.

In the derivation of dynamic link travel time functions, it is basically assumed that the delays are due to the signal control at the downstream intersection assuming isolated control. Since the travel time functions are developed to apply in dynamic network models, the interaction of upstream and downstream intersections should be taken into account. In addition to the filtering factor introduced in the travel time function, Rouphail and Akcelik (1991) observed that there is a critical queue length requirement for each link. The critical queue length, N_a, is defined as the longest downstream queue that allows upstream platoons to accelerate to and discharge at the full saturation flow rate. It follows that

$$x_{2a}(k) \leq N_a \qquad \forall a. \tag{13.26}$$

The critical queue length is assumed fixed for a given set of platoon speeds and speed change rates of the cruise inflow. This constraint reflects the reduction of inflow capacity as a result of downstream queue interaction effects, which in turn has an impact on the delays at downstream intersection; however, the functional relationship between the critical queue length and the cruise speed may complicate the formulation of dynamic network models. This difficulty needs further investigation.

13.4.3 Notes on Functions for Arterial Links

In the derivation of dynamic link travel time functions, it is basically assumed that delays are caused by the signal control at the downstream intersection assuming isolated control. Since these travel time functions are developed to apply in dynamic network models, the interaction of upstream and downstream intersections should be taken into account.

There is a critical queue length requirement for each link (Rouphail and Akcelik, 1991). The critical queue length, N_a, is defined as the longest downstream queue that allows upstream platoons to accelerate to and discharge at the full saturation flow rate. It follows that

$$x_{2a}(k) \leq N_a \qquad \forall a. \tag{13.27}$$

The critical queue length is assumed fixed for a given set of platoon speeds and speed change rates of the cruise inflow. This constraint reflects the reduction of inflow capacity as a result of downstream queue interaction effects, which in turn has an impact on delays at downstream intersection. However, the functional relationship between the critical queue length and the cruise speed may complicate the formulation of dynamic network models. This difficulty needs further investigation. In the following, we mainly discuss the factors affecting the stochastic delay term in the link travel time function for longer time horizons.

In the case of longer time horizons, the stochastic delay term is derived by assuming that the downstream intersection is an isolated intersection subject to Poisson arrivals, independent of the stochastic delay at other intersections. Newell (1990) noted that this assumption may grossly overestimate the stochastic delay on an arterial under certain conditions. This factor is especially important in dynamic network models since intersections in a network can no longer be considered isolated. Newell (1990) further pointed out that the cumulative stochastic delay on links over an arterial is mainly dependent on the stochastic delay at the critical intersection.

Also, van As (1991) noted that the randomness of arrivals in a traffic network is significantly reduced when the arrivals at upstream intersections are highly congested. Thus, delay is significantly overestimated in networks if the proposed formulae do not allow for the narrower distribution of arrivals, especially if the network is near saturation. The adjustment of dynamic link travel time functions to account for oversaturation effects in a longer time horizon case is another important issue in the application of dynamic network models.

13.5 Functions for Freeway Segments

Dynamic link travel time functions for freeway segments are simpler than those for arterials. In general, the average link travel time per vehicle $\tau_a(k)$ during time interval k can be expressed as the sum of two components: 1) a free-flow cruise time D_{a1} over the link; and 2) a flow-dependent congestion delay $D_{a2}(k)$. It follows that

$$\tau_a(k) = D_{a1} + D_{a2}(k) \qquad \forall a. \tag{13.28}$$

The free-flow cruise time (seconds) on link a is

$$D_{a1} = 3600 \frac{l_a}{w_{a0}} \qquad \forall a \tag{13.29}$$

where l_a is the link length (miles) and w_{a0} is the free-flow speed (miles/hour) on link a. The second delay term is caused by the traffic ahead of a vehicle when it enters link a. For simplicity, we assume that the average traffic density on link a determines this congestion delay. We note that a nonuniform distribution of traffic over link a may bring an error to this delay formula. However, as link a becomes shorter, the traffic is distributed more uniformly and this formula becomes more accurate.

The deterministic queue encountered by the first vehicle arriving at the beginning of time interval $[k, k+1]$ is $x_a(k)$. The deterministic queue encountered by the last vehicle arriving at the end of time interval $[k, k+1]$ is $\{[u_a(k) - v_a(k)]\Delta k + x_a(k)\}$. The average queue encountered by vehicles arriving during time interval $[k, k+1]$ is $\{[u_a(k) - v_a(k)]\Delta k/2 + x_a(k)\}$. Thus, the average traffic density (vehicles/mile) encountered by vehicles arriving during time interval $[k, k+1]$ is

$$e_a(k) = \frac{[u_a(k) - v_a(k)]\Delta k/2 + x_a(k)}{l_a} \tag{13.30}$$

We assume that the second delay term $D_{a2}(k)$ (seconds) can be expressed as

$$D_{a2}(k) = \alpha\, [e_a(k)]^m \tag{13.31}$$

where α is a real-valued parameter and m is an integer parameter. Both of them need to be calibrated.

In summary, we propose the following dynamic link travel time function for freeway segments.

$$\tau_a(k) = D_{a1} + D_{a2}(k) \qquad \forall a \tag{13.32}$$

where D_{a1} is given by equation (13.29) and $D_{a2}(k)$ by equation (13.31). This formula is different from the traditional BPR function because the second delay term depends on the average traffic density encountered by vehicles arriving during time interval $[k, k+1]$ instead of average traffic flow on the link. Sometimes, we use the link travel time function for vehicles entering link a at the beginning of interval $[k, k+1]$. In this situation, the average traffic density (vehicles/mile) encountered by vehicles arriving at the beginning of time interval $[k, k+1]$ is

$$e_a(k) = \frac{x_a(k)}{l_a} \tag{13.33}$$

Then, queuing delay $D_{a2}(k)$ in dynamic link travel time function (13.32) should be computed using equations (13.31) and (13.33) instead of equations (13.31) and (13.30).

13.6 Notes

This chapter has investigated different aspects of time-dependent link travel time functions for signal-controlled arterial and freeway links. For an arterial link, the following conclusions are emphasized.

1. Two sets of dynamic link travel time functions are proposed depending on the analysis time horizon. These functions can be applied to discrete-time dynamic network models.

2. For each link, six variables (three of which are independent) are necessary to describe the dynamics of traffic flow and calculate temporal link travel time on an arterial link with a signal controlled intersection.

3. Each physical link is decomposed into two dummy links in order to identify the queue length on the physical link. Thus, the link travel time functions for each physical link depend on the flow variables of two dummy links so that link interaction enters the dynamic link travel time functions. This interaction should be considered in dynamic network formulations.

There are many critical assumptions underlying the delay equations for arterial links. Those assumptions are especially important for the longer time interval delay model. The generalization of those assumptions will make the travel time functions more realistic. Among those assumptions, the most critical factors are the impact of signal coordination and the case when the flow is not zero after the peak ends. For the short time interval problem, we also need to investigate more realistic inflow arrival patterns at the exit of a link. Validation of the proposed dynamic link travel time functions is a major task for future research.

A dynamic link travel time function for a freeway segment link is also proposed. The major delay in the travel time function is dependent on the average traffic density on the freeway link. However, this function needs to be calibrated and validated.

The proposed link travel time functions for arterial and freeway segments are still subject to future challenges and validations from many sources. The first challenge is from traffic simulation models. Microscopic simulation models might provide detailed answers to many remaining questions for dynamic link travel time functions and also serve as validation tools.

The second challenge is from hydrodynamics theory (Newell, 1993) and its discrete form, the highway cell transmission model (Daganzo, 1993). The second challenge is more theory-oriented. Hydrodynamics theory and its discrete form may provide a basis for the derivation of dynamic link travel time functions, especially for freeway segments. We believe that with all these efforts plus a large amount of realistic data generated from ITS operational tests, we can produce appropriate link travel time functions for both freeway and arterial links. Consequently, those functions will ensure a good representation of travel times and traffic propagation on arterial and freeway links.

Problems

13.1 Discuss the requirements for link travel time functions for solving a dynamic network model.

13.2 Design a possible link travel time function for an arterial link with major/minor priority intersection.

13.3 Design a possible link travel time function for an arterial link with all-way-stop intersection.

13.4 Design a simple link travel time function for a local street link with no control.

13.5 In the proposed moving queue concept, the queue takes a physical length on the link. In the vertical queue concept, which has been used in many studies, the queue is assumed to have no physical length on the link and all vehicles entering the link have to traverse the entire link. Compare the differences of these two queue handling approaches. Discuss their advantages and disadvantages.

13.6 Suppose only three variables (u, v, x) are used to compute arterial link travel time. How can the link queuing time be approximately estimated?

13.7 To compute the cruise time on an arterial, derive a formula by using Greenshields' speed-density equation.

13.8 Consider a deterministic link travel time function for arterial. Suppose we have a link with length of 0.5 mile and free flow speed of 30 miles/hour. The discharging capacity at the exit of link is 1800 vehices/hour. The link has two lanes. The inflow for the link is 2000 vehicles/hour. Assume that no vehicles exist on link at time 0. Determine the link travel times for the first 20 minutes (using the moving queue concept) and draw the curve of $\tau_a(t)$ v.s. time t. Assume no flow dispersion; jam density is 240 vehicles/mile/lane.

13.9 Using the same parameters as in Problem 13.8, determine stochastic link travel times for the first 20 minutes.

13.10 Use Greenshields speed-density relationship to derive a link travel time function for freeway links. In such a link travel time function, if x achieves its maximum value, the link travel time function gives a travel time of infinity. This is inconsistent with reality. (Traffic does not stop indefinitely on a freeway even when it reaches its maximum density.) How can such an inconsistency be overcome?

Chapter 14

Implementation in Intelligent Transportation Systems

The rapid evolution of ITS technologies presents more and more specific requirements for dynamic network modeling. Conversely, implementation of dynamic models is becoming more and more important for the design and evaluation of ITS. In Section 14.1, several applications of dynamic models to various ITS components are investigated. We mainly consider technical aspects of applying these models. Then, practical considerations for applying dynamic network models are discussed in Section 14.2. Finally, we examine various data requirements for implementing these dynamic models in Section 14.3.

14.1 Implementation Issues

Dynamic transportation network models can be considered for various ITS applications, including both ATIS/ATMS and Automated Highway Systems (AHS). Among the various implementation possibilities that could be identified, we focus on the following items: 1) traffic prediction; 2) dynamic route guidance; 3) integrated traffic control and information system; 4) incident management; 5) congestion pricing; 6) automated highway systems analysis; and 7) transportation planning. In addition, how these models can serve operating and evaluation functions in ITS systems are examined. In the following, the application of dynamic network models to these problems is discussed.

14.1.1 Traffic Prediction

Dynamic network models can function as predictive models for these systems. Travelers' choice behavior determines which dynamic route choice model best fits into the predictive module of a realistic ATIS. In general, no single-class dynamic route choice model can represent the travel choices of the entire population. The most plausible model is a reasonable combination of several dynamic route choice models, including both deterministic and stochastic models.

For recurrent and non-recurrent congestion, both ideal and instantaneous DUO route choice criteria may apply. Note that in this application, the DUO state is defined using a general definition of travel disutility including fuel consumption, auto operating cost, etc., in addition to travel time. The population of travelers using either the instantaneous or ideal DUO route choice criterion could be determined by survey. Senior citizens or cautious travelers may prefer to choose routes based on traffic information from their past experience. In other words, they may choose routes using the ideal DUO criterion. On the other hand, young or aggressive travelers may prefer to choose routes based on current traffic information. That is, they may choose routes using the instantaneous DUO criterion.

For a traffic network, the multimodal problem should be handled explicitly. Conventional modes include HOV, bus, truck and passenger car. HOV lanes should be designated as separate links and may be subject to possible pricing or toll charges during congested periods. These will alter the travel cost and subsequently change the flow pattern of HOV lanes. Bus constitutes a special mode which should be handled carefully in the modeling. In general, buses move slowly and cause additional delays to other vehicles. For a road link with bus traffic, the travel time function needs to be adjusted for through traffic in the right lane and right turning traffic, since the impact of bus traffic is significant. For the left turn lane, buses only affect traffic flow at the time when making a left turn. Truck traffic flows also need to be transformed into equivalent passenger car flows. However, possible revisions of the travel time functions may be necessary, especially for turning movements because trucks make wide turns and take longer times for turning.

Since bus routes are fixed, there is no route choice for bus traffic itself. Thus, it belongs to the class with pre-specified routes. Truck traffic can be classified with the class with fewer alternative routes because some roads are closed to truck traffic. The trip chaining problem is not explicitly considered in this framework. However, a similar modeling framework is applicable to the situation of trip chaining when time-dependent trip information is available.

Concerning the route choice criterion, a more plausible model for dynamic traffic prediction may be a combination of instantaneous/ideal DUO/ SDUO route choice models. For travelers without route guidance devices or not complying with guidance information, instantaneous/ideal SDUO route choice models can be used to model this class of travelers. The population with prespecified routes may be quite large so that the dispersion parameter for this class is small. Table 14.1 presents route choice criteria for these two classes of travelers. Table 14.2 summarizes the stratification of travelers and their corresponding travel choice criteria. A general VI model for the above multi-class, multi-criteria route choice problem can be formulated by generalizing the VI models of Chapters 5, 7-9 and 10-11.

Furthermore, dynamic traffic prediction problems can be classified into several types based on time periods and travel purposes. Basically, we have:

Table 14.1: Route Choice Criteria for Different Travelers

Travelers	Route Choice Criteria	
Guided Travelers	Instantaneous DUO	Ideal DUO
Travelers with No Guidance	Instantaneous SDUO	Ideal SDUO

Table 14.2: Classes of Travelers and Travel Choice Criteria

Modes		Passenger Car	Trucks	HOV	Bus
Classes by Diversion	No	Y	Y	Y	Y
	Few	Y	Y	Y	N
	Many	Y	Y	N	N
Travel Choice Criterion	Instantaneous DUO	Y	Y	Y	N
	Ideal DUO	Y	Y	Y	N
	Instantaneous SDUO	Y	Y	Y	N
	Ideal SDUO	Y	Y	Y	N

Y – Yes; N – No.

1. morning home-to-work period (6:30 am - 9:30 am);
2. midday non-commuting period (9:30 am - 3:30 pm);
3. work-to-home period (3:30 pm - 7:30 pm);
4. evening period (7:30 pm - 10:30 pm);
5. night period (10:30 pm - 6:30 am).

The times indicated are illustrative and need to be determined by survey. For dynamic traffic prediction during the morning home-to-work period, the major concern is the travel time and the arrival time. The joint mode/departure time/route choice program might be revised as a joint mode/arrival time/route choice program. The arrival times for commuting travelers need to be specified and a large penalty charged for late arrival. Among the four modes specified in Table 14.2, the priority of bus and HOV should be guaranteed in terms of travel time reliability. For dynamic traffic prediction during other periods, the models should be revised accordingly to reflect characteristics of the period of interest.

14.1.2 Dynamic Route Guidance

We may soon see prototype in-vehicle route guidance systems in operation, where a subscribing driver is given a suggested route. There are a number ways of classifying different types of routing systems and many implications involved in representing such systems by dynamic network models, also known as a "dynamic traffic assignment" procedure, an implementation effort of the U.S. Federal Highway Administration (1994). These are classified as follows: 1) Historical Data-Based vs. Real-Time; and 2) Distributed vs. Centralized. In the following, each of the classifications is described and the applications of dynamic network models to these systems are discussed.

A historical data-based route guidance system operates with a self-contained historical information base, such as a map of a given area, digitized road network, and lists of streets or addresses on a high density storage medium such as a computer diskette or CD-ROM. There may be either a single set of information describing typical or free flow link characteristics, or a number of sets corresponding to various sets of peak and off-peak conditions. This information is used to route a driver independently of any other drivers. For this system, the traffic prediction procedure discussed in the previous section can be utilized to provide the historical information.

A real-time route guidance system will have access to real-time information on link characteristics via communication with an information system. The type of communication and information provided depends on whether the routing logic is centrally located or distributed.

A distributed route guidance system does not monitor the number of vehicles on each link. Vehicles route themselves using time-of-day and/or real-time link costs. Information is usually passed to on-board computers in the vehicles, which then route themselves. This saves the central computer having to compute specific information for individuals. More important still, only general information needs to be passed, so that less information needs to be transmitted; moreover, this type of information can be broadcast, so that the central system does not need to communicate directly with individual vehicles.

An individual in-vehicle computer can only route itself, independently of other self-routing vehicles. Therefore, vehicles which are in close proximity to one another, such as those leaving from the same origin, will tend to select the same links, especially if they have destinations in the same general direction, as they will all have the same information on relative link costs. This design may tend to overload one route for a short time causing the relative link costs to change. If the link costs used by the on-board computers can be made to fluctuate frequently, and thus cause oscillations between competitive alternate routes, the vehicles will tend to distribute themselves among the alternate routes on a virtually continuous basis, and avoid pulsating flow patterns.

A centralized route guidance system can do anything a distributed system can, and more. Rather than just broadcasting information, it calculates routes and passes specific information to each subscribing vehicle. It is able to monitor the subscribing vehicles it has routed to each link and use this information in subsequent routings. It can also seek to prevent the pulsating short term overloads on the parallel routes by distributing the trips among the different routes in a virtually continuous manner rather than in surges. Multi-route assignments can be created systematically, rather than relying on the randomization procedure used by distributed systems. This approach may provide better distribution of trips among the routes, and obviates the need for frequent re-building of shortest route trees.

In general, multi-class dynamic route choice models could be easily adopted for analysis in dynamic route guidance systems. To model a dynamic route guidance system in a simple way, travelers can be stratified into different classes based on the following route diversion behavior: 1) pre-specified routes; 2) a few alternative routes; 3) many alternative routes. Specifically, in-vehicle route guidance and changeable message signs (CMS) will be considered in a different modeling fashion. If the route guidance instruction is transmitted to drivers via in-vehicle devices, guided and unguided travelers can be represented by class 3 and class 2 travelers in the model, respectively.

Dynamic route guidance systems or ATIS will take information about the dynamic state of the road network, as provided by the multi-class dynamic route choice model, and transform it into the routes used by guided travelers. The transformation will be accomplished through a simulation of the various components of an ATIS system. The main components of the ATIS simulation will include traffic measurement devices (e.g. in-pavement loop detectors, cameras, probe vehicles), communication systems (e.g. cellular telephone, infrared, narrow-band FM), and computing resources (for tying information flows together). Thus, the model could be used to simulate the impact of ATIS market penetration on the accuracy of travel time information and traveler behavior changes. In addition, the benefits of dynamic route guidance systems can be evaluated by using this model. Furthermore, the effectiveness of various ATIS devices can be evaluated with the model as well.

14.1.3 Integrated Traffic Control/Information System

In an integrated traffic control and information system, route guidance information is provided to travelers, while traffic control is implemented to achieve specified objectives. To represent the interaction of the two modules, dynamic network models can be extended to incorporate traffic control measures, such as signal control and ramp metering. A bilevel program can usually be formulated utilizing the concepts of Section 2.3 to represent this combined dynamic traffic control and prediction problem (see Figure 14.1). The upper level is a dynamic traffic control model, which can be formulated using various objective functions as follows:

1. minimize total travel time;

2. minimize total travel cost or disutility;

3. minimize total number of vehicles during time period $[0, T]$;
4. minimize average congestion level during time period $[0, T]$;
5. minimize the length of the congested time period $[0, T]$;
6. minimize total emissions during time period $[0, T]$.

The lower level is a dynamic traffic prediction model, which determines dynamic traffic flows in the network by considering the control strategies provided by the upper level model. The lower level program can also provide route guidance instructions to the traveler information systems. This bilevel program is solved so that an optimal control strategy can be found while the traffic flows of various classes follow the desired DUO or SDUO route choice criteria. This bilevel program can be either hierarchical or non-hierarchical, depending on the nature of the problem.

Upper Level: Dynamic Traffic Controller:	Determine optimal traffic control strategies based on the predicted dynamic flows.

where the dynamic traffic flows are given by:

Lower Level: Dynamic Traffic Predictor:	Determine DUO/SDUO dynamic traffic flows using the provided dynamic traffic control strategy.

Figure 14.1: A Bilevel Program for Dynamic Traffic Prediction and Control

Control strategies can also be classified as centralized or decentralized, which can generate different versions of the objective function for the upper level control model. As an example, a freeway corridor model, which considers both ramp metering and route diversion, might be particularly interesting for ITS applications. For a general network including both freeways and arterials, signal control, ramp metering and diversion strategies can be considered in an integrated framework. For a traffic network with multimodal consideration, special control strategies could be designed for high-occupancy vehicles (HOV), such as bus. For example, bus priority signals could be considered in the objective function of the traffic control problem, and HOV lanes could be designated as separate links and may be subject to less pricing or toll charges during congested periods. These will alter the travel cost and subsequently change the flow pattern of HOV lanes.

Although conceptually straightforward, the formulation of the bilevel program involves numerous technical difficulties. One of these concerns the formulation of the upper level objective function, which we briefly consider now. We first consider the problem of minimizing the total travel time of all vehicles within a time period $[0, T]$. Using optimal control theory, the objective function for this dynamic system-optimal route choice problem is formulated as follows:

$$\min \quad \int_0^T \left\{ \sum_a u_a(t)\ \tau_a[x_a(t), u_a(t), v_a(t)] \right\} dt \tag{14.1}$$

In this objective function, if we replace the link travel time function $\tau_a(t)$ with the link travel *cost* function $\tilde{\tau}_a(t)$, then we obtain an objective function which minimizes the total travel cost during time period $[0, T]$:

$$\min \quad \int_0^T \left\{ \sum_a u_a(t)\ \tilde{\tau}_a[x_a(t), u_a(t), v_a(t)] \right\} dt \tag{14.2}$$

Equations (14.1) and (14.2) are different because the link travel cost function $\tilde{\tau}_a(t)$ includes other factors, such as automobile operating cost, link tolls, gasoline consumption, etc.

We now consider how to minimize the total number of vehicles traveling on the network during time period $[0, T]$. This objective function may be stated as follows:

$$\min \quad \int_0^T \sum_a [u_a(t) - v_a(t)] dt = \sum_a [x_a(T) - x_a(0)] \tag{14.3}$$

Since the initial value $x_a(0)$ is generally given, the above objective function is equivalent to

$$\min \quad \sum_a x_a(T) \tag{14.4}$$

Thus, we obtain an objective function which minimizes the total number of vehicles on the network at the final time T. This objective function is useful for reducing the average congestion level of peak-hour traffic when the final time T is set within the peak-hour.

Next, we consider a quantitative definition of the congestion level for the network. The congestion level $D(t)$ is defined as an indicator of average congestion in the network at any time t. There are several possible approaches to defining $D(t)$:

1. the ratio of mean travel time to mean free-flow travel time;

2. the ratio of mean free-flow speed to mean flow speed;

3. mean relative density.

where the calculation of the means is flow weighted. For example, we can use the mean relative density as a measure of the level of congestion. The relative density $D_a(t)$ for link a is then defined as

$$D_a(t) = \frac{x_a(t)/l_a}{e_{am}} \qquad \forall a \tag{14.5}$$

where l_a is the link length and e_{am} is the maximum density of traffic on link a. Therefore, the congestion level $D(t)$ for the network can be defined as

$$D(t) = \frac{\sum_a u_a(t)\; D_a(t)}{\sum_a u_a(t)} \tag{14.6}$$

where u_a is the inflow on link a. Then, the objective function that minimizes the mean relative density during time period $[0, T]$ is expressed as

$$\min \qquad \int_0^T \left\{ \frac{\sum_a u_a(t)\; D_a(t)}{\sum_a u_a(t)} \right\} dt \tag{14.7}$$

The above objective functions can be applied to situations either with or without elastic departure times.

The above objective functions can also be minimized subject to the constraint set (5.14)-(5.25) to determine the dynamic system-optimal route flows. Such a model is presented in Chapter 15 of Ran and Boyce (1994).

14.1.4 Incident Management

Incident management is another area in which dynamic network models can play an important role. In order to provide real-time incident management plans, the multi-class dynamic route choice model can be presented for predicting time-dependent travel times and routing strategies under major random incidents in a transportation network so that drivers could choose better routes and avoid nonrecurrent congestion. It could answer some important questions in incident management, such as when and where to divert the traffic and what proportion of vehicles should be diverted.

Traditional incident management strategies minimize incident congestion by clearing incidents as quickly as possible and diverting traffic before vehicles are trapped in the incident queue. However, such simple incident management plans cannot solve most incident congestion problems. Instead, a systematic and comprehensive incident management strategy is needed to tackle the incident congestion problem, as offered by advanced ATIS/ATMS/APTS technologies. Such an incident management strategy could use travelers' information on origins, destinations and departure times to develop a coordinated strategy to advise each person regarding a best mode and route to their destination on a real time basis. In this way, we can achieve either user-optimal or system-optimal objectives by appropriately integrating available information and control measures.

In the context of such an incident management strategy, dynamic network models are necessary for incident-related routing and travel time prediction. These models are both information and control oriented. The prediction procedure assumes either of two types of routing of vehicles: normal conditions and incident conditions. Routing for normal conditions provides time-dependent traffic flows and travel times under normal traffic conditions. Routing for incident conditions provides forecasts and suggestions following an incident. Both instantaneous and ideal DUO/SDUO route choice models can be used for this procedure. The system-optimal objective can also be applied to incident rerouting in conjunction with dynamic congestion pricing. Oversaturation and spillback can be serious problems when incidents happen. In using dynamic network models in response to incidents, corresponding constraints should be formulated to reflect these phenomena.

An incident can occur anywhere on a link. For simplicity, in the models we can specify that an incident is always located at the exit point of a link since a link with an incident can always be partitioned into two shorter links by placing a dummy node at the location of the incident. Thus, the location of the incident can be designated at the exit point of some shorter link. The models require that the start time and duration of an incident be reported by some source, such as the highway patrol. The duration of incident is defined as the period from the reported time through the clearance of the incident.

When incidents occur in a highway network, there are several ways for travelers to avoid congestion. Travelers already en route can shift to less congested routes which are less affected by the incident. If an incident happens in a central business district, some travelers may choose different parking places (destinations) and use other modes such as walking or buses to their final destinations. Travelers who have not yet departed may choose to delay their departures or shift to rapid transit. Thus, travelers' choices under incident conditions can be summarized as: route; departure time; mode; and destination. Instead of discussing all of these choices, we focus on route choice in the following.

A major difficulty of incident-related dynamic network models is that the continuity properties of most dynamic network models are destroyed. To overcome this difficulty, we consider a multi-period dynamic travel choice procedure in which each incident-related dynamic travel choice is treated separately and the continuity of traffic flow at the time and location of an incident is preserved. A detailed procedure is now described.

We consider two types of dynamic routing strategies under two route choice behavioral assumptions. The first type of route choice behavior is based on current or instantaneous travel time information. The second is based on projected or actual travel time information.

Instantaneous DUO/SDUO routing strategies pertain to drivers under incidents based on current traffic conditions. This framework is helpful in handling many unpredictable events that occur in traffic flow, such as accidents and illegal double parking on streets, etc. In case of such unforeseen events, optimal control models based on instantaneous travel times can provide improved results when feedback is taken into account and drivers adjust their routes enroute using updated traffic information.

Ideal DUO/SDUO routing strategies are appropriate for drivers' diversion under incidents based on projected travel time information. These strategies are useful when some future disturbance (incidents and other future events) are predictable, such as the increased traffic flow from a baseball stadium after a game. In general, if the future disturbance is more predictable, we can use the ideal DUO/SDUO route choice model. If the future disturbance is less predictable, we may prefer the instantaneous DUO/SDUO route choice model. Thus, the class stratification discussed in the dynamic traffic prediction module should be adjusted to the nature of the incident. An expert system can be designed to achieve this goal.

Under unpredictable incidents, our strategy is designed as a multi-period routing procedure. For simplicity, we first discuss the case of one incident only. We advise motorists of minimal travel cost routes based on current traffic information. Now, assume that an incident begins on link b at time t_1 and will be cleared at time T_1. We divide the period $[0, T]$ into three periods: $[0, t_1]$, $[t_1, T_1]$ and $[T_1, T]$. At the transition points t_1 and T_1, there are sudden capacity reductions and increases, respectively, resulting in discontinuities of traffic flows on incident link b. In our routing strategy, we assume that motorists choose routes based on current traffic information. Thus, for period $[0, t_1]$ (before the incident occurs), motorists' route choices are based on current travel cost information generated using an instantaneous DUO/SDUO route choice model over the whole time span $[0, T]$, as if no incident would occur. It covers the entire time period $[0, T]$. During the initial period $[0, t_1]$, rerouting is provided by a normal instantaneous DUO/SDUO route choice procedure.

During incident period $[t_1, T_1]$, rerouting advice must be rearranged. We consider the impact of the incident on the reduction of capacity. We assume that the capacity reduction is uniform during the incident duration in order to simplify the problem. During incident period $[t_1, T_1]$, rerouting is provided from time t_1 through T, as if the incident would not be cleared until time T_1. This strategy is consistent with the route choice criterion of current traffic information, because the clearance time of incident is not known at the initial moment of the incident. This process is called *incident-based routing.* During incident conditions, we reroute vehicle flows which are already on the network and route O-D departures for period $[t_1, T]$ at the same time. Regarding the vehicles already on the network, we consider each link as a dummy origin from which flows may have different destinations. Thus, those vehicle flows on each link a at time instant t_1 can be considered as dummy O-D flows. A routing procedure similar to the initial routing can be implemented for the same network with a reduced capacity on the incident link for period $[t_1, T]$.

We now consider the last period $[T_1, T]$, which can be called the *recovery routing* period. During period $[T_1, T]$ when the incident is cleared, the capacity of the incident link is recovered. Routing is provided as normal instantaneous

DUO/SDUO route choice. Still, the vehicle flows which are already on links at time instant T_1 must be considered as dummy O-D flows during the incident period. The above approach can be generalized to a general situation where there are multiple incidents occurring in the network during period $[0, T]$.

14.1.5 Congestion Pricing

Information and control have been identified as two major approaches to combat traffic congestion in an ATIS/ATMS system. Recently, congestion pricing has received increasing attention from policy makers as an effective measure of controlling congestion. With advances in ATIS/ATMS technology, real-time congestion pricing is becoming increasingly feasible.

Dynamic tolls can be collected with the application of automatic vehicle identification (AVI) technology. In conjunction with a dynamic route guidance system, effective traffic controls and congestion pricing can be implemented together to influence the routing strategies employed by either a central controller or individual drivers so that congestion levels in the transportation network can be controlled or adjusted. Real-time congestion pricing can be evaluated using dynamic network models.

In the evaluation of strategic ITS System Architectures, congestion pricing was identified as one of the most important control schemes in future ITS systems. What seemed impossible only a few years ago seems now possible with ATIS/ATMS technological advances, as well as changing attitudes regarding pricing policy, which in turn motivate the development of dynamic congestion pricing models for large scale transportation networks. We envision the application of such models as contributing to the evaluation of ATIS/ATMS. Eventually, such models may prove useful as well in the operation of such systems. In the short run, however, our principal objective is to improve understanding of the properties of dynamic congestion pricing models defined on large, complex road networks.

We anticipate that real-time congestion pricing can be evaluated using dynamic network models. Our focus is on technical aspects instead of policy issues. Traffic controls, including both surface street signal control and freeway ramp control, are assumed to be fixed in the current models. Their interaction with dynamic congestion pricing will be modeled as future extensions. In considering the impact of congestion pricing, therefore, we focus our attention on routing strategies instead of traffic signal controls. The typical toll strategies are: 1) bottleneck toll; 2) link or route toll; and 3) area toll. These toll strategies and their modeling issues are discussed in the following.

Bottleneck Tolls A typical example of a bottleneck toll is a bridge toll. The relationship between the congestion level and the toll level at the bottleneck can be represented using a time-dependent function. Thus, dynamic tolls can be combined with dynamic link travel cost functions, and dynamic prediction models can be applied to this situation.

Link and Route Tolls The link (or route) toll is charged for vehicles present on that link (or route) and is changing from link to link (or from route to route). A bilevel program can be constructed to determine dynamic link tolls. We use a leader-follower game to formulate such a dynamic congestion pricing problem. The objective of the upper-level problem (or the leader of the game) is to minimize the total travel time; the decision variable is the dynamic link toll. In the upper-level problem, we minimize the total travel *time* instead of total travel *cost*, because the link toll policy is designed to control the total congestion level. The DUO route choice problem (or the follower of the game) is formulated as the lower-level problem, in which the decision variables are the link flow variables. The upper level problem is a time-dependent minimization problem, whereas the lower-level problem is formulated as a variational inequality.

A bilevel program can also be constructed to determine dynamic *route* tolls. This program is similar to that for dynamic link tolls. However, solving this model requires explicit route enumeration.

Congested Area Tolls The area toll is charged for vehicles present within a congested area and is uniform across the area. It applies to any vehicle traveling in the network during the toll time period. The area tolls or congestion zone tolls, may be easier to implement than the link or route toll policy. This dynamic congestion pricing strategy is designed to control the congestion level within the central business district (CBD) area, such as central Los Angeles. We charge a uniform toll which may be time-varying and adjustable depending on the congestion level within the CBD area. In our continuous time model, we assume that the toll is continuous in time. This toll applies to any vehicle traveling within the CBD area during peak hour periods.

As with link tolls, this problem can also be formulated as a bilevel model. The upper level is the congestion control or toll policy decision level and the lower level is the description of responsive traffic status. The control authority keeps track of the congestion level in terms of average saturation degree in the CBD area and charges a congestion toll for each vehicle present in CBD areas during the toll period. However, such a bilevel model is computationally difficult for a large network. Thus, a single level model is needed, which endogenously considers the interaction of dynamic congestion toll policy and travelers' choice responses. This formulation is similar to that for the bottleneck toll problem.

A Link Toll Model

In the following, a link toll model is presented to illustrate the properties of dynamic congestion pricing models. We now consider the dynamic congestion pricing problem in which the dynamic system-optimal objective can be achieved while preserving the dynamic user-optimal route choice properties. Previous research on congestion pricing has concentrated on policy and practical implementation issues. A summary of recent changes in policy is given by Small (1992), emphasizing the changing political acceptance of congestion pricing. The theory of marginal cost pricing has been explored by many researchers in the context of static transportation networks; for example, see Beckmann et al (1956), Dafermos (1972) and Smith (1979). Few theoretical studies on congestion pricing are related to large transportation networks in real time. Recently, some models have been proposed by de Palma and Lindsey (1992) and Ghali and Smith (1993). Both of them use simple networks to explore the properties of dynamic tolls.

Consider an ideal situation with an AVI system installed and an automatic toll debiting system available. We also assume each vehicle is provided with perfect traffic information and travelers will comply with the user-optimal route guidance instructions. System optimal objectives can be defined in a variety of ways in dynamic transportation network problems. Thus, there are more toll strategies in dynamic problems than in their static counterpart.

After a link toll $\gamma_a(t)$ (dollars) is imposed, the final link travel cost on link a at time t is

$$\tilde{\tau}_a(t) = \alpha_a + \beta_a \tau_a(t) + \gamma_a(t) \tag{14.8}$$

where α_a is a term representing the fixed cost (dollars) on link a, β_a is a time-independent parameter to transform travel time (minutes) into travel cost (dollars), and $\tau_a(t)$ is travel time on link a (minutes). This problem is formulated in a bilevel structure. In the upper level problem, the total travel *time* is minimized. In the lower level problem, the variational inequality problem equilibrates the actual route travel *costs* based on travel times and link tolls.

We now discuss how we formulate a variational inequality for the lower-level ideal DUO route choice problem. The formulation is a simplified version of the link-based multi-class VI model for the ideal DUO route choice problem in Chapter 7. For simplicity, we only consider one class in this toll model. Denote $\tilde{\eta}_p^{rs}(t)$ as the actual route travel cost from origin r to destination s at time t. Also denote $\tilde{\pi}^{rs}(t)$ as the minimal actual route travel cost from origin r to destination s at time t, and $\bar{\pi}^{rs}(t)$ as the corresponding actual route travel time from origin r to destination s at time t. We also need to use a recursive formula to compute the route travel cost $\tilde{\eta}_p^{rs}(t)$ for all allowable routes. Assume route p consists of nodes $(r, 1, 2, \cdots, i, \cdots, s)$. Denote $\tilde{\eta}_p^{rj}(t)$ as the travel disutility *actually* experienced over route p from origin r to node j by vehicles departing origin r at time t. Then, a recursive formula for route travel cost $\tilde{\eta}_p^{rs}(t)$ is:

$$\tilde{\eta}_p^{rj}(t) = \tilde{\eta}_p^{r(j-1)}(t) + \tilde{\tau}_a[t + \eta_p^{r(j-1)}(t)] \qquad \forall p, r, j; j = 1, 2, \cdots, s;$$

where link $a = (j-1, j)$ and time $[t + \eta_p^{r(j-1)}(t)]$ is the arrival time instant at link a.

Recall the definition of the travel-cost-based ideal DUO state as follows.

Travel-Cost-Based Ideal DUO State: *If, for each class and each O-D pair at each instant of time, the actual travel* costs *for all routes that are being used equal the minimal actual route travel* cost, *the dynamic traffic flow over the network is in a travel-cost-based ideal dynamic user-optimal state.*

Denote $\tilde{\Omega}_a^{rj^*}(t)$ as the difference of the minimal travel cost from r to j and the travel cost from r to j via the minimal travel cost route from r to i and link a for vehicles departing from origin r at time t. It follows that

$$\tilde{\Omega}_a^{rj^*}(t) = \bar{\pi}^{ri^*}(t) + \tilde{\tau}_a[t + \bar{\pi}^{ri^*}(t)] - \bar{\pi}^{rj^*}(t) \qquad \forall a, r; a = (i, j). \tag{14.9}$$

We then rewrite the link-cost-based ideal DUO route choice conditions:

$$\tilde{\Omega}_a^{rj^*}(t) \geq 0 \qquad \forall a = (i, j), r; \tag{14.10}$$

$$u_a^{rs^*}[t + \bar{\pi}^{ri^*}(t)]\, \tilde{\Omega}_a^{rj^*}(t) = 0 \qquad \forall a = (i, j), r, s; \tag{14.11}$$

$$u_a^{rs}[t + \bar{\pi}^{ri^*}(t)] \geq 0 \qquad \forall a = (i, j), r, s. \tag{14.12}$$

The equivalent variational inequality formulation of link-cost-based ideal DUO route choice conditions (14.10)-(14.12) may be stated as follows.

Theorem 14.1. *The dynamic traffic flow pattern satisfying network constraints is in a link-cost-based ideal DUO route choice state if and only if it satisfies the variational inequality:*

$$\int_0^T \sum_{rs} \sum_a \tilde{\Omega}_a^{rj^*}(t) \left\{ u_a^{rs}[t + \bar{\pi}^{ri^*}(t)] - u_a^{rs^*}[t + \bar{\pi}^{ri^*}(t)] \right\} dt \geq 0 \tag{14.13}$$

The proof of the necessity and sufficiency of the variational inequality is similar to that in Chapter 5.

The dynamic link toll is designed so the total travel time on the network is minimized for period $[0, T]$. Thus, the bilevel program is formulated as

$$\min_{\gamma} \qquad \int_0^T \left\{ \sum_a u_a(t)\, \tau_a[u_a(t), v_a(t), x_a(t)] \right\} dt \tag{14.14}$$

where u, v and x solve the following variational inequality:

$$\int_0^T \sum_{rs} \sum_a \tilde{\Omega}_a^{rj^*}(t) \left\{ u_a^{rs}[t + \bar{\pi}^{ri^*}(t)] - u_a^{rs^*}[t + \bar{\pi}^{ri^*}(t)] \right\} dt \geq 0 \tag{14.15}$$

Because the optimality conditions for this bilevel program are very complex, no simple analytical link toll can be obtained for even a two-parallel-link network. Thus, the equivalence of the toll to the difference of marginal and unit link costs in the static network model does not exist in a dynamic congestion pricing problem.

14.1.6 Operations and Control for AHS

On fully automated highway systems, all driving tasks are performed by the vehicles and the roadside control system. To assign vehicles from on-ramps to appropriate routes and lanes so as to move the vehicles to their desired off-ramps, both route assignment and lane assignment must be performed dynamically by the roadside control system on a real-time basis. A primary objective of this type of model is to maximize AHS capacity. Since the AHS traffic is controlled and the control laws are known, the issues of driver behavior and driver compliance encountered in dynamic traffic management for conventional roadways no longer exist. This makes the dynamic network model a promising technique for AHS off-line capacity analysis and on-line operational strategies. This type of model could focus on system flow management and control, and seek to develop a robust system flow optimizer, where robustness refers to the ability to optimize the flow for various AHS operating scenarios. One of the fundamental characteristics of AHS is the conflicting relationship between longitudinal flow and lateral capacity. The robustness of the optimizer is achieved by enabling the representation of this relationship under various operating scenarios. A subsequent study could be focused on the merging/splitting and lane-changing effects of on-off ramp traffic on mainline traffic. Furthermore, the direct impact of AHS flow control and operations on vehicle emissions could be evaluated.

14.1.7 Transportation Planning

Dynamic network models can be generalized for application to long term planning purposes. Since dynamic models can predict traffic variations more accurately, they provide a good alternative to traditional planning models which are based on static equilibrium concepts. Furthermore, dynamic models can investigate disequilibrium aspects of traffic from day-to-day traffic variations.

With the rapid evolution of ITS systems, the concepts and principles of ITS are affecting the long-term transportation planning process. Thus, dynamic transportation network models can be extended as dynamic planning tools for a metropolitan area with various ITS field projects. Combining dynamic location choices into dynamic planning models could become another area of interest in order to evaluate the long-term impact of ITS projects on facility and residential locations.

14.2 Practical Considerations

Many practical considerations need to be examined in applying dynamic network models. Among these, the rolling horizon method which allows dynamic network models to be implemented, adjusted and calibrated periodically, is explored first. Furthermore, since dynamic route choice models may be regarded as constituting the core of dynamic network models, major implementation issues of dynamic route choice models are discussed. At the heart of dynamic route choice models are the assumptions about driver knowledge of traffic information and response to information. Some further stratification of drivers is also proposed.

14.2.1 Rolling Horizon Implementation

In a real-time implementation, the O-D matrices and network conditions are changing with time. Rather than assume that dynamic O-D matrices and network conditions are known a priori for the entire assignment horizon, a more realistic scenario would consider that the Traffic Management Center has deterministic information regarding dynamic O-D matrices and network conditions for a short period into the future, whereas information beyond this short period will not be available until some time later. In Figure 14.1, this short period is denoted as a roll period.

To assign vehicle flows to routes for the current time interval, dynamic network models could require knowledge of dynamic O-D matrices and network conditions for the rest of the total assignment period because those conditions are expected to influence the current assignment. Dynamic O-D matrices beyond the total assignment period are assumed to be zero or constant. Thus, the models need to be solved for the total assignment period. Alternately, dynamic O-D matrices could be assigned to routes only for the roll period because the dynamic O-D matrices and network conditions are available with reasonable certainty only for that period. Then, the time horizon is rolled forward by a length equal to the roll period and the above process is repeated. This rolling horizon implementation assumes that the current assigned vehicle flows will not be influenced by future vehicle flows assigned in the next period because the currently assigned vehicle flows will have exited the network by that time.

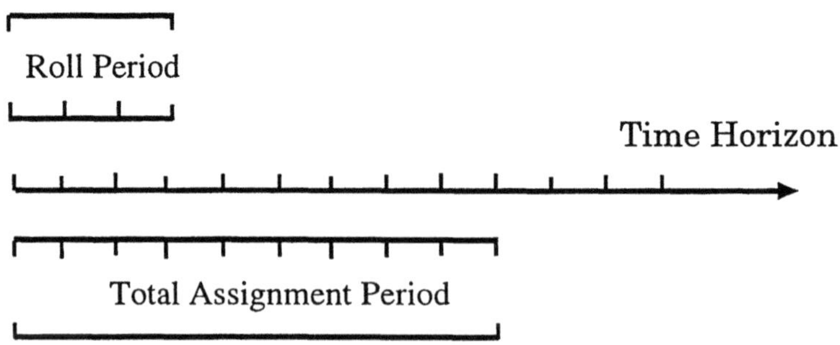

Figure 14.2: A Rolling Horizon Scheme for Dynamic Network Models

14.2.2 Traveler Knowledge of Information

Traveler knowledge is represented in the models by travel times, which are updated as new information becomes available. The computer database may either lead or lag the knowledge of travelers. Also, different travelers may respond differently to given information, which complicates dynamic network models considerably.

While the control system and simulation model have virtual real time access to changing conditions such as flows, queues, travel times, and even incidents, albeit with some reporting lag, the travelers only learn of these as they encounter these consequences, generally in the form of increased traffic density and delay. Regular or commuting drivers face the same conditions on a repetitive basis, and may be able to anticipate them even before they occur. For example, drivers traveling across town at the beginning of the peak period may know they will typically encounter congested conditions near the end of their trips, even though those facilities may be free-flowing at the present time. If drivers act on this experience, they would be considered pro-active. Any routing system, or simulated route selection, that is based on existing link conditions can not properly represent this pre-knowledge, and would be called reactive.

A distributed dynamic in-vehicle route guidance system, using real-time link conditions, can only be reactive. It will likely out-perform drivers after the occurrence of random incidents, when the system did not behave as usual, but not on a typical day, when the pro-active drivers can "out-guess" it. This type of system could be modeled as an instantaneous model. A model that has access to advance information on link travel times would be considered pro-active, and could be used to pre-test fixed-time network traffic management schemes prior to implementation. This type of system can be represented with an ideal model.

14.2.3 Response to Current and Anticipated Conditions

Different drivers place various relative values on the costs of travel on alternative routes. While one driver might prefer a shorter-distance route which saves fuel, another might prefer a shorter-time route or one with fewer stops or less queuing time. Some may accept longer travel times for safer or aesthetically pleasing routes.

Some drivers respond more to changes in network conditions than others. Route-sensitive drivers are more likely to alter their daily routes in response to traffic management schemes. Similarly, some are more likely to divert around incidents than others.

Information on an earlier portion of a trip can be more valuable to the driver than that on later portions. The anticipated costs of nearby links are more likely to be updated in drivers' minds in response to random effects, such as day-to-day variation or incidents, than those of downstream links, since the real-time information on the former is generally better. They are more likely to find out about events on nearby links which they can personally see, and to rely more on what they see than what they hear from a remote source. Furthermore, many drivers tend to optimize stochastically by being "short-term greedy", i.e.

taking any known advantage in the face of uncertainty downstream. Therefore, it is appropriate to temper a driver's pre-conceived anticipation with reality. The new information has more value upstream than downstream, where it may be re-updated before it is used. This phenomenon implies that instantaneous DUO/SDUO route choice models could be used to provide a good prediction under such scenarios.

Drivers that leave from their origins at different times will meet and interact on common links on their respective routes. A driver on a relatively long trip will merge downstream with drivers who enter the network later. A dynamic traffic assignment procedure must be able to blend these trips appropriately. As was discussed above, the driver who is just entering the network may be more sensitive to the costs on this shared link, tending to choose it only after the merging driver on a longer trip has tentatively pre-committed to it earlier. It is more likely for later departures to decide not to enter a common route, if recurrent local congestion favors an alternate route, than for earlier departures to decide to leave the route. Therefore, trips should tend to be assigned to the network in order of time of departure from their origins. Different drivers will interpret and/or comply differently with information and/or advice, such as that provided by changeable message signs. This implies that the ideal DUO route choice model can provide suitable predictions for this scenario.

A dynamic route choice model should be able to represent driver anticipation of downstream conditions that have not yet occurred, but typically will occur when the vehicle reaches the downstream link. This phenomenon was introduced into ideal DUO/SDUO route choice models in Chapters 5 and 9. Considering the above situations, the instantaneous DUO/SDUO route choice model and ideal DUO/SDUO route choice model should be applied for appropriate scenarios. Sometimes, they could be used in a combined fashion, or even interchangeably.

14.2.4 Flow-Based vs. Vehicle-Based Models

A flow-based dynamic network model provides a snapshot of the network for each time interval, and does not attempt to simulate a continuous record of network operation or equilibrium. It is feasible to achieve this level of precision by using the dynamic network models presented in this book. In contrast, a vehicle-based simulator requires a starting point from which to carry on in time on a virtually continuous basis. It can be run several times to refine its pre-estimates of time-dependent link travel times.

A flow-based dynamic network model can avoid solving an integer-based optimization program, which is usually a huge burden for a mid-sized network. A vehicle-based model usually simulates one dynamic route choice scenario without checking whether the desired solution property has been achieved. Using this type of simulation, one can avoid solving an integer-based optimization program. Therefore, an ideal scenario is to integrate flow-based model with vehicle-based model in one framework. The flow-based model is first used to ensure that the desired solution properties, such as the dynamic user-optimal conditions, have been achieved. Then, the vehicle-based model can be used to assign individual vehicles based on the solution results derived from flow-based models. Finally, the assigned vehicle flows are used to double check that flow-based model is consistent with the vehicle-based models.

14.2.5 Different Types of Travelers

As discussed earlier, a dynamic traffic assignment procedure needs to consider a variety of drivers who respond differently to events such as known traffic control measures or random incidents. They need different levels of information on these events and will even respond differently to the same information. Further, some will subscribe to in-vehicle route guidance systems. An individual vehicle simulator would typically apply the appropriate link information to each vehicle as required, while a flow-based model could blend users of different types by an ordered dynamic assignment by user type. While the former is more straightforward, the latter can be much more efficient.

In brief, a dynamic network model should also consider how to incorporate the following factors:

1. differing levels of driver sensitivity to traffic controls;

2. different perceptions of general information provided to drivers after an incident by radio or changeable message signs;

3. varying levels of driver response to the information and/or compliance with instructions;

4. variable numbers of drivers equipped with time-based and/or real time in-vehicle route guidance information.

Various types of drivers can be assigned differently by using alternative sets of link costs and shortest-route trees for each type of driver. To simplify the representation, travelers can be stratified into different classes based on the following route diversion behavior: pre-specified routes; a few alternative routes; many alternative routes. The population and characteristics of each class can be determined with surveys and updated periodically. For travelers with pre-specified routes, the route must be first generated exogenously. In general, this route includes a freeway segment and some surface streets. For networks with fewer alternative routes, such as the San Francisco Bay Area, the population with pre-specified routes may be large. On the other hand, for networks with many alternative routes, such as the Chicago Area, the population with pre-specified routes is relatively small. Similar arguments apply to travelers with few and many alternative routes.

14.3 Data Requirements

Inputs and parameter calibration for dynamic network models imply an extensive data collection effort. In general, most data required for dynamic models does not presently exist. Proposed and ongoing ITS field tests provide an excellent opportunity to collect such data. Associated with our dynamic transportation network models, the following data requirements are identified:

1. time-dependent O-D matrices;
2. network geometry and intersection/ramp control data;
3. traffic flow data for calibrating various dynamic link travel time functions;
4. traveler information for stratifying travelers into multi-classes and calibrating travel disutility functions.

14.3.1 Time-Dependent O-D Matrices

A time-dependent O-D matrix is considered as given in dynamic route choice models. For joint departure time/route choice models, the total number of departures between each O-D pair during a certain period of time has to be given. However, such a time-dependent O-D matrix does not exist in general. To overcome this difficulty, a tentative approach is to transform a 24 hour O-D matrix into an approximate time-dependent O-D matrix by using a prespecified transformation function.

In estimating a time-dependent O-D matrix, the maximum likelihood method can be used. Data collected from probe vehicles and traffic counts will serve this purpose. In addition, a time-dependent O-D matrix has to be estimated for different modes, such as passenger car, truck, HOV and bus.

The zone size for the time-dependent O-D matrix must be appropriate to the scale of the model. In general, it is smaller than the zones defined for static planning models. The time intervals for time-dependent O-D estimation can be 5, 10, 15 or 30 minutes, depending on the problem requirements. We note that zone size also depends on the length of a time interval. If a time interval is smaller, the zone can be larger and vice versa. During a time interval, the O-D departure flow rate is assumed to be constant. Since the O-D departure flow rate does not change substantially, as compared with the traffic flow during a short time interval, the interval for estimating a time-dependent O-D matrix can be larger than the interval used in dynamic route choice problems.

14.3.2 Network Geometry and Control Data

Network geometry and control data for dynamic problems must be more detailed than for static problems. For freeways, data are required for freeway segments, ramps and weaving areas. In general, network geometry and control data for freeways include:

1. link length;
2. free flow speed and speed limit on freeway segments and ramps;
3. length of weaving areas;
4. types of ramp control.

If a freeway has HOV lanes, these lanes should be treated as a set of special links. The length of HOV lanes and charges for single occupancy vehicles are additional data inputs. If there is main line metering on the freeway, a control algorithm is needed.

For arterials, the network geometry and control data include:

1. link length, and length and number of turning movement lanes;
2. number of midblock lanes;
3. number of bus stops on each link;
4. pedestrian activity;
5. number of bus/HOV lanes;
6. free flow speed and speed limit;
7. intersection control data.

The intersection controls can be classified more specifically as follows:

1. links with signalized intersections:
   (a) fixed signal control;
   (b) actuated signal control;
2. links with unsignalized intersections:
   (a) major/minor priority intersections;
   (b) all-way-stop intersections.

For local streets, the intersection controls can be classified as:

1. stop/yield control;
2. no control.

Intersection control data need to be collected for each of the above control types. In addition, offset data are necessary for signalized intersections.

14.3.3 Traffic Flow Data

Traffic flow data are needed to calibrate dynamic link travel time functions. Ideally, both detector data and probe data are desired. For a freeway segment, speed/occupancy and travel time data are required. When validating dynamic link travel time functions under highly congested conditions, information on queue length is also necessary.

For an arterial link, the following traffic flow data are required:

1. inflow/exit flow rates;

2. queue length;

3. link travel time;

4. intersection control parameters.

These traffic flow data can also be used to validate the FIFO constraints and oversaturation constraints.

14.3.4 Traveler Information

Since more realistic dynamic network models are based on travel costs or disutilities instead of travel times, it is necessary to stratify travelers into multiple classes according to their socio-economic characteristics. As discussed in Chapters 11-13 and in the dynamic prediction module in Section 14.1, traveler information should be collected based on: income and age; route diversion willingness; driving behavior; compliance degree with guidance information; compliance with current or predicted information.

In the following, we present some parameters which need to be calibrated and some travel characteristics data which need to be collected:

1. fixed travel disutilities (fuel consumption, automobile operating costs, etc.);

2. parameters to transform travel times into disutilities;

3. dispersion parameter for SDUO route choice models;

4. desired arrival time interval;

5. parameter for early arrival bonus;

6. parameter for late arrival penalty;

7. dispersion parameter for mode choice.

The above data must be collected for each class of travelers and parameters must be calibrated for each class of travelers as well.

14.4 Notes

The implementation issues discussed in Section 14.1 involve only a portion of the potential applications of dynamic network models. As the backbone of ATMS and ATIS, dynamic network models can be applied in different traffic control systems and various user services for travelers. Since ITS will definitely change travelers' behavior and the location of facilities, dynamic network models can be useful in both aspects. Another area is the integration with telecommunication systems, because transportation and telecommunication are more and more closely related in ITS. Some practical considerations of applying dynamic network models are presented in Section 14.2. In reality, there are more issues which should be identified, such as sensitivity and traffic control environment. The data requirements in Section 14.3 are generated based on the present level of development of dynamic network modeling. Further requirements are possible as the state of the art of dynamic network modeling evolves.

Successful deployment of dynamic network models in ATMS and ATIS requires extensive effort in the future. According to Yagar and Santiago (1993) and Solanki and Rathi (1993), ongoing research on dynamic network models should address the following issues:

1. integration of models of freeways and surface streets;
2. modeling dynamic route selection;
3. emulation of adaptive signal systems;
4. replication of surveillance and communication functions;
5. representation of driver behavior in ATMS/ATIS implementation;
6. incorporation of safety, energy and environmental impacts;
7. common databases and interfaces with other models;
8. interface with data reduction software at a Traffic Control Center;
9. real time capability;
10. simulation of Automated Vehicle Control Systems and Automated Highway Systems.

Problems

14.1 Identify ten key application areas of dynamic network models in ITS.

14.2 The changeable message sign has been widely used to provide real-time traffic information for drivers traveling on freeways and arterials. As travelers may or may not follow the instructions on changeable message signs, how can dynamic route choice models be used to evaluate the impact of changeable message signs on traffic performance?

14.3 Discuss the needs and obstacles of applying dynamic network models in urban transportation planning.

14.4 To evaluate the environmental impact of ITS, what kind of benefits can dynamic network models provide compared to static network models?

14.5 Travelers' compliance with route guidance instructions is very challenging. Propose one possible approach to recognize the interactions of travelers' compliance and traffic information provided in dynamic network models.

14.6 Explain how we can differentiate the degree of travelers' compliance and partial traffic information in a dynamic route choice model.

14.7 Consider the integration of traffic control and route guidance. What should be the role of a dynamic network model?

14.8 A dynamic route choice is capable of providing optimal route travel times and route flows. How can route guidance information be generated after solving an ideal DUO route choice model?

14.9 Define the goals of incident management and discuss the roles of dynamic network models in an incident management system.

14.10 Identify the major obstacles of implementing dynamic congestion pricing.

14.11 Discuss how dynamic network models can be used to help control and operate an AHS.

14.12 In a traffic management system, the reduction of air pollutants and gasoline consumption need to be considered together with travel time savings. Design an objective function for a dynamic route model which considers all three objectives.

14.13 In a rolling horizon implementation for an urban morning commuting problem, how long should the total assignment period and the roll period be?

14.14 Considering travelers' different response to traffic information, when should an instantaneous model be used and when should an ideal model be used? When can both models be integrated to provide a better representation of the real world?

References

Abkowitz M. 1981. Understanding the Effect of Transit Service Reliability on Work Travel Behavior. *Transportation Research Record*, **794**, 33-41.

Akcelik R. 1981. Traffic Signals: Capacity and Timing Analysis. *Australian Road Research Board Research Report*, **123**.

Akcelik R. 1988. The Highway Capacity Manual Formula for Signalized Intersections. *ITE Journal*, **58(3)**, 23-27.

Akcelik R. and Rouphail N.M. 1991. Estimation of Delays at Traffic Signals for Variable Demand Conditions. *Australian Road Research Board Report* **WD TE91/005**.

Alfa A.S. and Minh D.L. 1979. A Stochastic Model for the Temporal Distribution of Traffic Demand—the Peak Hour Problem. *Transportation Science*, **13**, 315-324.

Bazaraa M.S., Sherali H.D. and Shetty C.M. 1993. *Nonlinear Programming Theory and Algorithms*, 2nd edition, John Wiley & Sons, New York.

Beckmann M., McGuire C.B. and Winsten C.B. 1956. *Studies in the Economics of Transportation.* Yale University Press, New Haven.

Ben-Akiva M., Cyna M. and de Palma A. 1984. Dynamic Model of Peak Period Congestion. *Transportation Research*, **18B**, 339-355.

Ben-Akiva M. and Lerman S.R. 1985. *Discrete Choice Analysis*, MIT Press, Cambridge.

Ben-Akiva M., de Palma A. and Kanaroglu P. 1986. Dynamic Model of Peak Period Traffic Congestion with Elastic Arrival Rates. *Transportation Science*, **20**, 164-181.

Ben-Akiva M., Koutsopoulos H.N. and Mukundan A. 1994. A Dynamic Traffic Model System for ATMS/ATIS Operations, *IVHS Journal*, **2**(1).

Blum J. 1954. Multidimentional Stochastic Approximate Methods. *Annals of Mathematical Statistics*, **25**, 737-744.

Boyce D.E. 1978. Equilibrium Solution to Combined Urban Residential Location, Mode Choice and Trip Assignment Models. in *Competition Among Small Regions*, W. Buhr and P. Friedrich (eds), Nomos Verlagsgesellschaft, Baden-Baden, Germany, 246-264.

Boyce D.E. 1980. A Framework for Constructing Network Equilibrium Models of Urban Location. *Transportation Science*, **14**, 77-96.

Boyce D.E. 1989. Route Guidance Systems for Managing Urban Transportation Networks: Review and Prospects. Presented at *Tenth Italian Regional Science Conference*, Rome, Italy.

Boyce D.E., Day N.D. and McDonald C. 1970. *Metropolitan Plan Making.* Monograph Series No.4, Regional Science Research Institute, Philadelphia.

Boyce D.E., Kirson A.M. and Schofer J.L. 1994. ADVANCE - The Illinois Dynamic Navigation and Route Guidance Demonstration Program. in *Advanced Technology for Road Transport*, I. Catling (ed.), Artech House, Boston, MA, 247-270.

Boyce D.E., LeBlanc L.J. and Chon. K.S. 1988. Network Equilibrium Models of Urban Location and Travel Choices: A Retrospective Survey. *Journal of Regional Science*, **28**, 159-183.

Boyce D.E., Ran B. and LeBlanc L.J. 1991. Dynamic User-Optimal Traffic Assignment: A New Model and Solution Techniques. Presented at First Triennial Symposium on Transportation Analysis, Montreal, Canada.

Boyce D.E., Ran B. and LeBlanc L.J. 1995. Solving an Instantaneous Dynamic User-Optimal Route Choice Model. *Transportation Science*, **29**, 128-142.

Boyce D.E., Lee D.H., Janson B.N. and Berka S. 1996. Dynamic User-Optimal Route Choice Modeling of a Large-Scale Traffic Network. *Journal of Transportation Engineering*, ASCE, Forthcoming.

Bryson, Jr. A.E. and Ho Y.C. 1975. *Applied Optimal Control.* Hemisphere Publishing Corp., Washington, DC.

Burrow I.A. 1989. A Note on Traffic Delay Formulas. *ITE Journal*, **59(10)**, 29-32.

Canon M.D., Cullum, Jr., C.D. and Polak E. 1970. *Theory of Optimal Control and Mathematical Programming.* McGraw-Hill, New York.

Carey M. 1987. Optimal Time-Varying Flows on Congested Networks. *Operations Research*, **35**, 58-69.

Carey M. 1992. Nonconvexity of the Dynamic Traffic Assignment Problem. *Transportation Research*, **26B**, 127-133.

Carey M. and Srinivasan S. 1993. Externalities, Average and Marginal Costs and Tolls on Congested Networks With Time-Varying Flows. *Operations Research*, **41**, 217-231.

Carey M. 1995. Dynamic Congestion Pricing and The Price of FIFO. *Urban Transportation Networks: Dynamic Flow Modeling and Control*, Gartner N.H. and Improta G. (ed.), Springer-Verlag, 233-250.

Cascetta E. 1991. A Day-to-Day and Within-Day Dynamic Stochastic Assignment Model. *Transportation Research*, **25A**, 277-291.

Cascetta E., Inaudi D. and Marquis G. 1993. Dynamic Estimators of Origin-Destination Matrices Using Traffic Counts. *Transportation Science*, **27**, 363-373.

Cesari L. 1983. *Optimization - Theory and Applications.* Springer-Verlag, New York.

Chang G.L., Ho P.K. and Wei C.H. 1993. A Dynamic System-Optimum Control Model for Commuting Traffic Corridors. *Transportation Research*, **1C**, 3-22.

Chen M. and Alfa A.S. 1991. Algorithms for Solving Fisk's Stochastic Traffic Assignment Model. *Transportation Research*, **25B**, 405-412.

Codelli D.K., Niedringhaus W.P., Wang P.T.R. and Young B.S. 1993. *User's Manual for Traffic and Highway Object for REsearch Analysis, and Understanding (THOREAU)*, The MITRE Corporation, McLean, VA.

Codina E. and Barcelo J. 1991. An Algorithm for Dynamic Traffic Assignment Model Based on Optimal Control Approach. Presented at First Triennial Symposium on Transportation Analysis, Montreal, Canada.

Cruz, Jr. J.B. 1978. Leader-Follower Strategies for Multilevel Systems. *IEEE Transactions on Automatic Control* **AC-23**, **2**, 244-254.

Dafermos S.C. 1972. The Traffic Assignment Problem for Multiclass-User Transportation Networks. *Transportation Science*, **6**, 73-87.

Dafermos S.C. 1980. Traffic Equilibrium and Variational Inequalities. *Transportation Science*, **14**, 42-54.

Dafermos S.C. 1982. Relaxation Algorithms for The General Asymmetric Traffic Equilibrium Problem. *Transportation Science*, **16**, 231-240.

Dafermos S.C. and Nagurney A. 1984a. On Some Traffic Equilibrium Theory Paradoxes. *Transportation Research*, **18B**, 101-110.

Dafermos S.C. and Nagurney A. 1984b. Stability and Sensitivity Analysis for the General Network Equilibrium-Travel Choice Model. *Proceedings of 9th International Symposium on Transportation and Traffic Theory*, VNU Science Press, Utrecht, The Netherlands, 217-234.

Dafermos S.C. and Sparrow F.T. 1969. The Traffic Assignment Problem for a General Network. *Journal of Research of the National Bureau of Standards*, **72B**, 91-118.

Daganzo C.F. 1985. The Uniqueness of a Time-Dependent Equilibrium Distribution of Arrivals at a Single Bottleneck. *Transportation Science*, **19**, 29-37.

Daganzo C.F. 1993. The Cell Transmission Model. Part I: A Simple Dynamic Representation of Highway Traffic. *PATH Report*, **93-0409-3**, University of California, Berkeley.

Daganzo C.F. and Sheffi Y. 1977. On Stochastic Models of Traffic Assignment. *Transportation Science*, **11**, 253-274.

de Palma A., Ben-Akiva M., Lefevre C. and Litinas N. 1983. Stochastic Equilibrium Model of Peak Period Traffic Congestion. *Transportation Science*, **17**, 430-453.

de Palma A. and Lindsey R. 1992. The Potential Benefits of A Combined Route Guidance and Road Pricing System: An Economic Analysis. Presented at the Second Capri Seminar on Traffic Networks, Capri, Italy.

Dial R.B. 1971. Probabilistic Multipath Traffic Assignment Model Which Obviates Path Enumeration. *Transportation Research*, **5**, 83-111.

Drissi-Kaitouni, O. and Hameda-Benchekroun A. 1992. A Dynamic Traffic Assignment Model and A Solution Algorithm. *Transportation Science*, **26**, 119-128.

Fambro, D.B., Chang, E.C.P., and Messer, C.J. 1991. Effects of the Quality of Traffic Signal Progression on Delay, *National Cooperative Highway Research Program Report*, **339**, Transportation Research Board, Washington, DC.

Federal Highway Administration. 1994. Proceedings of IVHS Dynamic Traffic Assignment and Simulation Workshop.

Fisk C. 1980. Some Developments in Equilibrium Traffic Assignment. *Transportation Research*, **14B**, 243-255.

Fisk C. and Boyce D.E. 1983. Alternative Variational Inequality Formulations of Network Equilibrium Travel Choice Problem. *Transportation Science*, **17**, 454-463.

Florian M. 1986. Nonlinear Cost Network Models in Transportation Analysis. *Mathematical Programming Study*, **26**, 167-196.

Florian M. and Nguyen S. 1978. A Combined Trip Distribution Modal Split and Trip Assignment Model. *Transportation Research*, **12**, 241-246.

Florian M. and Spiess H. 1982. The Convergence of Diagonalization Algorithms for Asymmetric Network Equilibrium Problems. *Transportation Research*, **16B**, 447-483.

Frank M. and Wolfe P. 1956. An Algorithm for Quadratic Programming. *Naval Research Logistics Quarterly*, **3(1-2)**, 95-110.

Friesz T.L. 1985. Transportation Network Equilibrium, Design and Aggregation: Key Developments and Research Opportunities. *Transportation Research*, **19A**, 413-428.

Friesz T.L., Bernstein D., Smith T.E., Tobin R.L. and Wie B.-W. 1993. A Variational Inequality Formulation of the Dynamic Network User Equilibrium Problem. *Operations Research*, **41**, 179-191.

Friesz T.L., Luque F.J., Tobin R.L. and Wie B.-W. 1989. Dynamic Network Traffic Assignment Considered as a Continuous Time Optimal Control Problem. *Operations Research*, **37**, 893-901.

Ghali M. and Smith M.J. 1993. Traffic Assignment, Traffic Control and Road Pricing. *Proceedings of 12th International Symposium on Transportation and Traffic Theory*, Elsevier Science, Amsterdam, 147-170.

Greenshields B. 1933. A Study of Traffic Capacity. *Highway Research Board Proceedings*, **14**, 468-477.

Hendrickson C. and Plank E. 1984. The Flexibility of Departure Times for Work Trips. *Transportation Research*, **18A**, 25-36.

Ho J.K. 1980. A Successive Linear Optimization Approach to the Dynamic Traffic Assignment Problem. *Transportation Science*, **14**, 295-305.

Ho J.K. 1990. Solving the Dynamic Traffic Assignment Problem on a Hypercube Multicomputer. *Transportation Research*, **24B**, 443-451.

Hurdle V.F. 1974. The Effect of Queuing on Traffic Assignment in a Simple

Road Network. *Proceedings of 6th International Symposium on Transportation and Traffic Theory*, Elsevier Science, Amsterdam, 519-540.

Hurdle V.F. 1984. Signalized Intersection Delay Models – A Primer for the Uninitiated. *Transportation Research Record*, **971**, 96-105.

Janson B.N. 1991. Dynamic Traffic Assignment for Urban Road Networks. *Transportation Research*, **25B**, 143-161.

Janson B.N. 1993. Dynamic Traffic Assignment with Arrival Time Costs. *Proceedings of 12th International Symposium on Transportation and Traffic Theory*, Elsevier Science, Amsterdam, 127-146.

Janson B.N. and Robles J. 1995. A Quasi-Continuous Dynamic Traffic Assignment Model. *Transportation Research Record*, Forthcoming.

Jayakrishnan R., Tsai W.K. and Chen A. 1995. A Dynamic Traffic Assignment Model with Traffic Flow Relationships. *Transportation Research*, **3C**, 51-72.

Kamien M.L. and Schwartz N.L. 1981. *Dynamic Optimization: the Calculus of Variations and Optimal Control.* Elsevier North Holland, New York.

Kaysi I., Ben-Akiva M. and Koutsopoulos H. 1993 An Integrated Approach to Vehicle Routing and Congestion Prediction for Real-Time Driver Guidance. Presented at the 72nd Annual Transportation Research Board Meeting, Washington D.C.

Kimber R.M. and Hollis E.M. 1979. Traffic Queues and Delays at Road Junctions. *TRRL Laboratory Report*, **909**, U.K.

Kinderlehrer D. and Stampacchia G. 1980. *An Introduction to Variational Inequalities and Their Applications.* Academic Press, New York.

Kirk D.E. 1970. *Optimal Control Theory: An Introduction.* Prentice-Hall, Englewood Cliffs, NJ.

Kuhn H. and Tucker A.W. 1951. Nonlinear Programming. *Proceedings of Second Berkeley Symposium of Mathematical Statistics and Probability*, University of California Press, Berkeley, CA.

Lafortune S., Sengupta R., Kaufman D.E. and Smith R.L. 1991. A Dynamic System Model for Traffic Assignment in Networks. *Vehicle Navigation and Information Systems Conference Proceedings*, Society of Automotive Engineers, 701-708.

LeBlanc L.J. and Abdulaal M. 1982. Combined Mode Split-Assignment and Distribution-Model Split-Assignment Models with Multiple Groups of Travelers. *Transportation Science*, **16**, 430-442.

LeBlanc L.J., Morlok E.K. and Pierskalla W.P. 1975. An Efficient Approach to Solving the Road Network Equilibrium Traffic Assignment Problem. *Transportation Research*, **9**, 309-318.

LeBlanc L.J., Ran B. and Boyce D.E. 1992. Dynamic Travel Choice Models for Urban Transportation Networks. *Proceedings of 1992 IEEE International Conference on Systems Man and Cybernetics*, **1**, 225-231.

Li J. 1994. Enhanced Traffic Delay Models for Signalized Intersections and Congested Networks. *PhD Dissertation*, University of Illinois at Chicago.

Lieberman E.B. 1993. An Advanced Approach to Meeting Saturated Flow Requirements. Presented at the 72nd Annual Transportation Research Board Meeting, Washington, DC.

Lin F. 1989. Application of 1985 Highway Capacity Manual for Estimating Delays at Signalized Intersection, *Transportation Research Record*, **1225**, 18–23.

Lo H.K., Chatterjee A., Wegmann F., Richards S. and Rathi A. 1994. An Evaluation Framework for IVHS. To appear in *Journal of Transportation Engineering*, ASCE.

Lo H.K., Ran B. and Hongola B. 1996. A Multi-Class Dynamic Traffic Assignment Model: Formulation And Computational Experiences. *Transportation Research Record*, Forthcoming.

Luenberger D.G. 1984. *Linear and Nonlinear Programming*. Second Edition, Addison Wesley, Menlo Park, CA.

Mahmassani H.S. and Chang G.L. 1987. On Boundedly Rational User Equilibrium in Transportation Systems. *Transportation Science*, **21**, 89-99.

Mahmassani H.S. and Herman R. 1984. Dynamic User Equilibrium Departure Time and Route Choice on Idealized Traffic Arterials. *Transportation Science*, **18**, 362-384.

Mahmassani H.S., Hu T. and Jayakrishnan R. 1992. Dynamic Traffic Assignment and Simulation for Advanced Network Informatics (DYNASMART). Presented at The Second International Capri Seminar on Urban Traffic Networks, Capri, Italy.

Mahmassani H.S. and Mouskos K. 1988. Some Numerical Results on the Diagonalization Network Assignment Algorithm with Asymmetric Interactions Between Cars and Trucks. *Transportation Research*, **22B**, 275-290.

Mahmassani H.S., Peeta S., Hu T.Y. and Ziliaskopoulos A. 1993. Dynamic Traffic Assignment with Multiple User Classes for Real-Time ATIS/ATMS Applications. in *Large Urban Systems, Proceedings of the Advanced Traffic Management Conference*, Federal Highway Administration, US Department of Transportation, Washington, DC, 91-114.

McShane W.R. and Roess R.P. 1990. *Traffic Engineerıng*. Prentice Hall, Englewood Cliffs, NJ.

Merchant D.K. and Nemhauser G.L. 1978a. A Model and an Algorithm for the Dynamic Traffic Assignment Problems. *Transportation Science*, **12**, 183-199.

Merchant D.K. and Nemhauser G.L. 1978b. Optimality Conditions for a Dynamic Traffic Assignment Model. *Transportation Science*, **12**, 200-207.

Messer C.J. 1990. The 1985 Highway Capacity Manual Delay Equation. *ITE Compendium of Technical Papers*, 60th Annual Meeting, Orlando, FL, 205-209.

Mirchandani, P. and Soroush, H. 1987. Generalized Traffic Equilibrium with Probabilistic Travel Times and Perceptions *Transportation Science*, **21**(3).

Murchland J.D. 1966. Some Remarks on the Gravity Model of Traffic Distribution and an Equivalent Maximisation Formulation. Transport Network Theory Unit, London School of Economics, London.

Nagurney A. 1986. Computational Comparisons of Algorithms for General Asymmetric Traffic Equilibrium Problems with Fixed and Elastic Demands. *Transportation Research*, **20B**, 78-84.

Nagurney A. 1993. *Network Economics: A Variational Inequality Approach.* Kluwer Academic Publishers, Norwell, MA.

Newell G.F. 1990. Stochastic Delays on Signalized Arterial Highways. *Proceedings of 11th International Symposium on Transportation and Traffic Theory*, Elsevier Science, Amsterdam, 589-598.

Olszewski P.S. 1990. Traffic Signal Delay for Non-Uniform Arrivals, *Transportation Research Record*, **1287**, 42–53.

Powell W.B. and Sheffi Y. 1982. The Convergence of Equilibrium Algorithms with Predetermined Step Sizes. *Transportation Science*, **16**, 45-55.

Ran B. and Shimazaki T. 1989a. A General Model and Algorithm for the Dynamic Traffic Assignment Problems. *Transport Policy, Management and Technology Towards 2001, Proceedings of Fifth World Conference on Transport Research*, Yokohama, Japan, 463-477.

Ran B. and Shimazaki T. 1989b. Dynamic User Equilibrium Traffic Assignment for Congested Transportation Networks. Presented at the Fifth World Conference on Transport Research, Yokohama, Japan.

Ran B., Boyce D.E. and LeBlanc L.J. 1992. Dynamic User-Optimal Route Choice Models Based on Stochastic Route Travel Times. Presented at The Second International Capri Seminar on Urban Traffic Networks, Capri, Italy.

Ran B., Boyce D.E. and LeBlanc L.J. 1993. A New Class of Instantaneous Dynamic User-Optimal Traffic Assignment Models. *Operations Research*, **41**, 192-202.

Ran B. and Boyce D.E. 1994. Dynamic Urban Transportation Network Models - Theory and Implications for Intelligent Vehicle-Highway Systems. No.417 Lecture Notes in Economics and Mathematical Systems, Springer-Verlag, Heidelberg.

Rouphail N.M. and Akcelik R. 1991. Paired Intersections: Initial Development of Platooned Arrival and Queue Interaction Models. *Australian Road Research Board Report* **WD TE91/010**.

Russak B.I. 1970. On General Problems with Bounded State Variables. *Journal of Optimization Theory and Applications*, **6**, 424-452.

Sage A.P. and White, III C.C. 1977. *Optimum Systems Control.* Second Edition. Prentice-Hall, Englewood Cliffs, NJ.

Sheffi Y. 1985. *Urban Transportation Networks: Equilibrium Analysis with Mathematical Programming Methods.* Prentice-Hall, Englewood Cliffs, NJ.

Sheffi Y. and Powell W.B. 1982. An Algorithm for the Equilibrium Assignment Problem with Random Link Times. *Networks*, **12**, 191-207.

Singh M.G. and Titli A. 1978. *Systems: Decomposition, Optimization and Control.* Pergamon Press, Oxford, England.

Small K.A. Guest Editorial. 1992. Special Issue: Congestion Pricing. *Transportation*, **19**, 287-291.

Smith M.J. 1979. The Existence, Uniqueness and Stability of Traffic Equilibria. *Transportation Research*, **13B**, 295-304.

Smith M.J. 1993. A New Dynamic Traffic Model and the Existence and Calculation of Dynamic User Equilibria on Congested Capacity-Constrained Road Networks. *Transportation Research*, **27B**, 49-63.

Solanki R.S. and Rathi A.K. 1993. A Conceptual Framework for a Dynamic Traffic Flow Model. in *Large Urban Systems, Proceedings of the Advanced Traffic Management Conference*, Federal Highway Administration, US Department of Transportation, Washington, DC, 59-70.

von Stackelberg H. 1952. *The Theory of the Market Economy.* Oxford University Press, Oxford, England.

Tabak D. and Kuo B.C. 1971. *Optimal Control and Mathematical Programming.* Prentice-Hall, Englewood Cliffs, NJ.

Takaba S. 1991. Estimation and Measurement of Travel Time by Vehicle Detectors and License Plate Readers. *Vehicle Navigation and Information Systems Conference Proceedings*, Society of Automotive Engineers, 257-267.

Tarko A., Rouphail N. and Akcelik R. 1993. Overflow Delay at a Signalized Intersection Approach Influenced by an Upstream Signal: An Analytical Investigation. Presented at the 72nd TRB Annual Meeting, Washington, D.C.

Tarko A. and Rouphail N. 1994. A Distribution-Free Model for Estimating Random Queues in Signalized Networks. Presented at the 73nd TRB Annual Meeting, Washington, D.C.

Teply S. (ed.) 1984. *Canadian Capacity Guide for Signalized Intersections.* Institute of Transportation Engineers—District 7 and the University of Alberta, Edmonton, Alberta, Canada.

Transportation Research Board. 1985. Highway Capacity Manual. *Special Report*, **209**, Washington, DC.

Transportation Research Board. 1994. Highway Capacity Manual. Washington, D.C.

Van Aerde M. 1992. INTEGRATION: A Dynamic Traffic Simulation/Assignment Model. Presented at the IVHS Dynamic Traffic Assignment and Simulation Workshop, Federal Highway Administration, McLean, VA.

van As S.C. 1991. Overflow Delay at Signalized Networks. *Transportation Research*, **25A**, 1-7.

Vythoulkas P.C. 1990. Two Models for Predicting Dynamic Stochastic Equilibria in Urban Transportation Networks. *Proceedings of 11th International Symposium on Transportation and Traffic Theory*, Elsevier Science, Amsterdam, 253-272.

Wardrop J.G. 1952. Some Theoretical Aspects of Road Traffic Research, *Proceedings of the Institution of Civil Engineers, Part II*, **1**, 325-378.

Webster F.V. 1958. Traffic Signal Settings. *Road Research Laboratory Technical Paper*, **39**, HMSO, London.

Wie B.-W. 1989. Dynamic System Optimal Traffic Assignment on Congested Multidestination Networks. Presented at the Fifth World Conference on Transport Research, Yokohama, Japan.

Wie B.-W., Friesz T.L. and Tobin R.L. 1990. Dynamic User Optimal Traffic Assignment on Congested Multidestination Networks. *Transportation Research*, **24B**, 431-442.

Wie B.-W. 1993. A Differential Game Model of Nash Equilibrium on A Congested Traffic Network. *Network*, **23**, 557-565.

Wie B.-W., Tobin R.L. and Friesz T.L. 1994. The Augmented Lagrangian Method for Solving Dynamic Network Traffic Assignment Models in Discrete Time. *Transportation Science*, **28**, 204-220.

Wie B.-W., Tobin R.L., Friesz T.L. and Bernstein D. 1995. A Discrete Time, Nested Cost Operator Approach to the Dynamic Network User Equilibrium Problem. *Transportation Science*, **29**, 79-92.

Wilson A.G. 1969. The Use of Entropy Maximizing Models in the Theory of Trip Distribution, Mode Split and Route Split. *Journal of Transport Economics and Policy*, **3**, 108-126.

Yagar S. 1971. Dynamic Traffic Assignment by Individual Path Minimisation and Queuing. *Transportation Research*, **5**, 179-196.

Yagar S. and Santiago A.J. 1993. *Large Urban Systems. Proceedings of the Advanced Traffic Management Conference*, Federal Highway Administration, US Department of Transportation, Washington, DC.

Zawack D.J. and Thompson G.L. 1987 A Dynamic Space-Time Network Flow Model for City Traffic Congestion. *Transportation Science*, **21**, 153-162.

Subject Index

Author Index

Note: Citations are indexed by first author only.

List of Figures

List of Tables

MIX
Papier aus verantwortungsvollen Quellen
Paper from responsible sources
FSC® C105338

If you have any concerns about our products,
you can contact us on
ProductSafety@springernature.com

In case Publisher is established outside the EU,
the EU authorized representative is:
Springer Nature Customer Service Center GmbH
Europaplatz 3, 69115 Heidelberg, Germany

Printed by Libri Plureos GmbH
in Hamburg, Germany